Electrochemical Engineering: Emerging Technologies and Applications

Electrochemical Engineering: Emerging Technologies and Applications

Edited by **Heather Wright**

Willford Press
New York

Published by Willford Press,
118-35 Queens Blvd., Suite 400,
Forest Hills, NY 11375, USA
www.willfordpress.com

Electrochemical Engineering: Emerging Technologies and Applications
Edited by Heather Wright

International Standard Book Number: 978-1-68285-189-0 (Hardback)

Printed in China.

Contents

	Preface	**VII**
Chapter 1	**Preparation and characterization of RuO_2/polyaniline/polymer binder composite electrodes for supercapacitor applications** Suzana Sopčić, Marijana Kraljić Roković and Zoran Mandić	**1**
Chapter 2	**Corrosion inhibition of carbon steel by sodium metavanadate** Vijaya Gopal Sribharathy and Susai Rajendran	**13**
Chapter 3	**Silver ion recognition using potentiometric sensor based on recently synthesized isoquinoline-1,3-dione derivatives** Ajar Kamal, Rajiv Kumar Puri, Raghu Raj, Vipan Kumar and Rakesh Kumar Mahajan	**24**
Chapter 4	**Inhibition of corrosion of carbon steel in well water by arginine-Zn^{2+} system** Anthony Samy Sahaya Raja and Susai Rajendran	**36**
Chapter 5	**Characterization of an hrp-aox-polyaniline-graphite composite biosensor** Ana Carolina O. Santana, Erica F. Southgate, João Paulo B. G. Mendes, Jo Dweck, Eliana Mosse Alhadeff and Ninoska Isabel Bojorge Ramirez	**50**
Chapter 6	**Electrochemical extraction of oxygen using PEM electrolysis technology** Boulbaba Eladeb, Caroline Bonnet, Eric Favre and François Lapicque	**61**
Chapter 7	**Electrochemical degradation of reactive brilliant Red K-2BP on Ti/RuTiIrSnMn oxide anode in a batch cell** Luo Jiancheng, Yang Jie, Li Weishan, Huang Qiming and Xu Hongkang	**72**
Chapter 8	**Platinized titanium dioxide electrodes for methanol oxidation and photo-oxidation[§]** Svetlozar Ivanov, Ioanna Mintsouli, Jenia Georgieva, Stephan Armyanov, Eugenia Valova, Georgios Kokkinidis, Loannis Poulios and Sotiris Sotiropoulos	**85**
Chapter 9	**Electrochemical treatment of Acid Red 1 by electro-Fenton and photoelectro-Fenton processes** Camilo González-Vargas, Ricardo Salazar and Ignasi Sirés	**100**
Chapter 10	**Wastewater treatment by multi-stage batch adsorption and electrochemical regeneration** Fadhil M. Mohammed, Edward P. L. Roberts, Andrew K. Campen and Nigel W. Brown	**111**
Chapter 11	**Didanosine determination in diluted alkaline electrolyte by adsorptive stripping voltammetry at the mercury film electrode** Percio Augusto Mardini Farias, Arnaldo Aguiar Castro and Ana Isa Perez Cordoves	**125**
Chapter 12	**Mathematical modeling and reliability analysis of a 3D Li-ion battery** Richard Hong Peng Liang, Tangsheng Zou, Karthik Somasundaram, Wei Tong and Erik Birgersson	**135**

Chapter 13 **A review of liquid metal anode solid oxide fuel cells** **152**
Aliya Toleuova, Vladimir Yufit, Stefaan Simons, William C. Maskell and Daniel J. L. Brett

Chapter 14 **Development of an electrochemical sensor for the determination of the total antioxidant capacity in berries based on boron doped diamond** **167**
Bruna Pekec, Birgit Feketeföldi, Volker Ribitsch, Astrid Ortner and Kurt Kalcher

Chapter 15 **A multichannel frequency response analyser for impedance spectroscopy on power sources** **176**
Quentin Meyer, Simon Barass, Oliver Curnick, Tobias Reisch and Daniel J. L. Brett

Chapter 16 **Electrochemical Engineering - its appearance, evolution and present status. Approaching an anniversary** **184**
Velizar Stanković

Chapter 17 **Synthesis and electrochemical performance of hierarchical Sb_2S_3 nanorod-bundles for lithium-ion batteries** **198**
Xiaozhong Zhou, Shuhua Hua, Lianhua Bai and Dong Yu

Chapter 18 **Electrochemical combustion of indigo at ternary oxide coated titanium anodes** **207**
María I. León, Zaira G. Aguilar and José L. Nava

Chapter 19 **Enhanced electrocoagulation: New approaches to improve the electrochemical process** **219**
Carlos E. Barrera-Díaz, Gabriela Roa-Morales, Patricia Balderas Hernández, Carmen María Fernandez-Marchante and Manuel Andrés Rodrigo

Chapter 20 **Electrochemical studies on zirconium phosphoborate based heterogeneous membranes** **231**
Sandeep Kaushal, Pritpal Singh and Susheel K. Mittal

Permissions

List of Contributors

Preface

Electrochemical engineering is a significant branch of engineering which harnesses electrochemical phenomena for manufacturing substances. The material in this book discusses the latest theories and emerging technologies and applications of electrochemical engineering pertaining to topics such as energy conversion, storage, catalysis, electrochemical materials science, process engineering, physical and analytical electrochemistry, etc. For all readers who are interested in electrochemistry, the researches included in this book will serve as an excellent guide to develop a comprehensive understanding about the subject. It aims to facilitate the research in this field and contribute towards the progress of this discipline.

This book is a comprehensive compilation of works of different researchers from varied parts of the world. It includes valuable experiences of the researchers with the sole objective of providing the readers (learners) with a proper knowledge of the concerned field. This book will be beneficial in evoking inspiration and enhancing the knowledge of the interested readers.

In the end, I would like to extend my heartiest thanks to the authors who worked with great determination on their chapters. I also appreciate the publisher's support in the course of the book. I would also like to deeply acknowledge my family who stood by me as a source of inspiration during the project.

Editor

1

Preparation and characterization of RuO_2/polyaniline/polymer binder composite electrodes for supercapacitor applications

SUZANA SOPČIĆ, MARIJANA KRALJIĆ ROKOVIĆ and ZORAN MANDIĆ✉

University of Zagreb, Faculty of Chemical Engineering and Technology, Marulićev trg 19, HR-10000 Zagreb, Croatia

✉Corresponding Author: E-mail: zmandic@fkit.hr

Abstract

The composite electrodes consisting of amorphous and hydrous RuO_2, polyaniline and polymeric binder, Nafion® or poly(vinilydene fluoride) were prepared. The electrochemical and pseudocapacitive properties of the prepared electrodes were investigated by cyclic voltammetry and electrochemical impedance spectroscopy. The results show that the responses of composite electrodes are very sensitive to the presence of individual components and their respective ratio in the mixture. The difference in the electrochemical behavior was explained by the different physico-chemical properties of the polymeric binders.

Keywords

RuO_2; polyaniline; supercapacitors, composite electrodes, Nafion®.

Introduction

Transition metal oxides such as ruthenium and manganese oxides are increasingly investigated as active electrode materials for electrochemical supercapacitors. Due to variety of different solid forms and a range of oxidation states they usually show almost flat current profile during linear polarization experiments. Among transition metal oxides showing this type of behavior, ruthenium(IV) oxide, RuO_2, stands out since it exhibits very reversible and reproducible solid state reactions involving redox transitions between Ru(III) and Ru(IV) redox states. The redox response is stable enough allowing the oxide to undergo several thousands of charging/discharging cycles without significant deterioration of its charge storage ability and kinetics [1]. However, the gravimetric capacitance of RuO_2 is very sensitive to the degree of its hydration and crystallinity and in order to be used in widespread applications of Ru-based supercapacitors, its utilization should be maximized.

It has recently been demonstrated that Ru(III)-Ru(IV) solid state redox transition of RuO_2 consists at least of two redox reactions taking place at different rates during its charging and discharging [2]. Fast reaction includes the simultaneous proton/electron exchange according to the generally accepted mechanism:

$$RuO_x(OH)_y + ze^- + zH^+ \Leftrightarrow RuO_{x-z}(OH)_{y+z} \quad (1)$$

whereas additional reaction was proposed to proceed with the incorporation of oxygen into the hydrous RuO_2 matrix during oxidation (Equation 2). The latter reaction was recently defined as dissociative adsorption of water [3].

$$2RuO_2 + 2e^- + 2H^+ \Leftrightarrow Ru_2O_3 + H_2O. \quad (2)$$

Equations (1) and (2) stipulate that both electronic conductivity and proton availability are necessary for good performance of ruthenium oxide in supercapacitor applications. Since the hydration of ruthenium oxide promotes the proton conductivity, while at the same time it hinders the electron transport especially among the ruthenium oxide particles, it is obvious that the compromise between these two effects has to be made. Crystalline forms of RuO_2 give very low gravimetric capacitances due to insufficient hydration and limited rates of proton transport into the deeper layers of oxide, whereas on the other hand, the capacitances up to 720 F/g were achieved with hydrous and amorphous RuO_2 prepared by the sol-gel synthesis method [4]. It is now well established and confirmed by several authors that, in order to achieve the maximal utilization of ruthenium oxide and extract as much charge as possible, two parameters, electronic and protonic conductivity, should be optimized. This is usually done by annealing the amorphous $RuO_2 \cdot xH_2O$ at appropriate temperatures [5-13].

Several strategies were attempted to improve further the RuO_2 utilization and to increase its gravimetric capacitance. One strategy includes the different RuO_2 preparation methods [4, 8-11,14-23], whereas several authors recognized the importance of both intra- and inter-particle electronic conductivities and prepared the composite materials of RuO_2 with various forms of carbon [6,18,19,27-28]. It was observed that nano-structured RuO_2 particles were decisive factor in achieving the high gravimetric capacitance [29,30]. Carefully prepared electrodes with low RuO_2 loadings and good electric contact between RuO_2 and underlying conductive support gravimetric capacitances of RuO_2 as high as 1500 F/g and even higher can be achieved [6,7,13].

Another class of materials showing the promising pseudocapacitive properties is conducting polymers. In contrast to redox mechanism of the transition metal oxides, electrochemical reactions of conducting polymers are usually accompanied by the exchange of anions with surrounding medium. Among these materials, polyaniline (PANI) is the most frequently investigated due to simple synthesis, rich chemistry and electrochemistry along with the stable and reproducible behavior. The gravimetric capacitance of PANI can reach up to 400 F/g and higher comprising a combination of pseudocapacitance and double layer capacitance [31]. Although the potential range of pseudocapacitive behavior of PANI is limited between the two redox transitions, leucoemeraldine/emeraldine and emeraldine/pernigraniline, it could nevertheless be expected that the combination of PANI with RuO_2 might lead to the possible synergic effects resulting in the composite material with improved characteristics for supercapacitor applications.

Several routes for the preparation of RuO_2/PANI composite materials were attempted. Song *et al* [32] deposited electrochemically the hydrous RuO_2 from $RuCl_3$ solutions on the casted PANI/Nafion® support. Deposition of PANI by electropolymerization in the pores of IrO_x [33] was also carried out. The preparation of RuO_2–poly(3,4-ethylenedioxythiophene) nanocomposite for supercapacitor purpose was reported [34]. Recently, the procedure based on the chemical reaction between RuO_2 attached at the electrode surface and aniline from the solution has been developed [35]. This system showed good long-term stability and preserved electrochemical activity at higher pH values.

In this paper, we describe the preparation of the RuO_2/PANI composite by slightly different approach. Electrochemical deposition of PANI was carried out from aniline solution onto the glassy carbon (GC) electrode covered by casted hydrous RuO_2 particles. PANI was deposited in the pores of and between RuO_2 particles, what should lead to an increase in overall capacitance of the RuO_2 electrode by the contribution of the PANI pseudocapacitance. Equally important should be the expected effective increase in the inter-particle electronic conductivity which was shown in some cases to be the limiting factor in achieving high capacitance of high-power supercapacitors. To be as close as possible to practical applications, RuO_2 electrodes were made by mixing the oxide with the polymeric binder of variable amount. Two frequently encountered polymer binders, Nafion® and poly(vinylidene fluoride) (PVDF) were used. Due to their different physico-chemical properties, it is expected that the binders should affect the overall performance of the composite GC/RuO_2/binder/PANI electrodes differently.

Thus, the aim of this work was to prepare the composite electrodes with variable amounts of RuO_2, PANI and binders and to investigate their electrochemical and supercapacitor properties. The optimization of the composition of the composite electrode to achieve maximum capacitance and power was also attempted. The intention was to reveal the role of the individual components in the overall behavior of the composite electrode.

Experimental

Chemicals and materials

The following analytical grade chemicals were purchased: RuO_2 in hydrated and amorphous form, Nafion® 117 (5 % solution in isopropyl alcohol) and poly(vinylidene fluoride) (PVDF) from Aldrich Chemie, N-methyl pyrrolidone (NMP) from Merck, isopropyl alcohol from T.T.T. (Croatia), aniline and H_2SO_4 from Fluka. All solutions were prepared from bi-distilled water. X-ray diffraction and thermogravimetric analysis of the RuO_2 sample were published elsewhere [13].

Preparation of composite electrodes

Particles of the hydrous ruthenium oxide, $RuO_2 \cdot xH_2O$, were dispersed in corresponding binder solution using an ultrasonic bath. Two binders were used, Nafion® in isopropyl alcohol and PVDF in NMP. Homogenized suspensions were casted on the glassy carbon electrode (GC), which was previously polished by Al_2O_3 slurry, washed by bi-distilled water and degreased in ethanol. The electrodes were dried 24 h at room temperature in open air in the case of Nafion® and in vacuum in the case of PVDF. Various quantities of polyaniline (PANI) were electrodeposited onto the composite electrodes by cyclic voltammetry from the solution of 0.1 M aniline in 0.5 M H_2SO_4, in the potential range from -100 to 900 mV vs. Ag/AgCl reference electrode with the scan rate of v = 50 mV/s. The exact compositions of electrodes and their assignations are given in Tables 1 and 2. The same electrode compositions were achieved only in two cases (electrodes XX22 and XX23) since it was not possible to obtain compact composite layer with amount of PVDF higher than 0.05 mg and with amount of Nafion® smaller than 0.025 mg.

Table 1. *Composition of composite electrodes with Nafion® as a binder.*

Electrode	$m(RuO_2)$ / mg	m(Nafion®) / mg	Nafion®/RuO_2 ratio
RN14	0.04	0.10	2.50
RN15	0.04	0.40	10.0
RN22	0.10	0.025	0.25
RN23	0.10	0.05	0.50
RN24	0.10	0.10	1.00
RN25	0.10	0.40	4.00
RN35	0.46	0.40	0.86

Table 2. *Composition of composite electrodes with PVDF as a binder.*

Electrode	$m(RuO_2)$ / mg	m(PVDF) / mg	PVDF/RuO_2 ratio
RP11	0.04	0.01	0.25
RP21	0.10	0.01	0.10
RP22	0.10	0.025	0.25
RP23	0.10	0.05	0.50
RP31	0.46	0.01	0.02

Electrochemical characterization of composite electrodes

Two electrochemical methods, cyclic voltammetry (CV) and electrochemical impedance spectroscopy (EIS), were used to investigate the electrochemical and capacitive properties of prepared composite electrodes. The measurements were done in three-electrode system where the working electrode was as prepared composite on GC substrate, the counter electrode was a Pt-foil and the reference electrode was Ag/AgCl (3M KCl). CV experiments were carried out in an one-compartment cell by the Potentiostat/Galvanostat PAR, model 263A whereas EIS measurements were carried out with lock-in amplifier EG&G PAR, model 5210, connected to the potentiostat.

Electrochemical properties of the prepared composite electrodes were investigated in 0.5 M H_2SO_4 in the potential range from -100 to 1000 mV with the scan rate of v = 50 mV/s. EIS measurements were carried out at dc potential of E_{dc}=0.45 V *vs.* Ag/ AgCl, with the amplitude of 5 mV in the frequency range from 100 kHz to 10 mHz.

Gravimetric capacitances were determined on the basis of RuO_2 content.

Results and Discussion

The results obtained indicate that the responses of the composite electrodes are very sensitive not only to their constituents but also to the quantities of the individual components as well as to their ratios.

Figures 1a and 1b show cyclic voltammograms of RuO_2/Nafion® composite electrodes of different compositions, RN15 and RN23, respectively, during several cycles of PANI electrodeposition. A brief inspection of the Figures 1a and 1b reveals the importance of Nafion® on the characteristics of composite electrodes. The specific currents and consequently the charge extracted from RN23 electrode are at least an order of magnitude higher than currents obtained for RN15 electrode. In addition, RN15 composite electrode shows close to rectangular shape prior to the PANI deposition (Fig. 1a, black line), which is characteristic for capacitive behavior. As the amount of RuO_2 is increased in RN23 with respect to RN15 electrode, larger deviations from rectangular shape were observed (Figure 1b). These deviations might be the consequence of the poor interparticle contacts among oxide particles of the RN23 electrode since it contains lower Nafion® content comparing to the RN15 electrode (Table 1) leading to the increased overall ohmic resistance. PANI deposition is also influenced by the composition of the composite electrode. The PANI growth seems to be much faster in the case of RN15 compared to RN23 composite electrode with well resolved current peaks corresponding to the well known leucoemeraldine/emeraldine and emeraldine/pernigraniline redox transitions superimposed onto the RuO_2 capacitive current. In the case of RN23 electrode, the PANI deposition causes only negligible change in cyclic voltammograms (Figure 1b).

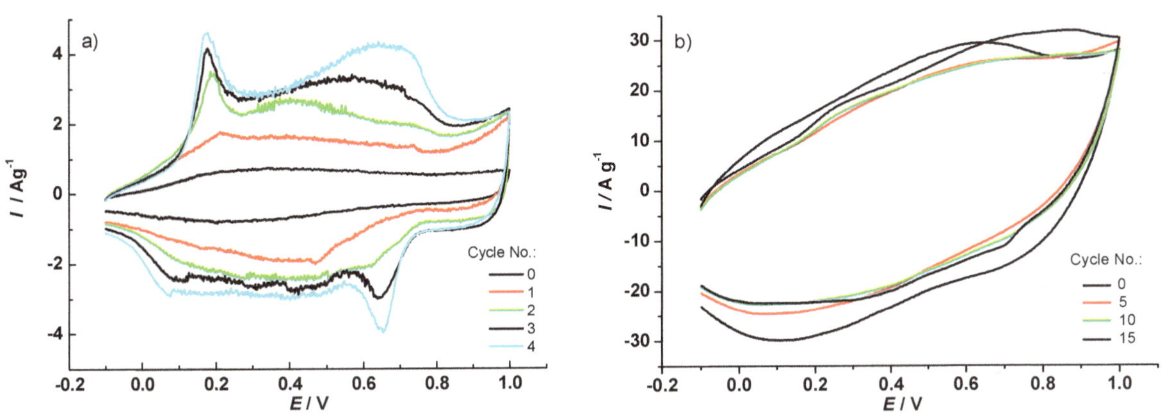

Figure 1. *Cyclic voltammograms of a) RN15 and b) RN23 composite electrode during several voltammetric cycles of the PANI electrodeposition. Scan rate = 50 mV/s.*

Cyclic voltammograms of the composite electrodes with PVDF (Figure 2, black lines) as a binder show very broad anodic peak at about 0.5 V, which is commonly observed in the cyclic voltammograms of hydrous $RuO_2{\cdot}xH_2O$ [6-8,25]. It is usually explained by the reversible processes which accompany the sorption of protons and water molecules [38,39]. In contrast to the composite electrodes prepared with Nafion® as a binder, PVDF influences the specific currents in the opposite way. With the decrease in RuO_2 content relative to the PVDF, an increase of specific currents was achieved.

Cyclic voltammograms recorded during the deposition of PANI through several voltammetric cycles are also shown in Figures 2a and 2b. In the first few cycles of PANI deposition the specific current was somewhat decreased, while at later stages the continuous specific current increase was observed. The effect of PANI deposition is even more pronounced for RP21 electrode since high currents characteristic for RuO_2 disappear at the potentials higher than 850 mV, at which the conductive emeraldine form of PANI transforms into the non-conductive pernigraniline state. It seems that in this case non-conductive pernigraniline form increases inter-particle resistance and resistance between particles and the current collector resulting in the sudden current decrease in the pernigraniline potential region. It is interesting to note that such behavior was not obtained in the case when Nafion® was used as a binder (Figure 1). Further synthesis of PANI leads for all compositions of PVDF electrodes to the overall current increase due to the contribution of PANI redox reactions. Obviously, the polymerization process as well as the resulting properties of the electrodes with PVDF binder is different from those obtained when Nafion® is used.

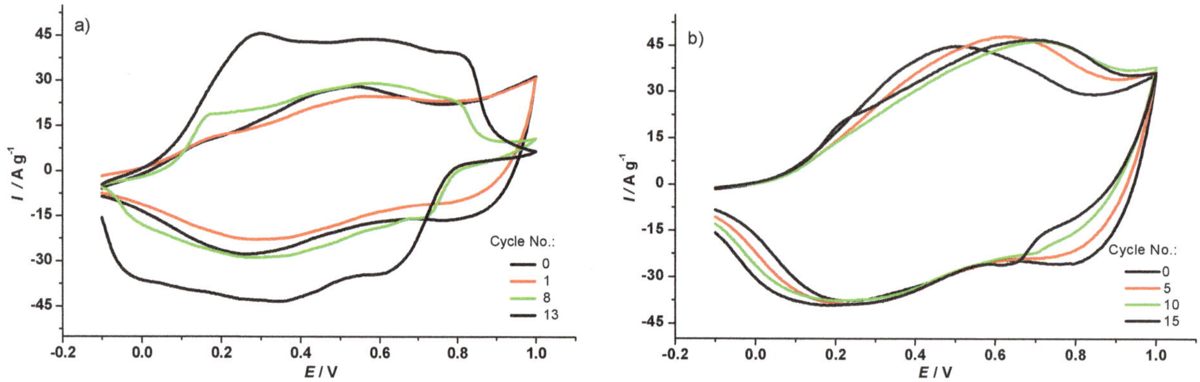

Figure 2*. Cyclic voltammograms of (a) RP21 and (b) RP23 composite electrodes during several cycles of PANI deposition. Scan rate = 50 mV/s.*

Although the shape of cyclic voltammograms and current heights can reveal important information about the system studied, in the investigations aimed to the development of electrodes for supercapacitor applications more important quantity is the gravimetric capacitance, C_s. The gravimetric capacitance for the investigated composite electrodes could be determined from the cyclic voltammograms according to the Equation 1. The results obtained for the Nafion® and PVDF composite electrodes are shown in Figures 3a and 3b, respectively, as plots of gravimetric capacitance versus voltammetric cycles of PANI deposition.

$$C_s = \frac{Q}{(2\Delta E m)} \qquad (1)$$

Q is the sum of anodic and catodic charges obtained by the integration of cyclic voltammograms in the potential interval ΔE, and m are the mass of casted RuO_2.

The data shown on Figure 3 could be analyzed from two different standpoints. The data values belonging to y-axis correspond to the gravimetric capacitances of electrodes composed of various RuO_2/Nafion® (Figure 3a) and RuO_2/PVDF (Figure 3b) ratios. The curves representing the dependence of gravimetric capacitance on the number of voltammetric cycles reveal the influence of PANI on the overall capacitance of the particular composite electrode. The analysis of the obtained results indicates the significant influence of the binder, its chemical structure or/and its physico-chemical properties on the overall behavior of RuO_2 composite electrode. While Nafion® exerts a negative influence, *i.e.* gravimetric capacitance decreases with the increase of Nafion® content relative to RuO_2, the gravimetric capacitance of RuO_2/PVDF composite electrodes shows completely opposite behavior (Figure 3c). With the increase of PVDF content relative to RuO_2 the overall capacitance increases. The highest RuO_2 gravimetric capacitance of 526 F g^{-1} was obtained for the RP11 composite electrode.

The formation of PANI on the surface and inside the pores of RuO_2/Nafion® composite electrodes contributes to the gravimetric capacitances although for some composite electrodes (RN22, RN23, RN24, RN35) slight decrease of the overall charge was observed after a few voltammetric cycles. The trend depends on Nafion®/RuO_2 ratio and decrease in charge is obtained in the cases when this ratio was less than 1.

In contrast to the RuO_2/Nafion® composite electrodes, deposition of PANI on RuO_2/PVDF electrode leads to the capacitance decrease in the first few voltammetric cycles but later on the capacitances steadily increase indicating the contribution of PANI redox reactions to the total charge of the composite electrodes.

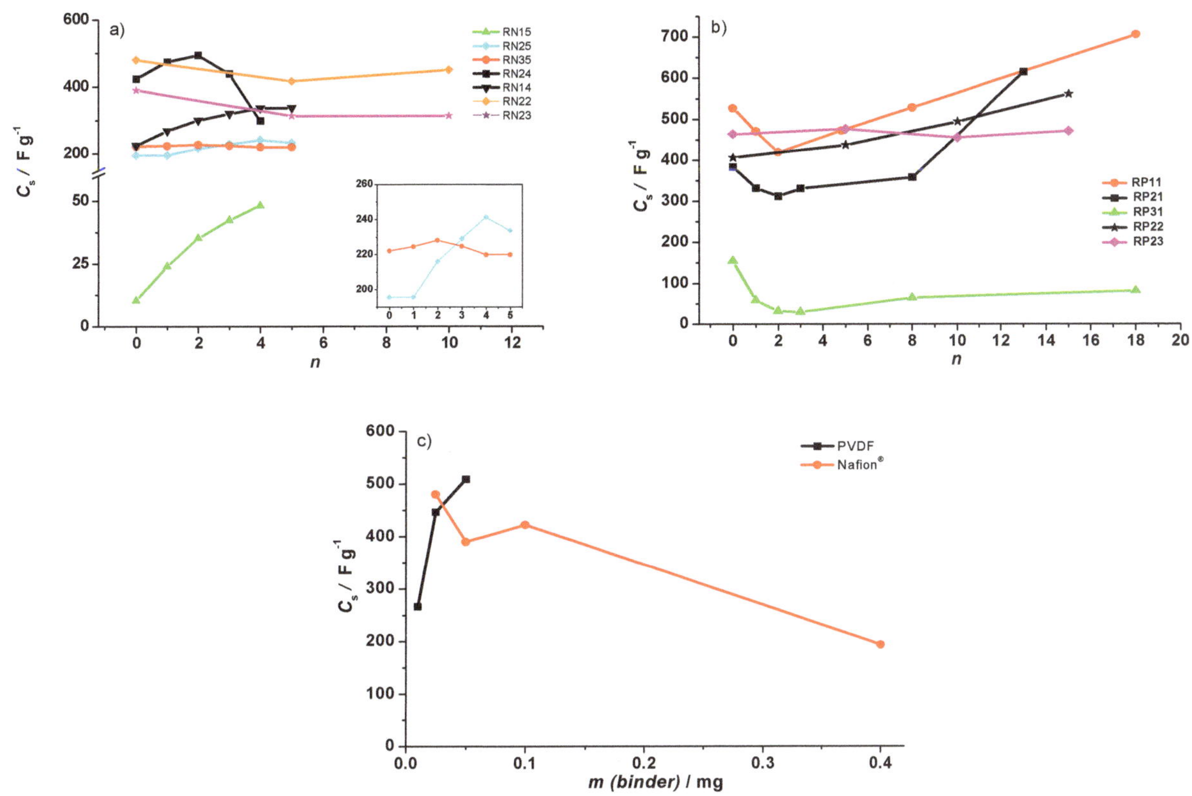

***Figure 3**. Dependence of the specific capacitance,* C_s, *on the number of voltammetric cycles of PANI deposition,* n, *for a)* RuO_2*/Nafion® composite electrodes, b)* RuO_2*/PVDF composite electrodes, c) the comparison of* C_s *dependence for the two composite electrodes* (n = 0), $m(RuO_2)$ *= 0.1 mg*

In order to get a deeper insight into the origins of such a peculiar behavior of the RuO_2 electrode and the role individual components play in the electrochemical properties of the composite electrodes, further investigations by the electrochemical impedance spectroscopy (EIS) were carried out. In Figures 4a and 4b Bode plots for the various compositions of RuO_2/Nafion® and RuO_2/PVDF electrodes are shown, respectively. Depending on the frequency range three regions of impedance response could be distinguished, resistive behavior at high frequencies, capacitive-like response at low frequencies and, in some cases, diffusion-like behavior in the intermediate frequency range. These impedance responses were usually observed for the porous electrodes involving RuO_2 pseudocapacitive behavior [9,38,39]. Diffusion characteristics are frequently explained by the rate-limiting transport of ions, mainly protons, into the deeper layers of porous and hydrous RuO_2.

Figure 4a displays impedance spectra of three composite electrodes prepared with Nafion® as a binder with the same amount of Nafion® and various amounts of RuO_2. Depending on the composite, various processes dominate the impedance spectra. All three composites show capacitive response at low frequencies. However, for the electrodes with relatively low RuO_2 content, the onset of the capacitive responses falls in the range of 1-10 Hz. For RN35 electrode, the electrode with very high RuO_2 and low Nafion® content, in the same frequency range a process controlled by diffusion appears and the onset of the capacitive response is shifted toward lower frequencies. On the other hand, the electrodes with very low RuO_2 content (RN15) exhibit in the high frequency range a behavior which could be modeled with parallel resistive/capacitive combination. This is probably due to increased inter-particle resistance caused by their separation by Nafion® chains or poor contact between oxide particles and GC support.

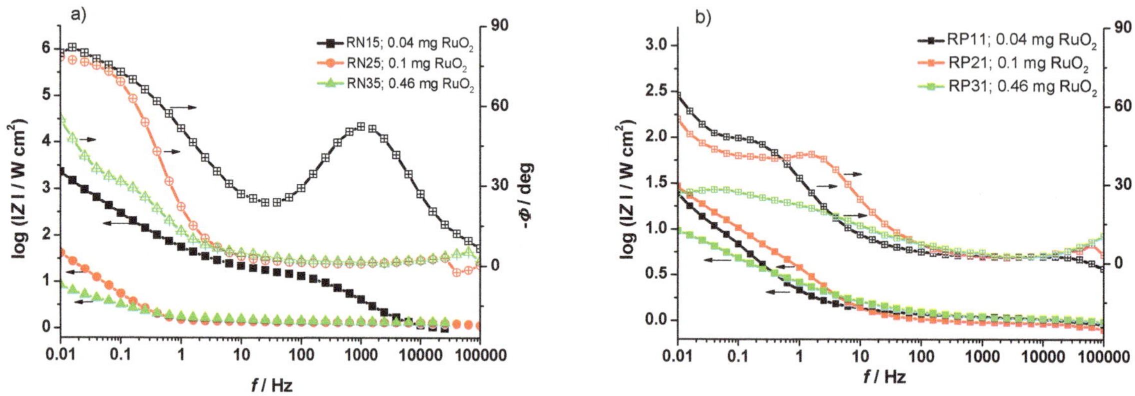

***Figure 4**. Bode plots of a) RuO_2/Nafion® and b) RuO_2/PVDF composite electrodes*

As impedance spectra were concerned, composite electrodes with PVDF as a binder (Figure 4b) behaved differently from the RuO_2/Nafion® composites. None of the RuO_2/PVDF composites showed resistive/parallel combination in high-frequency region of impedance spectra similar to RN15 electrode. However, diffusional process in the intermediate frequency range was observed for almost all composites.

PANI deposition causes the EIS spectra of RuO_2/Nafion® and RuO_2/PVDF composite electrodes to change. The changes in the impedance spectra during each successive voltammetric cycle are shown in Figures 5a and 5b, respectively. The changes depicted in the Figure 5 are mostly located in the medium, diffusion region of the frequency range. However, surprising observation was that in the case of RuO_2/Nafion® composite, each successive cycle of PANI deposition promotes diffusion process (Figure 5a), while on the other hand, diffusion process usually observed for RuO_2/PVDF electrode is hindered and starts to disappear with PANI deposition (Figure 5b).

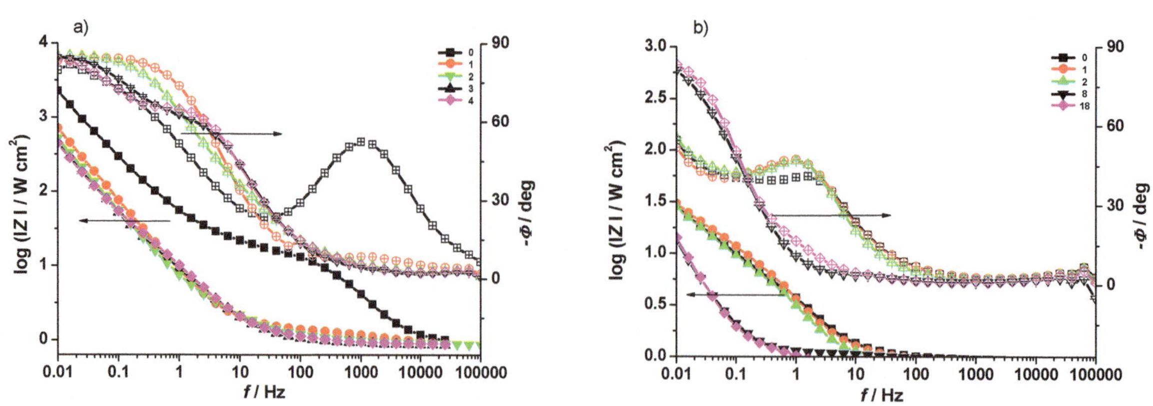

***Figure 5.** Bode plots of the composite electrodes, a) RN15 and b) RP21 at different voltammetric cycles of PANI deposition*

Specific capacitance values obtained from EIS (Table 3) are in a very good agreement and indicate the same trend as C_s values obtained from cyclic voltammetry. From the Table 3 it is also evident that C_s values obtained from EIS are higher compared to cyclic voltammetry values. These values are obtained using complex non-linear least square (CNLS) fitting of data in ZsimpWin program. The electrical equivalent circuit used in modeling procedure consisted of the parallel combination of Warburg diffusion impedance, Z_W, and constant phase element, Q, all in series with the solution resistance, R_S. Some C_S are estimated from $C=1/\omega Z''$ at 0.01 Hz, and for the two results (RN35 and RP11) it was not possible to obtain good fit or to estimate C_S. This could be explained by the fact that diffusion behavior which dominates in wide frequency range influences capacitive response in the low frequency range.

Table 3. *C_s values obtained from CV and EIS of the RuO_2/binder electrodes for various compositions*

Electrode	C_s / F g^{-1} (CV)	C_s / F g^{-1} (EIS)
RN14	224	275
RN15	10.4	12.4
RN22	482	687
RN23	391	446
RN24	423	519
RN25	199	219
RP21*	381	671
RP22	409	526
RP23	437	770
RP31*	155	330

*C_s (EIS) values are estimated from $C = 1/\omega Z''$ at 0.01 Hz.

The reason for such an unusual behavior of the RuO_2/binder/PANI electrodes is obviously due to the specific inter-molecular interactions among various components. Although at the first glance two binders have similar chemical structure, there is, in fact, a tremendous difference in physico-chemical properties between them. Nafion® consists of two distinct domains, hydrophilic one around sulphonic groups, which exist at the end of the side chains of the hydrophobic polymer backbone. This is in contrast to the PVDF composed of completely hydrophobic chains. The sulphonic groups in Nafion® are able to interact with charged species and hydrophilic environment. When the RuO_2/Nafion® composite is formed, it is reasonable to assume that the sulphonic groups are oriented toward the hydrous RuO_2 surface effectively enveloping and blocking the solid particles with Nafion® chains, as depicted in Figure 6. The blocking effect of Nafion® is demonstrated by the decrease of gravimetric capacitance (Figures 3a and 3c) as well as by the disappearance of processes related to the diffusion of ions into the deeper layers and pores of RuO_2 (Figure 4a). The deposition of PANI on the RuO_2/Nafion® composite electrode removes Nafion® from the surface of RuO_2 due to energetically more favored ionic attraction between positive PANI backbone and sulphonic groups. In fact, sulphonic groups might play a role of counter-ions during PANI synthesis as well. The result is the liberation of the RuO_2 surface by PANI deposition enabling its unhindered oxidation and reduction including the transport of species into deeper layers what is manifested as a diffusion appearing in the middle frequency range during PANI deposition (Figure 5a).

PVDF binder, due to its hydrophobicity, forms separate isolated domains within its composite with RuO_2 and does not hinder the reactions of RuO_2 active sites by blocking the oxide surface (Figure 6). PVDF has no affinity toward PANI either and the deposition of PANI leads to the coverage of RuO_2 surface interfering with the RuO_2 redox reactions, especially those requiring transport of ionic species. As a result, the disappearance of diffusion characteristic was observed (Figure 5b) and total charge extracted after the initial decrease continuously increases due to the contribution of PANI redox reactions.

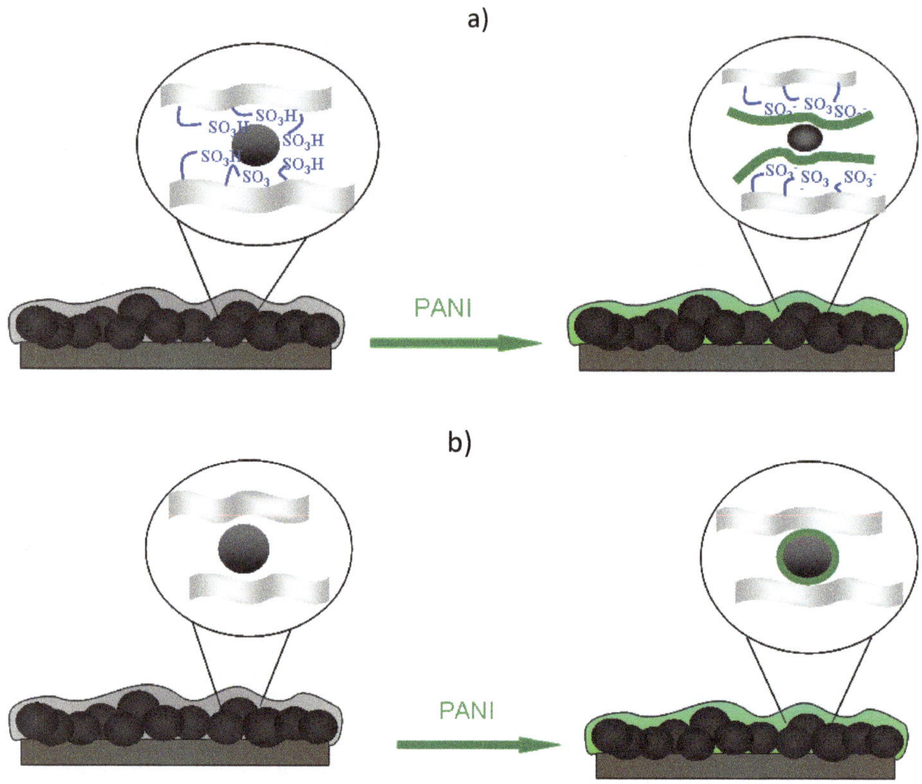

***Figure 6**. Schematic view of PANI growth within the composite electrodes of: a) RuO_2/Nafion® and b) RuO_2/PVDF.*

The data presented in this paper reveal the importance of understanding of all properties of the components involved and interactions they undergo with each other in construction of the RuO_2 electrode with desired properties for supercapacitor applications. In order to prepare efficient capacitive electrode, one has to keep in mind its particular application. If high power supercapacitors are aimed, the electrodes exhibiting no diffusion processes, such as RN15 and RN25, are favorable. On the other hand, if high energy devices are needed, RuO_2/PDVF/PANI composite electrodes are preferable. Finally, in high energy devices for the applications with intermittent high current demand, the electrodes with the low Nafion® contents would be the composition of choice.

Conclusions

For the construction of active electrodes for supercapacitor applications consisting of hydrous ruthenium oxide, with or without polyaniline, one has to pay particular attention to the exact composition and amount of additives used in the manufacturing process. Different physico-chemical properties of polymeric binders influence the overall behavior of composite electrodes differently. Nafion consisting of perfluorinated chains having hydrophilic sulphonic groups attached on side-chains tends to block the surface of RuO_2, preventing deeper layers charging/discharging reactions. This decreases the overall capacitance of the RuO_2 but on the other hand enables its applications in the high-power devices.

Poly(vinlydene fluoride) forms separate domains within the composite electrode leaving the surface of RuO_2 free for the ionic transport which accompanies charging/discharging reactions. As a result, higher amount of RuO_2 material is utilized and consequently higher energy electrodes can be constructed, but with limited ability for high power applications.

Acknowledgements: *This work is supported by the national research project No 125-1252973-2576 from the Ministry of Science, Education and Sports of the Republic of Croatia. The authors wish to thank Višnja Horvat Radošević and Krešimir Kvastek for fruitful discussion related to this work.*

References

[1] K.H. Chang, C.C. Hu, *Appl. Phys. Lett.* **88** (2006) 193102
[2] S. Sopčić, M. Kraljić Roković, Z. Mandić, A. Róka, G. Inzelt, *Electrochim. Acta* **56** (2011) 3543-3548
[3] P. Kurzweil, *J. Power Sources* **190** (2009) 189-200
[4] J. P. Zheng , T.R. Jow, *J. Electrochem. Soc.* **142** (1995) L6-L8
[5] J. P. Zheng , P.J. Cygan, T. R. Jow, *J. Electrochem. Soc.* **142** (1995) 2699-2703
[6] C.C. Hu, W. C. Chen, K. H. Chang, *J. Electrochem. Soc.* **151** (2004) A281-A290
[7] O. Barbieri, M. Hahn, A. Foelske, R. Kötz, *J. Electrochem. Soc.* **153** (2006) A2049-A2054
[8] K. H. Chang, C. C. Hu, *J. Electrochem. Soc.* **151** (2004) A958-A964
[9] C. C. Hu, Y. H. Huang, K.H. Chang, *J. Power Sources* **108** (2002) 117-127
[10] I. H. Kim, K. B. Kim, *J. Electrochem. Soc.* **153** (2006) A383-A389
[11] J.H. Jang, A. Kato, K. Machida, K. Naoi, *J. Electrochem. Soc.* **153** (2006) A321-A328
[12] A. Foelske, O. Barbieri, M. Hahn, R. Kötz, *Electrochem. Solid St. Lett.* **9** (2006) A268-A272
[13] S. Sopčić, Z. Mandić, G. Inzelt, M. Kraljić Roković, E. Meštrović, *J. Power Sources* **196** (2011) 4849-4858
[14] V. Horvat Radošević, K. Kvastek, M. Vuković, D. Čukman, *J. Electroanal. Chem.* **482** (2000) 188-201
[15] C. C. Hu, Y. H. Huang, *J. Electrochem. Soc.* **146** (1999) 2465-2471
[16] K. Kvastek, V. Horvat-Radošević, *J. Electroanal. Chem* **511** (2001) 65-78
[17] T.R. Jow, J.P. Zheng, *J. Electrochem. Soc.* **145** (1998) 49-52
[18] J. Zhang, D. Jiang, B. Chen, J. Zhu, L. Jiang, H. Fang, *J. Electrochem. Soc.* **148** (2001) A1362-A1367
[19] V. Panić, T. Vidaković, S. Gojković, A. Dekanski, S. Milonjić, B. Nikolić, *Electrochim. Acta* **48** (2003) 3805-3813
[20] V. Panić, A. Dekanski, S. Gojković, V. Mišković-Stanković, B. Nikolić, *Mater. Sci. Forum* **453-454** (2004) 133-138
[21] V. Panić, A. Dekanski, S. Milonjić, R. Atanasoski, B. Nikolić, *Electrochim. Acta* **46** (2000) 415-421
[22] I. H. Kim, K. B. Kim, *Electrochem. Solid St. Lett* **4** (2001) A62-A64
[23] K. H. Chang, C. C. Hu, *Electrochem. Solid St. Lett* **7** (2004) A466-A469
[24] V. Panić, A. Dekanski, S. Gojković, S. Milonjić, V. Mišković-Stanković, B. Nikolić, *Mater. Sci. Forum* **494** (2005) 235-240
[25] I.H. Kim, J.H. Kim, Y.H. Lee, K.B. Kim, *J. Electrochem. Soc.* **152** (2005) A2170-A2178
[26] M. Min, K. Machida, J.H. Jang, K. Naoi, *J. Electrochem. Soc.* **153** (2006) A334-A338
[27] M. Ramani, B. Haran, R. White, B. Popov, *J. Electrochem. Soc.* **148** (2001) A374-A380
[28] V.V. Panić, A.B. Dekanski, R.M. Stevanović, *J. Power Sources* **195** (2010) 3969-3976
[29] V. Barranco, F. Pico, J. Ibanez, M.A. Lillo-Rodenas, A. Linares-Solano, M. Kimura, A. Oya, R.M. Rojas, J.M. Amarilla, J.M. Rojo, *Electrochim. Acta* **54** (2009) 7452–7457.

[30] F. Pico, J. Ibanez, M.A. Lillo-Rodenas, A. Linares-Solano, R.M. Rojas, J.M. Amarilla, J.M. Rojo, *J. Power Sources* **176** (2008) 417–425.

[31] Z. Mandić, M. Kraljić Roković, T. Pokupčić, *Electrochim. Acta*, **54** (2009) 2941-2950

[32] R.Y. Song, J.H. Park, S.R. Sivakkumar, S.H.Kim, J.M. Ko, D.-Y Park, S.M. Jo, D.Y. Kim, *J. Power Sources* **166** (2007) 297- 301

[33] K.-M. Lin, K.-H. Chang, C.-C Hu, Y.-Y Li, *Electrochim. Acta* **54** (2009) 4574-4581

[34] L. Chen, C. Yuan, B. Gao, S. Chen, X. Zhang, *J. Solid State Electr.* **13** (2009) 1925-1933

[35] S. Sopčić, M. Kraljić Roković Z. Mandić, G. Inzelt, *J. Solid State Electr.* **14** (2010) 2021-2026

[36] W. Sugimooto, T. Kizaki, K. Yokoshima, Y. Murakami, Y. Takasu, *Electrochim. Acta* **49** (2004) 313-320

[37] S. Trasatti, G. Buzzanca, *J. Electroanal. Chem.* **29** (1971) A1-A5

[38] R. Kötz, M. Carlen, *Electrochim. Acta* **45** (2000) 2483-2498

[39] X. Liu, P. G. Pickup, *J. Power Sources* **176** (2008) 410-416

2

Corrosion inhibition of carbon steel by sodium metavanadate

VIJAYA GOPAL SRIBHARATHY✉ and SUSAI RAJENDRAN*

Corrosion Research Centre, PG and Research Department of Chemistry, GTN Arts College, Dindigul-624005, Tamil Nadu, India

**Department of Chemistry, RVS School of Engineering and Technology, Dindigul-624005, Tamil Nadu, India*

✉Corresponding Author

Abstract

The inhibition efficiency of sodium metavanadate (SMV)-adipic acid (AA) system in controlling corrosion of carbon steel in an aqueous solution containing 60 ppm of Cl^- has been evaluated by weight-loss method; 250 ppm of SMV exhibits inhibition efficiency of 56 %. Addition of adipic acid to SMV improves the inhibition efficiency of the system. The formulation consisting of 250 ppm of SMV and 250 ppm of adipic acid has inhibition efficiency of 98 %. A synergistic effect exists between SMV and adipic acid with the synergism parameters greater than 1. Mechanistic aspects of corrosion inhibition have been studied by electrochemical methods like potentiodynamic polarization and electrochemical impedance spectroscopy. FTIR spectra reveal that the protective film consists of Fe^{2+}-SMV complex and Fe^{2+}-adipic acid complex. The protective film has been analyzed by fluorescence spectra, SEM and EDAX.

Keywords

Carbon steel, corrosion inhibition, fluorescence, synergism parameters, SEM, adipic acid, sodium metavanadate.

Introduction

Vanadium based oxy-anion, also referred to as vanadate, has been investigated as corrosion inhibitor for Al alloys [1-4]. Smith *et al.* explored the release kinetics and protection performance of vanadate based pigments in epoxy-coated AA 2014-T6 panels [5]. Cook *et al.* investigated several corrosion inhibitors including vanadates, molybdates and ions of rare earth elements like Ce, Y and La [6]. Blanc *et al.* developed a surface conversion process for aluminium alloy based on acidic vanadium formulations which imparted protection to AA 2024-T3 coated panel [7]. These coating can be characterized by optical microscopy and scanning electron microscopy. Schumz and Frankel performed several AFM Scraching experiments on pure Al and AA 2024-T3 [8-10]. Buchheit *et al.* noticed that vanadates from the coating can impart active corrosion protection [11]. ^{51}V nuclear magnetic resonance spectroscopy (NMR) has been used to prove that the inhibition efficiency of vanadate strongly depends on the speciation [12-14]. It was shown that the clear meta vanadate solutions inhibited AA 2024-T3 corrosion to an extent similar to chromate while orange decavanadate solution was poor inhibitor

Experimental

Preparation of the specimens

Carbon steel specimens, containing 0.02 - 0.03 % S, 0.03 - 0.08 % P, 0.4 - 0.5 % Mn, 0.1 - 0.2 % C, of the dimensions 40×10×2 mm were polished to a mirror finish, degreased with trichloroethylene, and used for weight-loss and surface examination studies. For potentiodynamic polarization studies, carbon steel encapsulated in Teflon, with an exposed cross section of 5 mm diameter, was used as the working electrode. The working electrode surface was polished to a mirror finish and it was degreased with trichloroethylene.

Solution preparation

Clear metavanadate solution was prepared by dissolving 1 g $NaVO_3$ in 100 ml double distilled water. The adipic acid solution was prepared by dissolving 1 g $COOH\text{-}CH_2\text{-}CH_2\text{-}CH_2\text{-}CH_2\text{-}COOH$ in 100 ml double distilled water.

Weight-loss method

Three mild steel specimens were immersed for a period of three days in 100 ml of the neutral aqueous test solutions (with a chloride content of 60 ppm) containing various concentrations of adipic acid as the inhibitor in the absence and presence of SMV. After the exposure, corrosion products were removed with Clarke's solution [15], and the weights of the specimens before and after the immersion were determined using a SHIMADZU AY62 electronic balance.

The percentage of inhibition efficiency (*IE*) was calculated using the following equation:

$$IE = \frac{W_1 - W_2}{W_1} 100 \tag{1}$$

W_1 = corrosion rate (CR) in the absence of inhibitor and
W_2 = corrosion rate (CR) in the presence of inhibitor
The corrosion rate was calculated using the following formula:

$$\text{Corrosion rate, mm year}^{-1} = \frac{\text{Loss in weight, mg}}{\text{Surface area of specimen, dm}^2 \times \text{period of imersion, day}} \times \frac{0.0365}{\rho} \tag{2}$$

ρ-density of the metal in g cm^{-2} (7.86).

Potentiodynamic polarization studies

The polarization studies were carried out in a three-electrode cell assembly, using carbon steel as the working electrode, platinum as the counter electrode and saturated calomel electrode (SCE) as the reference electrode. Electrochemical impedance and polarization curve measurements were achieved using H & CH model CHI 660A provided with *iR* compensation option. Polarization curve measurements were carried out at scan rate of 0.01 V s^{-1}. The exposed area (1 cm^2) was mechanically polished with a series of emery sheets of variable grade. The samples were washed thoroughly with double distilled water before insertion in the cell. During the polarization study, the scan rate was 0.01 V s^{-1}; hold time at *Ef* was 0 s and quiet time was 2 s.

AC impedance spectra

AC impedance spectra were recorded using the same instrument as for polarization study and the same type of three electrode cell assembly. The real part Z' and imaginary part Z'' of the cell impedance were measured in Ω for various frequencies. The charge transfer resistance (R_t) and double layer capacitance (C_{dl}) values were calculated.

$$R_t = (R_s + R_t) - R_s \quad (3)$$

$$C_{dl} = ½ \pi R_t f_{max} \quad (4)$$

where R_s = solution resistance, f_{max} = maximum frequency.

AC impedance spectra were recorded with dc potential E = 0 V, in the frequency range $1x10^5$ - 10 Hz, amplitude = 0.005 V, and quiet time = 2 s.

Analysis of the protective film

The carbon steel samples were immersed in various test solutions for a period of three days. After three days samples were taken out and dried up. The film formed on the metal surface was carefully removed, mixed thoroughly with KBr and made as pellets samples which were recorded using Perkin Elmer FTIR 1600 spectrometer. The UV and Fluorescence spectra were recorded using Hitachi F-4500 fluorescence spectrophotometer. Surface analysis by SEM was recorded with JOEL MODEL 6390 to study the surface smoothness of the metal.

Synergism parameter

The synergism parameter can be calculated by using the equation indicating the synergistic effect existing between the inhibitors [16-18]:

$$S_i = \frac{1-\theta_{1+2}}{1-\theta'_{1+2}}$$

where $\theta_{1+2} = (\theta_1 + \theta_2) - (\theta_1\theta_2)$,

θ_1 = Surface coverage by substance 1

θ_2 = Surface coverage by substance 2

θ'_{1+2} = combined surface coverage for substance 1 and substance 2

Let the surface coverage be represented by θ; $\theta = IE/100$.

If synergistic effect exists between the inhibitors, S_I value will be greater than 1.

Analysis of variance (F-test)

An F-test was carried out to investigate whether the synergistic effect existing between inhibitor systems is statistically significant [19]. If *F*-value is greater than 5.32 for 1,8 degrees of freedom, the synergistic effects proves to be statistically significant. If it is less than 5.32 for 1,8 degrees of freedom, it was statistically insignificant at a 0.05 level of significance.

Results and Discussion

Weight-loss method

Corrosion rate of carbon steel in an aqueous solution containing 60 ppm of Cl^- in the presence and absence of inhibitor was obtained by weight loss method.

The inhibition efficiency is given in Table 1. It is observed that sodium metavanadate (SMV) has good inhibition efficiency (IE), *i.e.* as the concentration of SMV increases the inhibition efficiency also increases: 250 ppm of SMV has *IE* = 56 % and 250 ppm of adipic acid (AA) has *IE* = 61 %, but the combination of 250 ppm of SMV and 250 ppm of adipic acids show *IE* = 98 %. This suggests a synergistic effect between the binary inhibitor formulation of SMV and AA.

Table 1. *Corrosion rates (CR) of carbon steel in aqueous solution containing 60 ppm Cl^- in the presence of inhibitor obtained by weight-loss method. Inhibitor: SMV – AA, immersion period: 3 days*

Amount of SMV, ppm	Amount of AA, ppm	*IE* / %	*CR* / mm year^{-1}
0	0	-	1.0374
50	0	8	0.9544
100	0	12	0.9544
150	0	29	0.6933
200	0	32	0.5765
250	0	56	0.4564
0	50	33	0.7261
0	100	40	0.6221
0	150	48	0.5390
0	200	56	0.4564
0	250	61	0.4044
250	50	64	0.3697
250	100	92	0.0826
250	150	96	0.0411
250	200	96	0.0411
250	250	98	0.0310

Synergism parameter

The value of synergism parameter is shown in Table 2, here values of S_I are greater than 1 suggesting a synergistic effect. S_I approaches 1 when no interaction exists between the inhibitor compounds. If $S_I > 1$ it shows that the synergistic effect exists between the inhibitor compounds. In the case of $S_I < 1$ the negative interaction of inhibitor prevails (*i.e.* the corrosion rate increases).

Table 2. *Inhibition efficiencies and synergism parameters for various concentrations of AA-SMV (250 ppm) system, when carbon steel is immersed in aqueous solution containing 60 ppm of Cl^-. Immersion period: 3 days, pH 7*

Amount of AA, ppm	*IE* / %	θ_1	Amount of SMV, ppm	*IE* / %	θ_2	I'_{1+2}	θ_{1+2}	S_I
50	33	0.33	250	56	0.56	64	0.64	0.75
100	40	0.40	250	56	0.56	92	0.92	3.3
150	48	0.48	250	56	0.56	96	0.96	5.72
200	56	0.56	250	56	0.56	96	0.96	4.84
250	61	0.61	250	56	0.56	98	0.98	8.8

Analysis of variance (ANOVA)

F-test is used to confirm if the synergistic effect existing between SMV and AA is statistically significant. The results are given in Table 3. The influence of various concentrations of adipic acid (50, 100, 150, 200, 250 ppm) on the inhibition efficiency 250 ppm of SMV is tested in Table 3 for the statistical significance. The observed F-value 6.17 is greater than the critical F-value (5.32) for 1,8 degrees of freedom at 0.05 level of significance so it was concluded that the synergistic effect existing between sodium metavanadate and AA is statistically significant.

This is in agreement with conclusion derived from the calculation of synergism parameter.

Table 3. *Distribution of F-Value between the inhibition efficiencies of the influence of SMV-AA*

Source of variance	Sum of squares	Degree of freedom	Mean square	F	Level of significance
Between the sample	1708	1	1708	6.17	$p < 0.05$
With the sample	521	8	278		

Analysis of FTIR spectra

The FTIR spectrum of pure SMV is shown in Fig 1a. The vanadate stretching frequency of the sodium metavanadate appears at 1385 cm^{-1}. The FTIR spectrum of pure adipic acid (AA) is shown in Fig 1b. The -C=O stretching frequency appears at 1650 cm^{-1}, $-OH^-$ stretching frequency at 3413 cm^{-1}, and -CH stretching frequency appears at 2900 cm^{-1}. The FTIR spectrum of film formed on the carbon steel surface after immersion in solution containing 250 ppm of SMV and 250 ppm of adipic acid is shown in Fig. 1c. The vanadate frequency of sodium metavanadate increased from 1385 cm^{-1} to 1389 cm^{-1}. The $-OH^-$ stretching frequency shifted from 3413 cm^{-1} to 3434 cm^{-1}. The stretching frequency of carboxylic acid shifted from 1650 cm^{-1} to 1639 cm^{-1}. This indicates that the oxygen atom of carboxyl group has coordinated with Fe^{2+}-AA and Fe^{2+}-SMV complex on the metal surface [20].

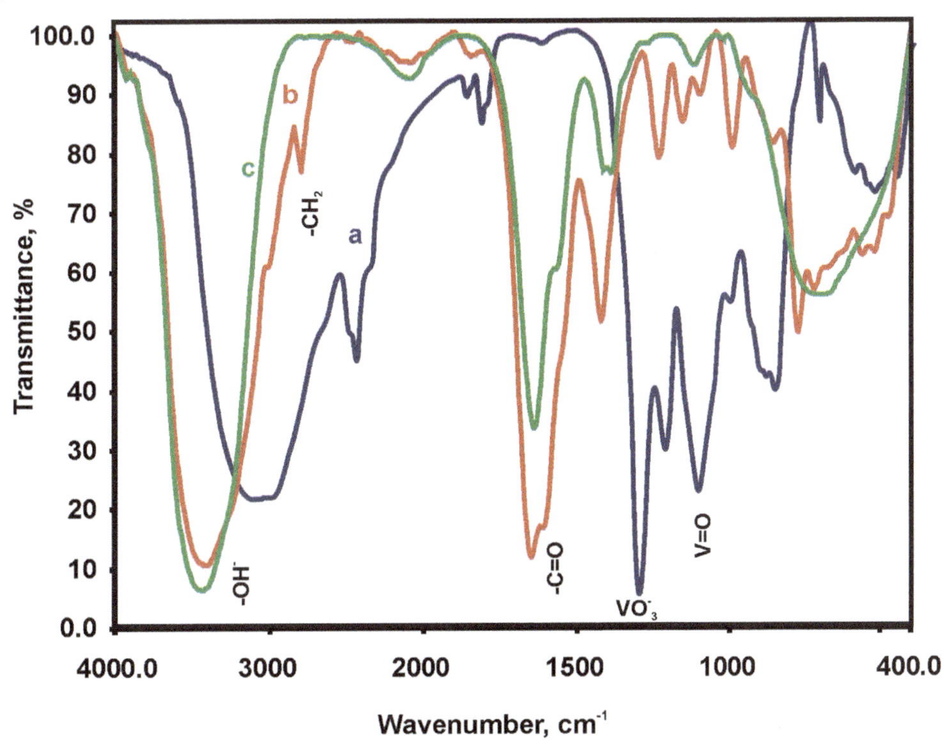

Figure 1. *FTIR spectra of* ***(a)*** *Pure SMV.* ***(b)*** *Adipic Acid,* ***(c)*** *Film formed on the metal surface after immersion in the solution containing 50 ppm of Cl^- + 250 ppm of SMV + 250 ppm of AA*

Analysis of fluorescence spectra

The analysed formulation contains 250 ppm of SMV and 250 ppm of adipic acid. Its emission spectrum (λ_{ex} = 300 nm) is shown in Fig 2a. The peak appears at 400 nm.

Solution containing 250 ppm of SMV and 250 ppm of adipic acid were mixed with a few drops of freshly prepared Fe^{2+} ions (ferrous sulphate) solution. After drying, its emission spectrum (λ_{ex} = 300 nm) is shown in Fig 2b. The intensity of the peak shifted from 400 to 403 nm [21]. The increase in intensity of the peak is due to the fact that the metal surface, after the formation of the protective film, is very bright and the film is very thin. Only one peak is obtained. Hence it is inferred that the complex is somewhat highly symmetric in nature.

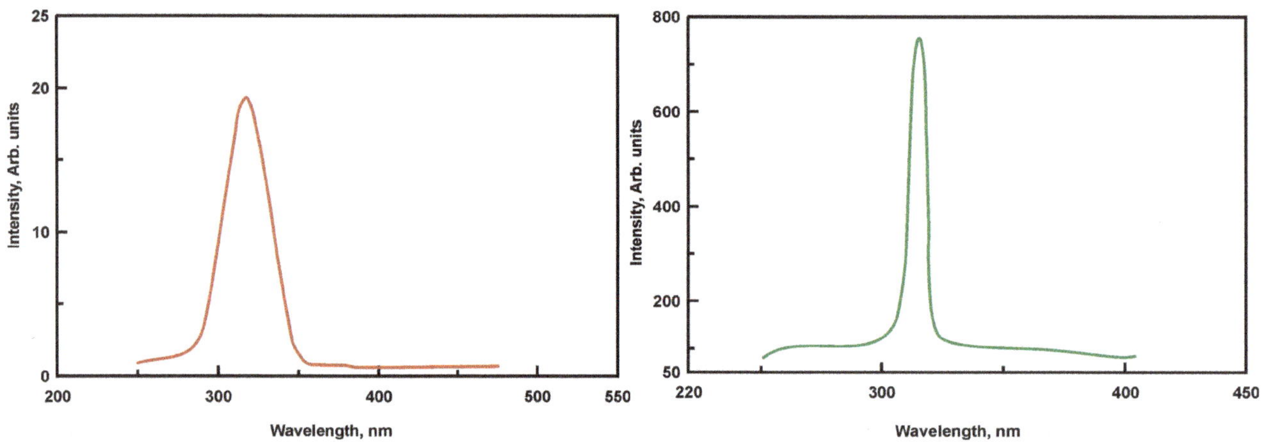

Figure 2. *Fluorescence spectra of: (a) SMV + Fe^{2+}, (b) SMV + AA + Fe^{2+}*

Analysis of potentiodynamic polarization curve

Polarization study has been carried out to confirm the formation of protective film formed on the metal surface during the corrosion inhibition process. If a protective film is formed on the metal surface, the linear polarization resistance value (LPR) increases and the corrosion current value (I_{corr}) decreases.

The potentiodynamic polarization curves of carbon steel immersed in aqueous solution containing 60 ppm of Cl^- are shown in Fig. 3 and the corrosion parameters are given in Table 4. When sodium metavanadate and adipic acid were added to the above system, the corrosion potential shifted to the anodic sites of the metal surface. This film controls the anodic reaction by forming Fe^{2+}-SMV and Fe^{2+}-AA complex on the anodic sites of the metal surface.

The formation of a protective film on the metal surface is further supported by the fact that the anodic Tafel slope (b_a) increases from 198 to 288 mV. Furthermore, the linear polarization resistance (LPR) value increases from $2.0157\times10^4\ \Omega\ cm^2$ to $4.0990\times10^4\ \Omega\ cm^2$; the corrosion current decreases from 2.025×10^{-6} A cm^{-2} to 1.120×10^{-6} A cm^{-2}. Thus the polarization study confirms the formation of the protective film on the metal surface [22-23].

Table 4. *Corrosion parameters of carbon steel immersed in an aqueous solution containing 60 ppm of Cl^- in the absence and presence of inhibitors obtained by polarization method*

Amount of		E_{corr}	b_c	b_a	I_{corr}	LPR / Ω cm^2
AA, ppm	SMV, ppm	mV *vs.* SCE	mV $decade^{-1}$	mV decade-1	A cm^{-2}	
0	0	-542	193	198	2.025×10^{-6}	2.0157×10^4
250	250	-554	166	288	1.120×10^{-6}	4.0990×10^4

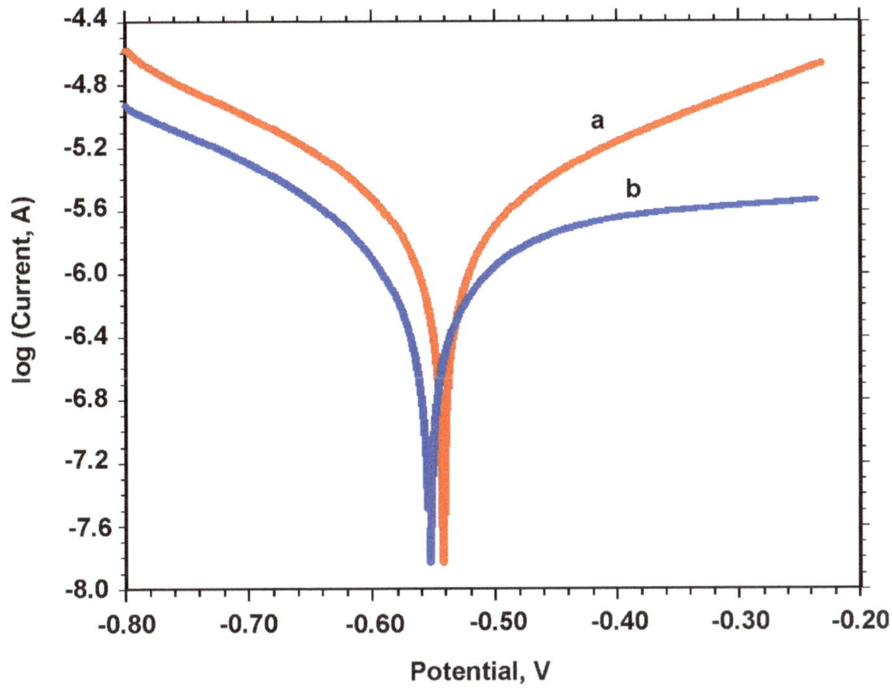

Fig.3. *Polarization curves of carbon steel immersed in various test solutions (a) 60 ppm of Cl⁻. (b) 60 ppm of Cl⁻ + 250 ppm of SMV + 250 ppm of AA*

Analysis of AC impedance spectra

AC impedance spectra (electrochemical impedance spectra) have been carried out to confirm the formation of the protective film on the metal surface since charge transfer resistance (R_{ct}) increases, double layer capacitance value (C_{dl}) decreases and the impedance log (Z/Ω) value increases. The AC impedance spectra of carbon steel immersed in aqueous solution containing 60 ppm of Cl^- in the absence and presence of inhibitors are shown in Figs. 4a,b (Nyquist plots) and in Figs. 5a,b (Bode plots). The impedance log (Z/Ω) values derived from Bode plots are also given in Table 4.

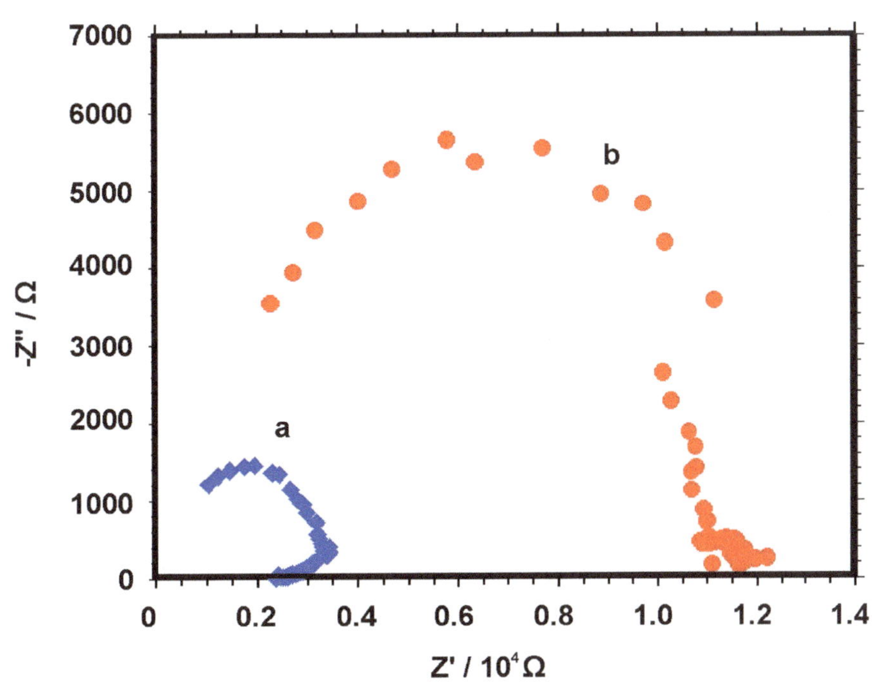

Figure 4. *AC impedance spectra (Nyquist plot) of carbon steel immersed in various test solutions:* ***(a)*** *60 ppm of Cl⁻.* ***(b)*** *60 ppm of Cl⁻ + 250 ppm of SMV + 250 ppm of AA*

AC impedance parameters, namely, charge transfer resistance (R_{ct}) and double layer capacitance (C_{dl}) are given in Table 5 for the aqueous solution containing 60 ppm of Cl^- . The obtained values are: R_{ct} 2500 Ω cm^2 and C_{dl} value of 2.04×10^{-8}. When adipic acid and SMV are added to aqueous solution containing 60 ppm Cl^-, R_{ct} value increases from 2500 Ω cm^2 to 3500 Ω cm^2. The C_{dl} value decreased from 2.04×10^{-8} to 1.45×10^{-8} F cm^{-2}. This suggests that the protective film is formed on the metal surface.

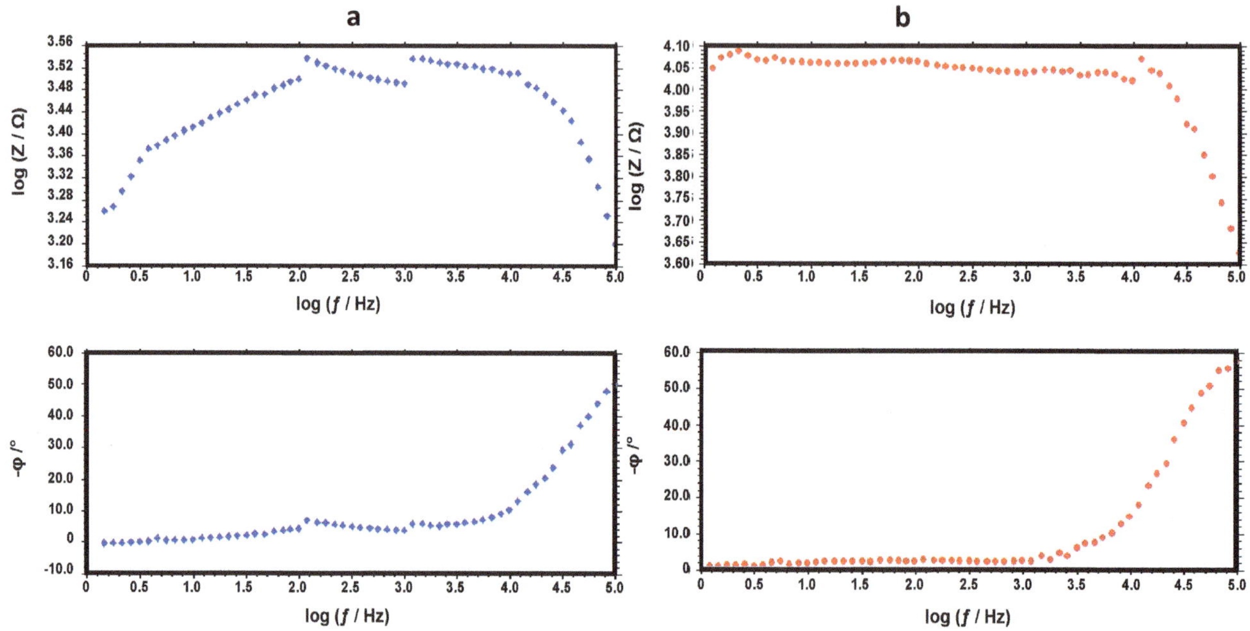

Figure 5. *AC impedance spectra (Bode plots) of carbon steel immersed in various test solutions:* ***(a)*** *60 ppm of Cl^-* ***(b)*** *60 ppm of Cl^- + 250 ppm of SMV + 250 ppm of AA*

Table 5 : *Corrosion parameters of carbon steel immersed in an aqueous solution containing 60 ppm Cl^- in the absence and presence of inhibitors obtained by AC impedance spectra*

Amount of AA, ppm	Amount of SMV ppm	Nyquist plot		Bode plot
		R_t /Ω cm^2	C_{dl} / F cm^{-2}	log (Z / Ω)
0	0	2500	2.04×10^{-8}	3.125
250	250	3500	1.45×10^{-8}	3.44

Scanning electron microscopy (SEM)

The SEM images of different magnification (×500, ×1000) of carbon steel specimen immersed in an aqueous solution containing 60 ppm Cl^- for 3 days in the absence and presence of inhibitor systems are shown in Figs. 6(b,d) and 6(c,f), respectively.

The SEM micrographs of polished carbon steel in Figs. 6(a,b) show the smooth surface of the metal. The carbon steel surface immersed in aqueous solution containing 60 ppm Cl^-, Figs. 6(b,d), shows the roughness of the metal surface that indicates the corrosion of carbon steel. Figs. 6(c,f) indicate that in the presence of inhibitor the rate of corrosion is suppressed, as it can be seen from the decrease of the corroded area. In the presence of SMV + AA, the surface is covered by a thin layer of inhibitors, which effectively controls the dissolutions of carbon steel [24].

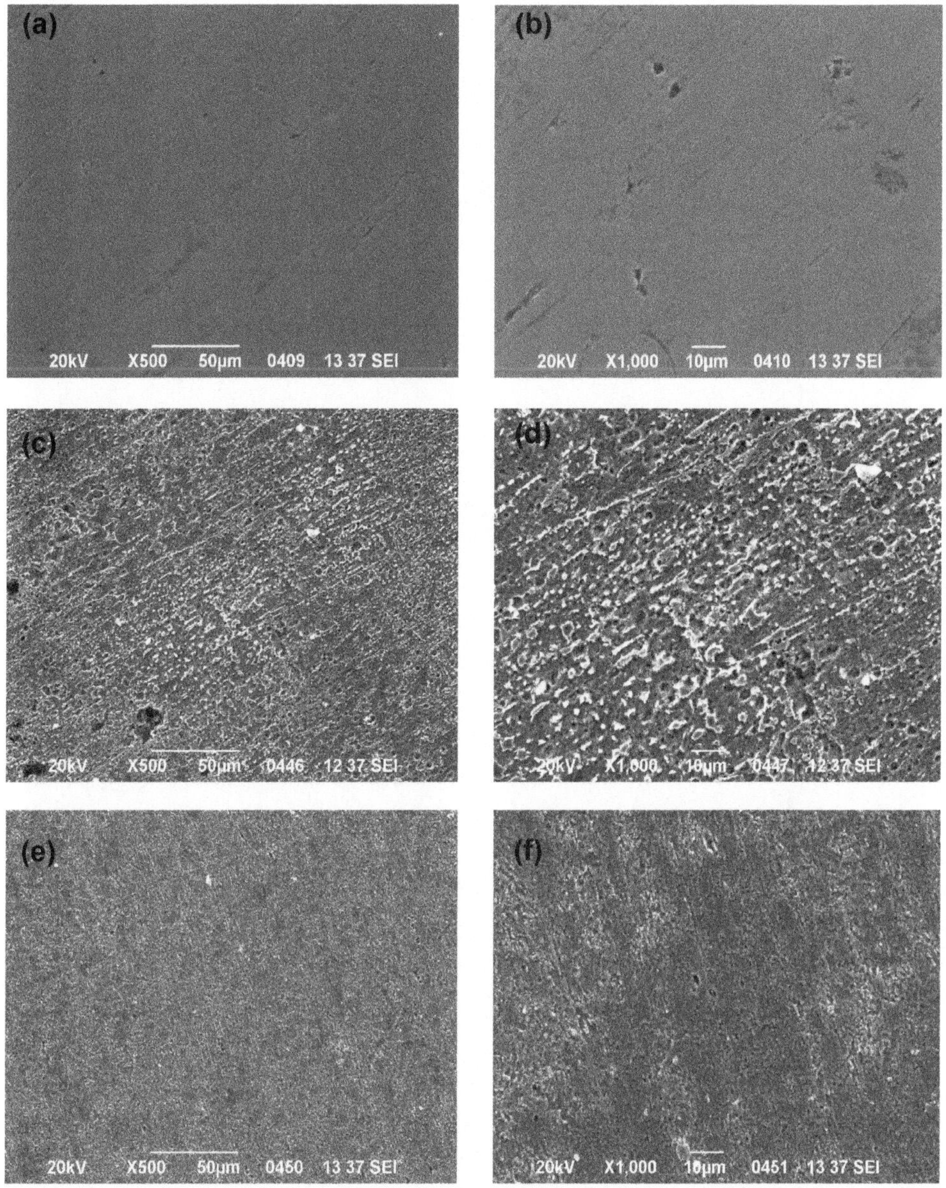

Figure 6. *SEM micrographs of **(a)** Polished carbon steel (Control) (magnification ×500);*
***(b)** Polished carbon steel (Control) (magnification ×1000),*
***(c)** 60 ppm of Cl^- (magnification ×500), **(d)** 60 ppm of Cl^- (magnification ×1000),*
***(e)** Cl^- 60 ppm + AA 250 ppm + SMV 250 ppm (magnification ×500),*
***(f)** Cl^- 60 ppm + AA 250 ppm + SMV 250 ppm (magnification ×1000)*

EDAX analysis

EDAX analysis of various surfaces is shown in Fig. 7. The EDAX analysis of polished carbon steel is shown in Fig.7a.

The EDAX analysis of polished carbon steel immersed in an aqueous solution containing 60 ppm of Cl^-, 250 ppm of AA and 250 ppm of SMV is shown in Fig. 7(b). Due to the formation of Fe-inhibitor complexes formed on the metal surface in presence of AA and SMV, the observed Fe signal is high in Fig 7a and low in Fig 7b. Thus the presence of vanadate on the metal surface is confirmed. It is also found that the intensity due to oxygen increases in Fig. 7b. This is due to the adsorption of carboxyl group of adipic acid of the inhibitor system. Thus it is concluded that the protective film consists of Fe^{2+}-adipic acid complex and Fe^{2+}-vanadate complex.

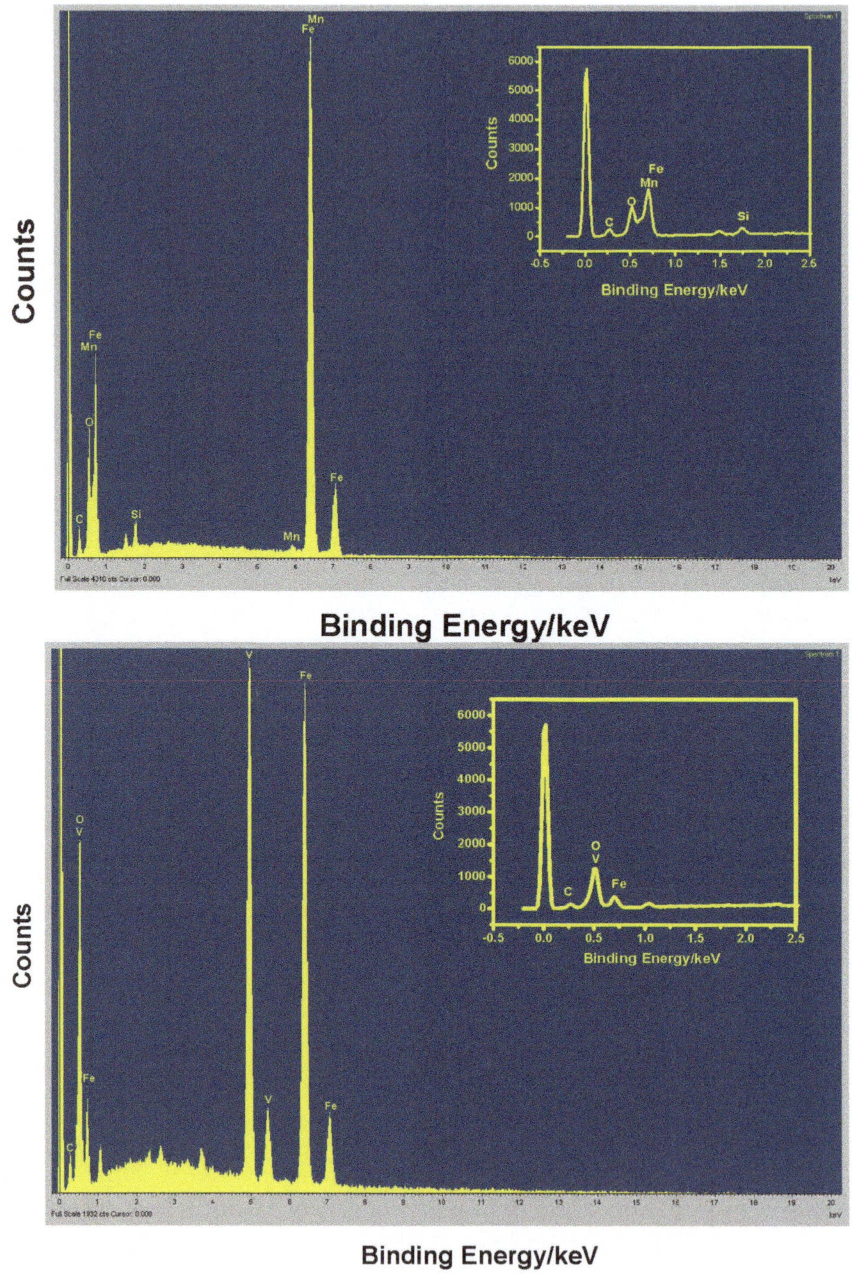

Figure 7. *EDAX **(a)** Polished carbon steel (blank)*
***(b)** Carbon steel in the solution containing 60 ppm Cl^- + 250 ppm of AA + 250 ppm of SMV*

Conclusions

The corrosion inhibition by sodium metavanadate and adipic acid in aqueous solution containing 60 ppm of Cl^- in the presence and absence of inhibitor was studied by weight-loss study and electrochemical measurements. The results show that inhibitor has the ability of reducing the corrosion rate of carbon steel in aqueous solution containing 60 ppm of Cl^-, and it acts as an anodic inhibitor. This effectiveness is confirmed by electrochemical impedance spectra and potential polarization analysis. The SEM image in the presence of the inhibitor shows the formulation of the inhibitor layer. EDAX analysis revealed that vanadium is adsorbed on the metal surface. The formulation consisting of 250 ppm of SMV and 250 ppm adipic acid shows more than 96 % inhibition efficiency and no pitting tendency.

Acknowledgements: *The authors are thankful to their respective Management and UGC for their encouragement and help.*

References

[1] M. Kending, R.G. Buchheit, *Corrosion* (*Houston*), **59** (2003) 379-400.
[2] G.S. Frankel, H.S. Isaacs, J.R. Scully and J.D. Sinclair, eds., Corrosion Science : A Retrospective and Current Status in Honour of Robert P. Frankenthal, PV 2002-13; p.430, The Electrochemical Society, Proceeding Series NJ (2002).
[3] H. Gaun,R.G. Buchheit, *Corrosion* (*Houston*), **60** (2004) 284-296.
[4] M. Trypuc and G. Lyjak, *Polish J. App. Chem.*, **XLI** (**Z3**) (1997) 187-192.
[5] C.J.E. Smith, M.A.H. Hewins, P.L.Lane, K.R.Baldwin, M.C.Gibson in Proceeding of Eurocorr 98, Utrecht, The Netherlands, Paper N5 (1998).
[6] R.L. Cook, S.R. Taylor, *Corrosion* (*Houston*), **56** (2000) 321-333.
[7] C. Blanc, B. Lavelle, G. Mankowski, *Corros. Sci.*, **39** (1997) 495-510.
[8] P. Schmutz, G.S. Frankel, *J. Electrochem. Soc.*, **145** (1998) 2285-2295.
[9] P. Schmutz, G.S.Frankel, *J. Electrochem. Soc.*, **145** (1998) 2295-2306.
[10] P. Schmutz, G.S.Frankel, *J. Electrochem. Soc.*, **146** (1999) 4461-4472.
[11] R.G. Buchheit, H.Gaun, S. Mahajanam, F. Wong, *Prog. Org. Coat.*, **47** (2003) 174-182.
[12] M. Iannuzzi, T. Young, G.S. Frankel, *J. Electrochem. Soc.*, **153** (2006) B533-B541.
[13] M. Iannuzzi, *PhD. Dissertation*, The Ohio State University, 2006.
[14] M. Iannuzzi, G.S. Frankel, *Corros. Sci.*, **49** (2007) 2371-2391.
[15] G. Wrangle, *Introduction to corrosion and protection of metals*.Chapman and Hall, London, 1986, p. 236.
[16] S. Rajendran, S. Vaibhavi, N. Anthony, D.C. Trivedi, *Corrosion,* **59** (2003) 529-534.
[17] S. Rajendran, A. John Amalraj, M.J. Joice, N. Antony, D.C. Trivedi, N. Sundaravadivelu, *Corros. Rev.,* **22** (2004) 233-248.
[18] S. Rajendran, A. John Amalraj, B.R.E. Peter, P.P. Regis, A.M. Sundaravadivelu, *Trans SAEST,* **40** (2005) 35-39.
[19] S. Rajendran, A. Raji, A. Arokia Selvi, J. Rosaly, A. Thangaswamy, *Edutracks,* **6** (2007) 30-33.
[20] S. Rajendran, B.V. Apparao, N. Palaniswamy, *Anti–Corros. Method. M.,* **47** (2000) 83-87.
[21] S. Rajendran, N. Palaniswamy, B.V. Apparao, in proceeding of the 8th European symposium on corrosion inhibitors, *University of Ferrara, Italy,* (1996) 465.
[22] S. Rajendran, R.M. Joany, B.V. Apparao, N. Palaniswamy, *Indian Journal Of Chemical Technology,* **9** (2002) 197-200.
[23] R. Hariharaputhran, A. Subramanian, A.A. Antony, P. Manishanker, T. Vasudevan, S.V. Iyer, *Anti–Corros. Method. M.,* **46** (1999) 35-39.
[24] R. Touir, N. Dkhireche, M. Ebn Touhami, M. Sfaira, O. Senhaji, J.J. Robin, B. Boutevin, M. Cherkaoui, *Mater. Chem. Phys.,* 122 (2010) 1-9.

3

Silver ion recognition using potentiometric sensor based on recently synthesized isoquinoline-1,3-dione derivatives

AJAR KAMAL, RAJIV KUMAR PURI, RAGHU RAJ, VIPAN KUMAR and RAKESH KUMAR MAHAJAN✉

Department of Chemistry, UGC-Centre for Advanced Studies, Guru Nanak Dev University, Amritsar-143005, India

✉Corresponding Author

Abstract

The four derivatives of isoquinoline-1,3-dione based on β-lactum (I-IV), have been explored as neutral ionophores for preparing poly(vinylchloride) based polymeric membrane electrodes (PME) selective to silver(I) ions. The addition of sodium tetraphenylborate (NaTPB) and dioctylsebacate (DOS) as a plasticizer was found to improve the performance of ion selective electrodes. The best performance was obtained with PME-1 based on ionophore I having composition: ionophore (9.2 mg), PVC (100.1 mg), DOS (201.1 mg) and NaTPB (1.5 mg) in 5 mL tetrahydrofuran. The electrode response was linear with Nernstian slope of 58.44 mV/decade in the concentration range of 1.0 x 10^{-1} M to 5.0 x 10^{-6} M and detection limit of 5.83 x 10^{-6} M. It performs satisfactorily over wide pH range of 1.0-5.5. The proposed sensor can be used over a period of more than three months without any significant drift in potential and shows good selectivity to silver(I) ion over a number of cations especially with no interference of mercury(II) ions. Sharp end point was obtained when the sensor was used as an indicator electrode for the potentiometric titration of silver(I) ions with chloride ions and therefore this electrode (PME-1) could be used for quantitative determination of silver(I) ion in synthetic water, silver foil and dental amalgam samples.

Keywords

Isoquinoline-1, 3-diones, Silver(I) ion, Potentiometric sensors, Selectivity, Indicator Electrode.

1. Introduction

Silver has wide commercial importance in photography, medicine, optics, electronics, and production of currency, jewelry and silverware as well as high capacity silver-zinc and silver-cadmium batteries [1,2]. Because of its antibacterial properties, silver compounds have been used to disinfect drinking water and water for recreational purposes, in dental, pharmaceutical, antibacterial, anti-HIV preparations and implanted prostheses [3]. Soluble silver salts are lethal and slowly absorbed by body tissues, with consequent bluish or blackish skin pigmentation (argyria), and causes severe corneal injury if liquid comes in contact with the eyes. It may produce other toxic effects including kidney, eye, lung, liver and brain damage as well as change in blood cells. Metallic silver, however, appears to pose a minimal risk to health [4]. It has been found that an excess of silver is toxic to fish and micro-organisms at a concentration as low as 0.17 $\mu g\ L^{-1}$ [3]. It is therefore crucial to closely monitor the activity of silver in the environment. There exist several techniques for silver metal ion determination, such as inductively coupled plasma atomic emission spectrometry [5,6], inductively coupled plasma mass spectrometry [7,8] and flame absorption spectrometry [9,10], but these techniques are normally time consuming, involving sample manipulation, and they are relatively expensive. Ion-selective electrodes (ISEs) have been established for many years as powerful analytical tools. Carrier-based ISEs have found widespread use for the direct determination of various ionic species in complex samples [11-15]. The ISE is an ion analysis technique that provides unique characteristics such as selectivity, good precision, simplicity and low cost [16-19]. ISEs have been successfully applied for trace analysis of ions of environmental and physiological importance [20-22]. The determination of silver(I) ions in presence of mercury(II) ions is often a challenging job. Silver selective potentiometric sensors based on ion carriers like podands, crown ethers and calixarenes derivatives have been reported [2,23-32]. In addition, bis-pyridinetetramide macrocyclic [33], bis-dialkyldithiophosphate derivative [34] and O,O,O-trialkylphosphorothioates [35] have been reported as effective ionophores to construct silver(I) ion selective electrodes. These ionophores are generally hampered with mercury(II) ions interference. In this work, we report the use of isoquinoline-1,3-dione derivatives based on β-lactum as a highly selective neutral ion carrier for detection of silver(I) ions using membrane ion selective electrodes. The need for highly sensitive and selective determination of silver(I) ions arises from silver's economic value and the toxicity toward human beings. In present work, an attempt has been made to improve the sensitivity and selectivity of silver(I) ions ISE towards various secondary ions particularly mercury(II) ions. The electrochemical selectivity for various metal ions, the effect of membrane components, internal solution and pH of sample solution on the electrode response was investigated. Further, the electrode was also used as indicator electrode in titration experiments.

2. Experimental

2.1. Materials

All reagents used were of analytical grade, and doubly distilled deionised water was used to prepare solutions. All isoquinoline-1,3-dione derivatives *i.e.* (4RS,4aSR)-4-[(SR)phenyl(p-tolylamino)methyl)-4a,7,8,8a-tetrahydro-4H-isoquinoline-1,3-dione [I], (4RS,4aSR)-4-[(SR)p-tolyl(p-tolylamino)methyl)-4a,7,8,8a-tetrahydro-4H-isoquinoline-1,3-dione [II], (4RS,4aSR)-4-[(SR)4-chloro-phenyl(p-tolylamino)methyl]-4a,7,8,8a-tetrahydro-4H-isoquinoline-1,3-dione [III], (4RS,4aSR)-4-[(SR)4-chloro-phenylamino(p-tolyl)methyl]-4a,7,8,8a-tetrahydro-4H-isoquinoline 1,3-dione [IV] were synthesied as reported in literature [36] and they are shown in Fig. 1. The plasticizers bis-2-ethylhexylsebacate (DOS), bis-2-ethylhexylphthalate (DOP), bis-2-ethylhexyladipate (DOA), dibutylphthalate (DBP) and tributylphosphate (TBP) and high molecular weight poly(vinylchloride)(PVC) were used as received from Fluka. Anion excluder sodium tetraphenylborate (NaTPB) was obtained from Aldrich while silver nitrate and other metal nitrates received from Merck were used without any further purification.

2.2. Preparation of PVC membrane

The general procedure for preparation of PVC membrane was to mix thoroughly PVC, plasticizer, additive and varying amount of isoquinoline-1,3-dione based ionophore in about 5 mL of tetrahydrofuran. Mixture was shaken vigorously and clear solution was poured into petri-dish (50 mm diameter). Solvent was allowed to evaporate at room temperature. The resulting membrane of 0.4 mm thickness was cut to size, attached to PVC tube with help of PVC glue and conditioned with metal ion solution (10^{-2} M) for about 24 hours till it gave reproducible and stable potential.

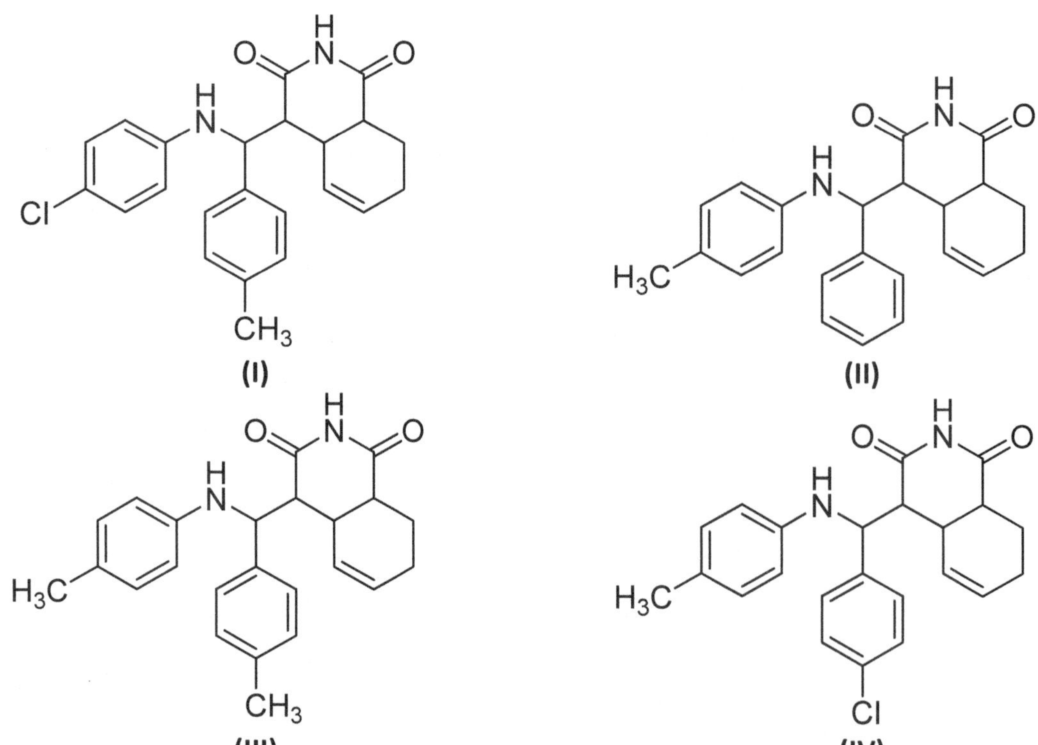

Figure 1. *Structures of isoquinoline-1,3-dione receptors I-IV*

2.3. EMF measurements

Silver/silver chloride electrodes with 3 M KCl as salt bridge were used as internal and external reference electrodes. The electrochemical cell assembly used for the study was:

Ag-AgCl	3 M KCl	1.0 x 10^{-2} M AgNO3	PVC membrane	Test Solution	3 M KCl	Ag-AgCl

All the measurements of electrode potential were made with EQUIPTRONICS MODEL EQ-602 potentiometer. The pH measurements were made using Elico LI MODEL-120 pH meter.

3. Results and Discussion

3.1. Potentiometric response for multiple-ions using polymeric membrane electrodes

In preliminary experiments, different PVC-based ion selective membrane sensors, using iso-quinoline-1,3-dione based receptor I as ion carrier, were prepared by mixing 100.1 mg PVC, 201.1 mg DOS and 9.2 mg receptor I. Membranes so prepared were conditioned with 10^{-2} M solution of different ions for 24 hours. Their potential responses were noted against 10^{-1} M to 10^{-8} M solutions of respective analyte ions to determine their potentiometric response for different metal ions (Figure 2). Silver(I) ions were found to give better response in terms of slope, working concentration range and lower detection limit as compared to other metal ions. Results obtained indicate that electrode based on receptor I has preferential affinity for silver(I) ions and it can be employed as an ionophore for the preparation of silver(I) ion selective electrodes. The preferential affinity of receptor I for silver(I) ions can be due to the presence of nitrogen and oxygen as donor atoms.

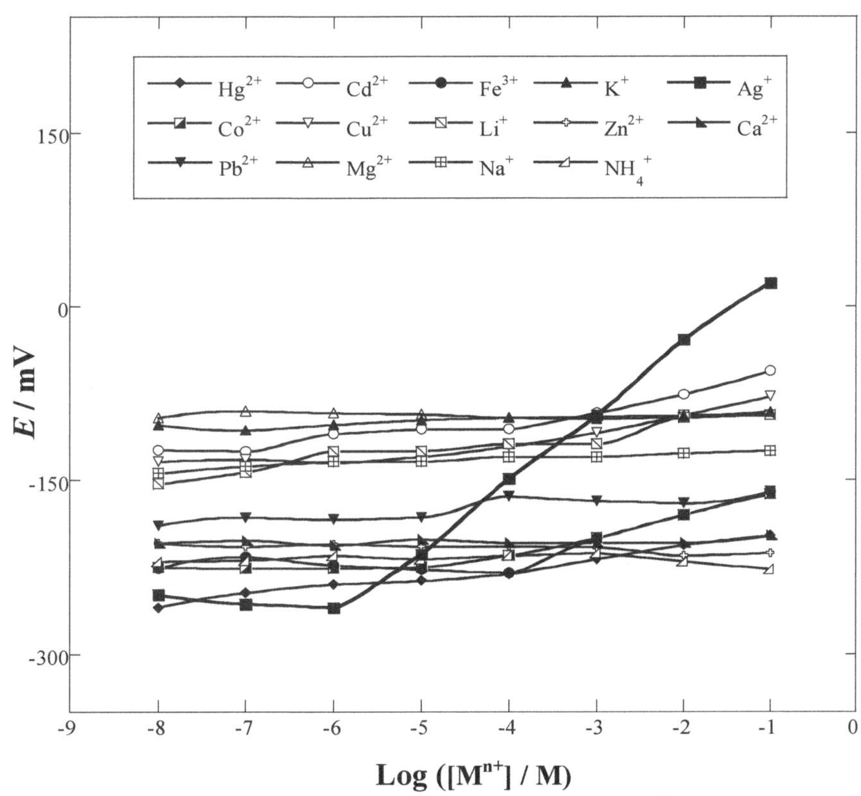

Figure 2. *Potentiometric response of polymeric membrane electrodes based on isoquinoline-1 3-dione receptor I for various metal ions*

3.2 Potential response for silver(I) ions using polymeric membrane electrodes

Based on the results obtained from the preliminary investigation with receptor I, it was decided to employ isoquinoline-1,3-dione based receptors I-IV as ion carrier for the preparation of silver(I) ion selective electrodes. Response curves of polymeric membrane electrodes PMEs 1-4 incorporating receptors I-IV, respectively, as ion carriers are shown in Fig. 3. The composition and the response characteristics obtained for the polymeric membrane electrodes PMEs 1-4 are shown in Table 1.

Table 1. *Composition and response characteristics of silver(I) ion selective electrodes based on isoquinoline-1,3-dione receptors I-IV as ionophores*

S. No.	PVC, mg	Plasticizer, mg	NaTPB, mg	Ionophore, mg	Internal Solution, M	Linear Range, M	Detection Limit , M	Slope, mV dec^{-1}
PME-1	100.1	201.1 **DOS**	1.5	9.2 I	1.0x10^{-2}	1.0x10^{-1}-5.0x10^{-6}	5.83x10^{-6}	58.44
PME-2	100.6	199.8 **DOS**	1.5	9.0 II	1.0x10^{-2}	1.0x10^{-1}-1.0x10^{-5}	9.54x10^{-6}	52.72
PME-3	100.0	201.7 **DOS**	1.6	9.2 III	1.0x10^{-2}	1.0x10^{-1}-1.0x10^{-5}	1.15x10^{-5}	49.62
PME-4	100.3	199.6 **DOS**	1.6	8.9 IV	1.0x10^{-2}	1.0x10^{-1}-1.0x10^{-5}	1.07x10^{-5}	67.47
PME-5	100.2	201.0 **DOS**	1.5	9.2 I	1.0x10^{-3}	1.0x10^{-1}-1.0x10^{-5}	1.01x10^{-5}	53.20
PME-6	100.0	201.2 **DOS**	1.5	9.2 I	1.0x10^{-1}	1.0x10^{-1}-5.0x10^{-5}	8.12x10^{-5}	44.96
PME-7	100.0	201.0 **DOS**	1.5	7.0 I	1.0x10^{-2}	1.0x10^{-1}-1.0x10^{-5}	3.16x10^{-5}	44.50
PME-8	100.5	200.2 **DOS**	1.6	5.2 I	1.0x10^{-2}	1.0x10^{-1}-1.0x10^{-5}	1.07x10^{-5}	49.09
PME-9	101.0	201.5 **DOS**	1.5	3.2 I	1.0x10^{-2}	1.0x10^{-1}-5.0x10^{-6}	6.45x10^{-6}	27.52
PME-10	100.4	200.7**DOS**	3.0	9.0 I	1.0x10^{-2}	1.0x10^{-1}-1.0x10^{-5}	1.05x10^{-5}	50.31
PME-11	100.8	200.3 **DOS**	-	9.1 I	1.0x10^{-2}	1.0x10^{-1}-5.0x10^{-6}	4.36x10^{-6}	38.09
PME-12	100.0	200.0 **DBP**	1.5	9.2 I	1.0x10^{-2}	1.0x10^{-1}-1.0x10^{-5}	1.90x10^{-5}	29.25
PME-13	100.0	200.3 **DOA**	1.5	9.2 I	1.0x10^{-2}	1.0x10^{-1}-5.0x10^{-6}	8.91x10^{-5}	71.86
PME-14	99.9	200.1 **TBP**	1.6	9.0 I	1.0x10^{-2}	1.0x10^{-1}-1.0x10^{-5}	5.01x10^{-6}	77.56
PME-15	100.6	201.4 **DOP**	1.6	9.1 I	1.0x10^{-2}	1.0x10^{-1}-1.0x10^{-5}	1.77x10^{-5}	17.39

The silver(I) ion ISEs PME-2, PME-3 and PME-4 exhibited Nernstian slopes of 52.72, 49.62 and 67.47 mV dec^{-1} , respectively, in the concentration range of 1.0×10^{-1} M to 1.0×10^{-5} M silver(I) ions and produced lower detection limits of the order of 9.54×10^{-6} M, 1.15×10^{-5} M and 1.07×10^{-5} M, respectively. Electrode PME-1 based on receptor I shows Nernstian slope of 58.44 mV dec^{-1} over concentration range of 1.0×10^{-1} M to 5.0×10^{-6} M silver(I) ions with 5.83×10^{-6} M as lower detection limit.

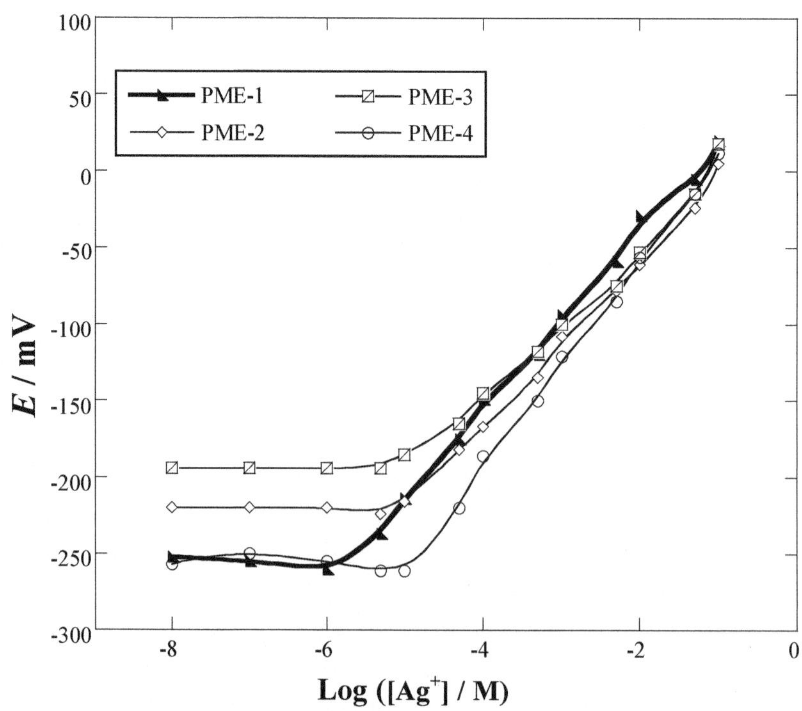

Figure 3. *Potentiometric responses of silver(I) selective polymeric membrane electrodes based on isoquinoline-1,3-dione receptors I-IV as ionophores*

3.3. Effect of internal solution

The effect of the concentration of internal solution on the potential response of the PME-1, PME-5 and PME-6 for silver(I) ion based on receptor I was studied. The internal solution concentration of silver(I) ion was varied from 1.0×10^{-1} M to 1.0×10^{-3} M and the potential response of PME-1 was observed as shown in Table 1. The concentration of internal solution has been found to affect the slope and the working range of the electrode. Further, it was found that the best results, in terms of Nernstian slope and the working concentration range , were obtained with the internal solution of concentration 1.0×10^{-2} M. Thus, 1.0×10^{-2} M solution was found to be quite appropriate for the smooth functioning of the proposed polymeric electrode PME-1.

3.4. Effect of ionophore content

The influence of the ionophore content on the response behavior of silver(I) ion selective electrode incorporating receptor I as an ion carrier was studied by preparing membrane electrodes PME-1, PME-7, PME-8 and PME-9, using varying amount of receptor I (Table 1). Polymeric membrane electrode PME-1 containing 9.2 mg of receptor I exhibited Nernstian slope of 58.44 mV dec^{-1} over the wide concentration range of 1.0×10^{-1} M to 5.0×10^{-6} M with lower detection limit of 5.83×10^{-6} M, whereas, polymeric membrane electrodes PME-7, PME-8 and PME-9, containing 7.0 mg, 5.2 mg and 3.2 mg of receptor I, exhibited sub-Nernstian slopes over comparatively narrow concentration range with lower detection limit of 3.16×10^{-5} M, 1.07×10^{-5} M and 6.45×10^{-6} M, respectively (Table 1). The best result was obtained in case of PME-1 containing 9.2 mg content of receptor I.

3.5. Effect of additive content

The presence of lipophilic negatively charged additive improve the potentiometric behavior of cation-selective electrodes by reducing ohmic resistance, increasing the membrane sensitivity in case of ionophore whose extraction capability is poor [37] and by catalyzing the exchange kinetics [38]. Thus the presence of lipophilic negatively charged additive improve the response behavior and selectivity of ion selective electrodes [37,39,40]. In our study sodium tetra phenyl borate (NaTPB) was incorporated as an additional membrane component (Table 1). Polymeric membrane electrodes PME-1 and PME-10 were prepared by using 1.5 mg and 3.0 mg of NaTPB and PME-11 was prepared without any additive (NaTPB). Response characteristics for the polymeric membrane electrodes PME-1, PME-10 and PME-11 incorporating varying amount of NaTPB are given in Table 1. It is clear from Table 1 that 1.5 mg NaTPB is the optimal content required for proper functioning of receptor I based silver(I) ion selective electrode. It may be due to the fact that 1.5 mg of NaTPB is the appropriate amount for the charge compensation of counter ion in the membrane and thus facilitates the process of the ion charge transduction. On the other hand, PME-11 which contains no lipophilic additive shows sub-Nernstian slope. That may be due to the presence of anionic impurities within the polymer matrix [41].

3.6. Effect of nature of plasticizer

The dielectric constant of the membrane phase, the mobility of ionophore molecules and the state of ligands are dependent on the nature of the plasticizer incorporated in the membrane [14,42]. The selectivity and the sensitivity of the selective electrode strongly depend on the membrane composition and the nature of plasticizer used [11,43]. The effect of plasticizer on the potential response of silver(I) ion ISE incorporating isoquinoline-1,3-dione receptor I as an ion carrier was studied by preparing different membrane sensors containing plasticizer such as DOS,

DBP, DOA, TBP, DOP and the results are summarized in Table 1. The electrodes PME-13 and PME-14, prepared by incorporating DOA and TBP, exhibited super-Nernstian slopes (71.86, 77.56 mV dec^{-1}), while DBP and DOP incorporating electrodes PME-12 and PME-15 exhibited sub-Nernstian response (29.25, 17.39 mV dec^{-1}). The response of membrane sensor PME-1 incorporating DOS as a plasticizer was found to be better in terms of Nernstian slope of 58.44 mV dec^{-1} and wide linear concentration range of 1.0×10^{-1} M to 5.0×10^{-6} M.

3.7. Effect of pH

The potential response of an ion selective electrode is affected by pH of the analyte ion solution. The pH dependence of the polymeric membrane electrodes PMEs 1-4 containing receptors I-IV as ionophores were examined at 1.0×10^{-2} M concentration of silver(I) ions as shown in Fig. 4. The pH values were adjusted using conc. nitric acid and hexamine and the potential was measured after each addition. The electrode potential for all four electrodes remained constant over a pH range 1.0 to 5.5, which was the working pH range of the proposed silver(I) ion electrode. At lower pH<1 deviation in potential may be due to the hydrogen error of the indicator electrode, and the sharp change at higher pH values may be due to the formation of some hydroxyl complexes of the charge transport process by the membrane, thereby causing interference.

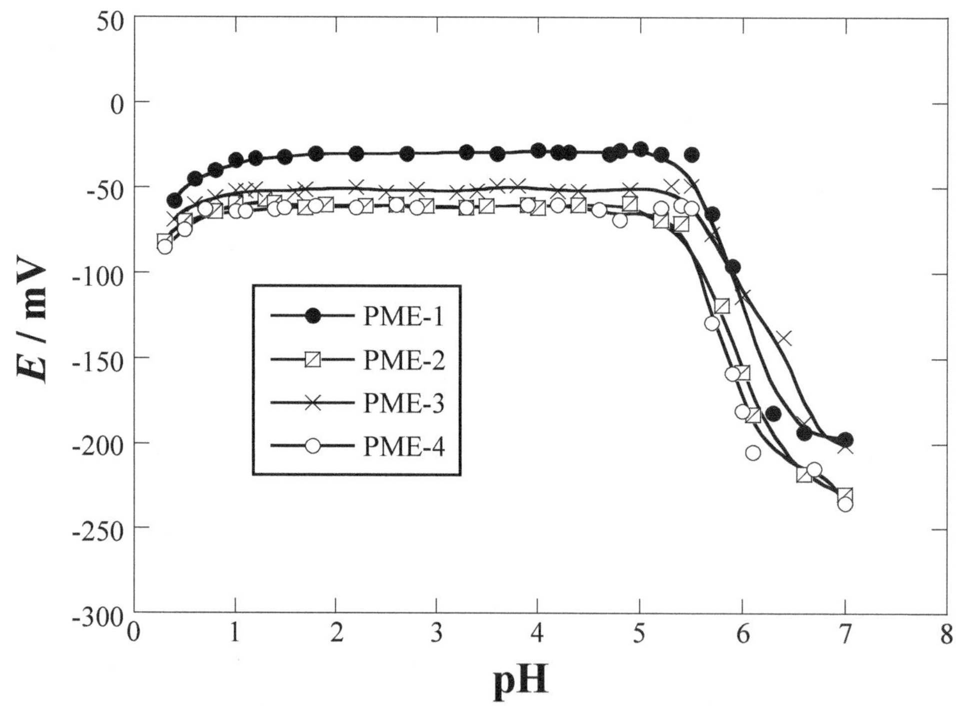

Figure 4. *Effect of pH on potential response of silver(I) ion selective polymeric membrane electrodes PME 1-4 based on isoquinoline-1,3-dione receptors I-IV as ionophores*

3.8. Response time and life time of proposed electrode

The response time of an ion selective electrode is the average time required for attaining the equilibrium value of the potential. In this study the practical response time was recorded from lower to higher successive concentrations of silver(I) ions solutions (10^{-5} M to 10^{-1} M). A potential versus time trace for electrode PME-1 is shown in Fig. 5. It is evident that the electrode reached the equilibrium response in a very short time of less than 12 seconds and no change was observed up to 5 minutes. The life time of the present electrode was more than three months and electrode potentials were reproducible. After more than three months the selectivity and sensitivity of polymeric membrane started degrading. It may be due to the leaching out of ionophores from the membrane. The response time of the proposed sensor PME-1 is compared with the best silver(I) ion-selective electrodes reported earlier in literature (Table 2). It is clear from Table 2 that proposed electrode is better as compared to other reported electrodes in regard to Nernstian slope, lower detection limit, pH range and working concentration range.

Table 2. *A comparative study of the proposed silver(I) ion selective electrodes with previously reported ion selective electrodes for silver(I) ion*

Ionophore	Linear Range, M	Detection Limit, M	pH Range	Slope, mV dec^{-1}	Response time, s	Ref. No.
Schiff-base ($N^2E,N^{2'}E$)-$N^2,N^{2'}$-bis(thiophen-2-ylmethylene)-1,1'-biNaphthyl-2,2'-diamine.	1.0×10^{-7}- 1.0×10^{-2}	-[a]	2.0-6.0	57.40	9	[44]
Calix[4]arene.	1.0×10^{-5}- 1.0×10^{-2}	-[a]	>2	56.00	-[a]	[45]
Schiff-base–p-*tert* butyl-calix[4]arene.	1.0×10^{-5}- 1.0×10^{-1}	6.3×10^{-6}	1.0-6.0	58.90	30	[32]
Calix[4]arene.	1.0×10^{-1} - 1.0×10^{-4}	1.0×10^{-4}	-[a]	50.01	3	[46]
Schiff-baseN,N′-Bis(pyridine-2-Ylmethylene)benzene-1,2-diamine.	1.0×10^{-7}- 1.0×10^{-1}	-[a]	3.0-8.0	58.60	10	[47]
Isoquinoline-1, 3-diones	5.0×10^{-6}- 1.0×10^{-1}	5.83×10^{-6}	1.0-5.5	58.44	<12	This work

[a] not available in the literature

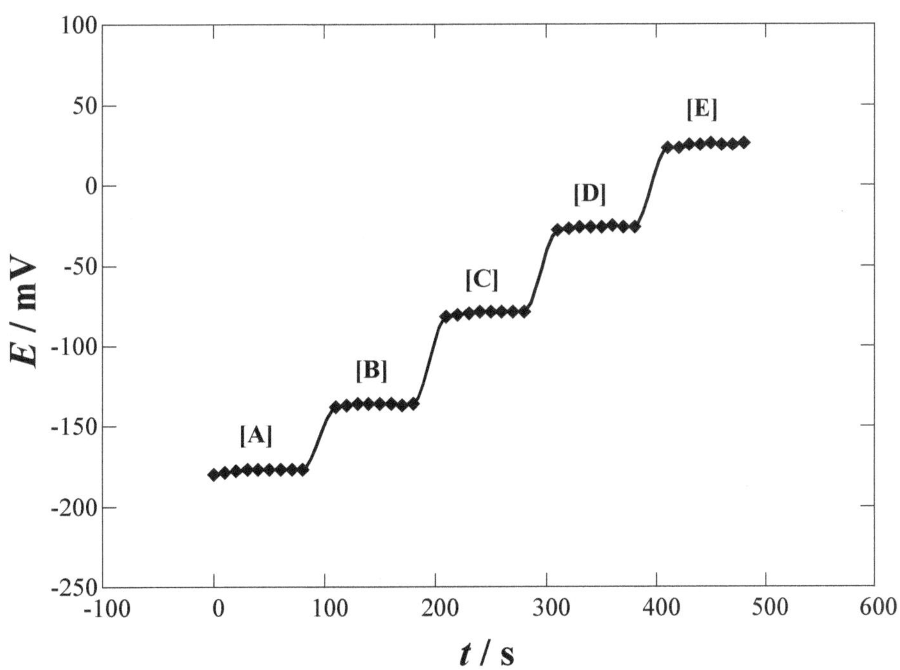

Figure 5. *Response time of silver(I) ion selective electrode PME-1, based on isoquinoline-1,3-dione receptor I for step changes in concentration of silver(I) ion. (A) 1.0×10^{-5} M (B) 1.0×10^{-4} M (C) 1.0×10^{-3} M (D) 1.0×10^{-2} M (E) 1.0×10^{-1}M.*

3.9. Electrode selectivity

The most important characteristic of any ion-selective sensor is its response to the primary ions (A) in the presence of the secondary ions (B) in the solution, which is expressed in terms of the potentiometric selectivity coefficients ($K^{pot}_{A,B}$). In the present work the selectivity coefficients ($LogK^{pot}_{A,B}$) were determined using fixed interference method (FIM) based on the semi empirical Nikolsky-Eisenman equation [48]. In FIM, the selectivity coefficient was evaluated from potential measurements in solutions containing a fixed concentration of interfering cation (1.0×10^{-2} M) and varying amount of silver(I) ions. The potential values obtained are plotted versus the activity of the primary silver(I) ions. The potentiometric selectivity coefficients for proposed silver(I) ion selective electrodes PMEs 1-4 were evaluated by FIM as shown in Figure 6. Comparing selectivity data of polymeric membranes PMEs 1-4, it has been found that the $LogK^{pot}_{Ag^+,B}$ values for most of the secondary ions are better in case corresponding to PME-1. As seen from Fig. 6 the silver(I) ion electrode PME-1 is significantly more selective to silver(I) ions over all the interfering ions. Although, the selectivity coefficient value for mercury(II) ion is higher than the values for other secondary metal ions ($LogK^{pot}_{Ag^+,Hg^{2+}}$ = -2.81) still, it has been found that a mercury(II) ion does not interfere in the normal functioning of the proposed silver(I) ion selective electrode PME-1. Moreover, it is found that most of the studied cations such as K^+, Zn^{2+}, Fe^{3+}, Cu^{2+}, Co^{2+}, Pb^{2+}, NH_4^+, Al^{3+}, Mg^{2+}, Na^+, Li^+ and Cd^{2+} would not cause any interference in determination of silver ions, even if present in large amount.

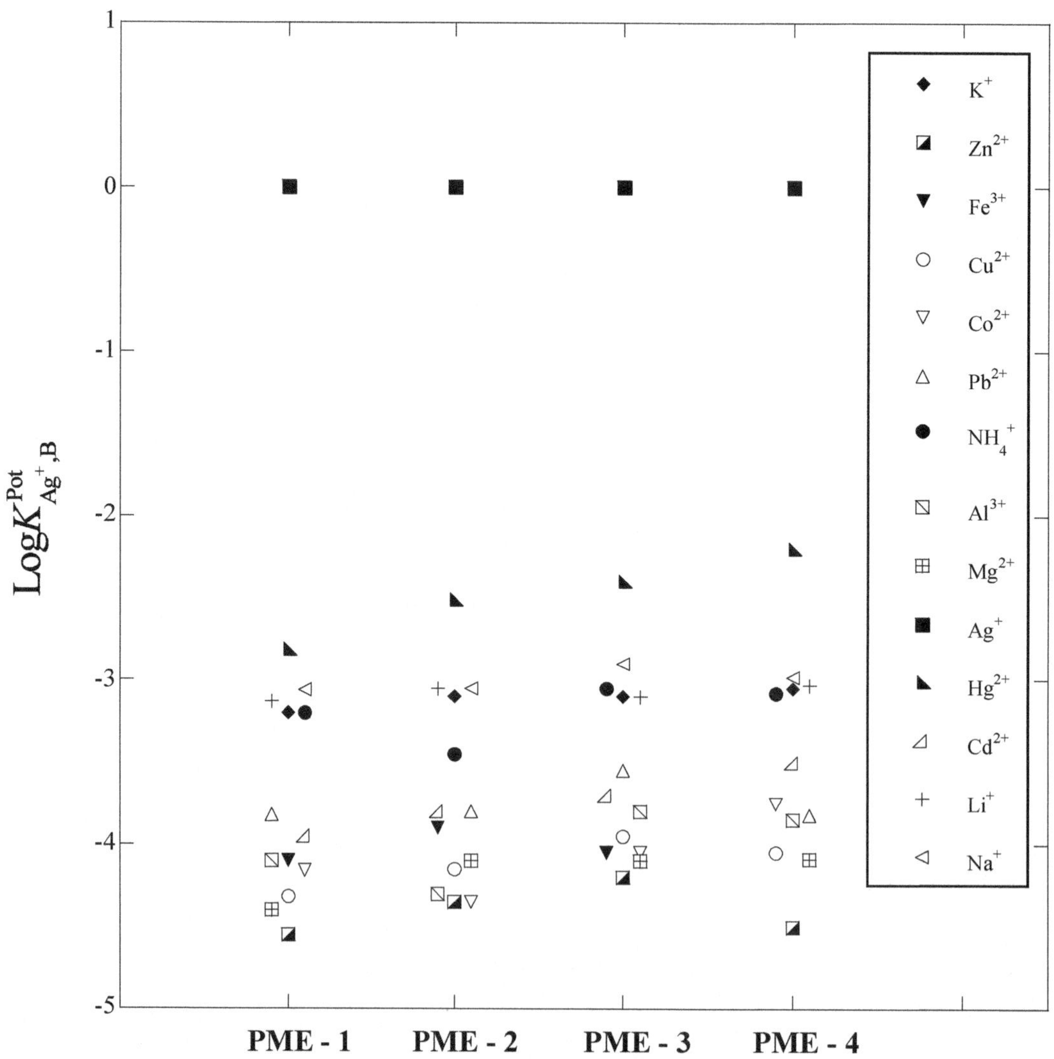

Figure 6. *Selectivity of silver(I) ion sensors based on isoquinoline-1,3-dione receptors I-IV*

4. Analytical application

The optimized proposed polymeric membrane sensor PME-1 was found to work well under laboratory conditions. Practical utility of the proposed sensors was tested by using them as indicator electrode for potentiometric titration of silver solution (1.0×10^{-2} M) with sodium chloride solution (1.0×10^{-2} M) as shown in Fig.7. The plot is of sigmoidal shape and the inflexion point of the plot corresponds to 1:1 stoichiometry of the AgCl precipitation. Therefore, the end point and the amount of silver(I) ions in the solution can be accurately determined by extrapolation of the three linear portions of the titration plot. Also, PME-1 was employed for the quantitative determination of silver(I) ion content in synthetic water (1 and 2), silver foil (3) and dental amalgam (4) samples. In case of silver foil, the sample was prepared by desolving it in diluted nitric acid (3 ml of double distilled water and 5 ml of conc. HNO_3), and then diluted to 100 ml by adding double distilled water. Silver(I) ion content in these samples was also determined volumetrically. The results summarized in Table 3 make it clear that both of these methods of analysis give compatible results providing the practical significance of proposed electrode.

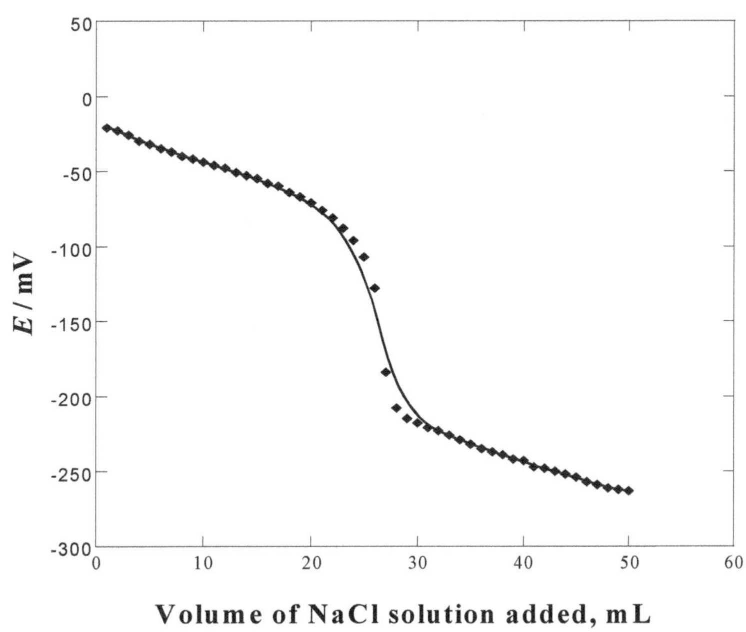

Figure 7. *Potentiometric titration of 1.0 x 10^{-2} M silver(I) solution with 1.0 x 10^{-2} M NaCl using PME-1, based on isoquinoline-1,3-dione receptor I*

Table 3. *Estimation of silver(I) ion content in synthetic water, silver foil and dental amalgam samples*

Sample	Silver content, M		
	Volumetric method	Estimated using PME-I	Compatibility, %
Synthetic Water (1)	3.72×10^{-3} M	3.86×10^{-3} M	103.76
Synthetic Water (2)	6.69×10^{-4} M	6.58×10^{-4} M	98.36
Silver foil (3)	9.13×10^{-3} M	9.23×10^{-3} M	98.91
Dental amalgam(4)	1.67×10^{-2} M	1.61×10^{-2} M	96.41

5. Conclusion

Isoquinoline-1,3-dione based receptor I has been found to be novel carrier of silver(I) ions in the PVC based polymeric membrane PME-1. The main advantage of the electrode is the simplicity of construction. It has good response properties and performance, with Nernstian response (58.44 mV/decade) in the concentration range of the 1.0×10^{-1} M to 5.0×10^{-6} M and detection limit of 5.83×10^{-6} M. The sensors have working pH range of 1.0 to 5.5 with fast response time of less than 12 seconds. The selectivity coefficient indicated that PME-1 sensor is significantly selective to silver(I) ion over many different secondary ions. Even mercury(II) ions were not found to interfere in normal functioning of the proposed PVC membrane PME-1. Additionally, the life time of this electrode longer than three months could be of importance as well. The electrode was successfully applied as an indicator electrode for potentiometric titration of silver(I) ion solution against sodium chloride solution and it can be used for determination of silver(I) ion content in different real samples.

Acknowledgements: *This paper was financially supported by the Council of Scientific and Industrial Research (CSIR), New Delhi as a part of project (01)2469/11/EMR-II. One of us AK is thankful to CSIR for the award of Junior Research Fellow (JRF).*

References

[1] E. Malinowska, Z. Brzozka, K. Kasiura, R. J. M. Egberink, D. N.Reinhoudt, *Anal. Chim. Acta* **298** (2004) 810-817.
[2] L. Chen, H. Ju, X. Zeng, X. He, Z. Zhang, *Anal. Chim. Acta* **437** (2001) 191-197.
[3] C. Z. Lai, M. A. Fierke, R. C. D. Costa, J. A. Gladysz, A. Stein, P. Buhlmann. *Anal. Chem.* **82** (2010) 7634-7640.
[4] P. L. Drake, K. J. Hazelwood, *Ann. Occup. Hyg.* **49** (2005) 575-585.
[5] T. Wang, X. Jai, J. Wu, *J. Pharma. Biomed. Anal.* **33** (2003) 639-646.
[6] X. J. Yang, R. Foley, G. K. C. Low, *Analyst* **127** (2001) 505-507.
[7] X. Liu, W. Li, Q. Shen, Z. Nie, M. Guo, Y. Han, W. Liu, S. Yao, *Talanta* **85** (2011) 1603-1608.
[8] M. Krachler, C. Mohl, H. Emons, W. Shotyk, *Spectrochim. Acta* **57** (2002) 1277-1289.
[9] P. B. Barrera, A. M. Pineiro, A. B. Barrera, *Talanta* **43** (1996) 35-44.
[10] S. Dadfarnia, A. M. H. Shabani, M. Gohari, *Talanta* **64** (2007) 682-687.
[11] E. Bakker, P. Buhlmann, E. Pretsch, *Electroanalysis* **11** (1999) 915-933
[12] E. Bakker, D. Diamond, A. Lewenstam, E. Pretsch, *Anal. Chim. Acta* **393** (2001) 191-197
[13] E. Bakker, E. Pretsch, P. Buhlmann, *Anal. Chem.* **72** (2000) 1127-1133.
[14] E. Bakker, P. Buhlmann, E. Pretsch, *Chem. Rev.* **97** (1997) 3083-3132.
[15] S. Shahrokhian, A. Hamzehloei, M. Bagherzadeh, *Anal. Chem.* **74** (2002) 3312-3320.
[16] Z. Szigeti, A. Malon, T. Vigassy, V. Csokai, A. Grun, K. Wygladacz, N. Y. C. Xu, V. J. Chebny, I. Bitter, R. Rathore, E. Bakker, E. Pretsch, *Anal. Chim. Acta* **572** (2006) 1-10.
[17] A. Ceresa, A. Radu, S. Peper, E. Bakker, E. Pretsch, *Anal. Chem.* **67** (1995) 1654-1660.
[18] A. Malon, T. Vigassy, E. Bakker, E. Pretsch, *J. Am. Chem. Soc.* **128** (2006) 8154-8155.
[19] R. S. Hutchins, L. G. Bachas, *Anal. Chem.* **67** (1995) 1654-1660.
[20] A. Ceresa, E. Bakker, B. Hattendorf, D. Gunther, E. Pretsch, *Anal. Chem.* **73** **(**2001) 343-351.
[21] A. Radu, M. T. Diaz, E. Bakker, *Anal. Chem.* **75** (2003) 6922-6931.
[22] M. Shamsipur, M. Javanbakht, V. Lippolis, A. Garau, G. D. Filippo, M. R. Ganjali, A. Yari, *Anal. Chim. Acta* **462** (2002) 225-234.
[23] Y. Liu, B. T. Zhao, L. Chen, X. He, *Microchem. J.* **65** (2000) 75-79.
[24] L. Chen, X. He, B. T. Zhao, Y. Liu, *Anal. Chim. Acta* **417** (2000) 51-56.

[25] S. M. Lim, H. J. Chung, K. J. Paeng, C. H. Lee, H. N. Choi, W. Y. Lee, *Anal. Chim. Acta* **453** (2002) 81-88.
[26] A. Malon, A. Radu, W. Qin, Y. Qin, A. Ceresa, M. M. Zurawska, E. Bakker, E. Pretsch, *Anal. Chem.* **75** (2003) 3865-3871.
[27] S. S. Lee, M. K. Ahn, S. B. Park, *Analyst* **123** (1998) 383-386.
[28] R. K. Mahajan, M. Kumar, V. Sharma, I. Kaur, *Analyst* **126** (2001) 505-507.
[29] M. Kumar, R. K. Mahajan, V. Sharma, H. Singh, N. Sharma, I. Kaur, *Tetrahedron Lett.* **42** (2001) 5315-5318.
[30] R. K. Mahajan, I. Kaur, V. Sharma, M. Kumar, *Sensors* **2** (2002) 417-423.
[31] R. K. Mahajan, I. Kaur, M. Kumar, *Sen. Actuators* **91** (2003) 26-31.
[32] R. K. Mahajan, I. Kaur, R. Kaur, V. Bhalla, M. Kumar, *Bull. Chem. Soc. Jap.* **78** (2005) 1635-1640.
[33] R. K. Mahajan, O. Parkash, *Talanta* **52** (2000) 691-693.
[34] D. Liu, J. Liu, D. Tian, W. Hong, X. Zhou, J. C. Yu, *Anal. Chim. Acta* **416** (2000) 139-144.
[35] D. Xu, T. Katsu, *Anal. Chim. Acta* **443** (2001) 235-240.
[36] R. Raj, V. Mehra, P. Singh, V. Kumar, G. Bhargava, M. P. Mahajan, S. Honda, L. G. M. Slaughter, *Eur. J. Org. Chem.* **2011** (2011) 2697-2704.
[37] T. Rostazin, E. Bakker, K. Suzuki, W. Simon, *Anal. Chim. Acta* **280** (1993) 197-206.
[38] P. Gehrig, W. E. Morf, M. Welti, E. Pretsch, W. Simon, *Helv. Chim. Acta* **73** (1990) 203-212.
[39] R. Eugster, P. M. Gehrig, W. E. Morf, U. E. Spichiger, W. Simon, *Anal. Chem.* **69** (1991) 2285-2289.
[40] U. Schaller, E. Bakker, U. E. Spichiger, E. Pretsch, *Anal. Chem.* **66** (1994) 391-398.
[41] P. Anker, E. Wieland, D. Ammann, R. E. Dohner, R. Asper, W. Simon, *Anal. Chem.* **53** (1981) 1970-1974.
[42] X. Yang, N. Kumar, H. Chi, D. B. Hibbert, P. W. Alexander, *Electroanalysis* **9** (1997) 549-553.
[43] D. Ammann, E. Pretsch, W. Simon, E. Lindner, A. Bezegh, E. Pungor, *Anal. Chim. Acta* **171** (1985) 119-129.
[44] E. Jeong, M. S. Ahmed, H. Jeong, E. Lee, S. Jeon, *Bull. Korean Chem. Soc.* **32** (2011) 800-804.
[45] L. Chen, X. He, H. Zhang, Y. Liu, X. Hu, Hu, Y. Sheng, *Anal. Lett.* **34** (2001) 2237-2248.
[46] K. M. O'Connor, G. Svehla, S. J. Harris, M. A. Mckervey, *Talanta* **39** (1992) 1549-1554.
[47] H. Seo, E. Jeong, M. S. Ahmed, H. K. Lee, S. Jeon, *Bull. Korean Chem. Soc.* **31** (2010) 1699-1703.
[48] D. G. Hall, *J. Phys. Chem.* **100** (1996) 7230-7236.

4

Inhibition of corrosion of carbon steel in well water by arginine-Zn^{2+} system

ANTHONY SAMY SAHAYA RAJA✉ and SUSAI RAJENDRAN*

PG and Research Department of Chemistry, GTN Arts College, Dindigul-624005, Tamil Nadu, India
**Department of Chemistry, RVS School of Engineering and Technology, Dindigul-624005, Tamil Nadu, India*

✉Corresponding Author

Abstract
The environmental friendly inhibitor system arginine-Zn^{2+}, has been investigated by weight-loss method. A synergistic effect exists between arginine and Zn^{2+} system. The formulation consisting of 250 ppm of arginine and 5 ppm of Zn^{2+} offers good inhibition efficiency of 98 %. Polarization study reveals that this formulation functions as an anodic inhibitor. AC impedance spectra reveal that a protective film is formed on the metal surface. The FTIR spectral study leads to the conclusion that the Fe^{2+}- DL-arginine complex, formed on anodic sites of the metal surface, controls the anodic reaction. $Zn(OH)_2$ formed on the cathodic sites of the metal surface controls the cathodic reaction. The surface morphology and the roughness of the metal surface were analyzed with Atomic Force Microscope. A suitable mechanism of corrosion inhibition is proposed based on the results obtained from weight loss study and surface analysis technique.

Keywords
Carbon steel; arginine; corrosion inhibition; synergistic effect; SEM; AFM.

Introduction

The principles and practices of corrosion inhibition have begun in recent years to take into account the health and safety considerations. The use of hazardous chemicals has been restricted to no contact with the environment. Hence, there is a search for non-toxic, eco-friendly corrosion inhibitors. The use of inhibitors is one of the most practical methods to protect metals from corrosion. Corrosion inhibitor is a chemical substance which, when added to the corrosive environment at an optimum concentration, significantly decreases the corrosion rate of metals (or) alloys. Unfortunately, many common corrosion inhibitors are highly toxic and health-hazardable, such as chromates [1], nitrite [2] aromatic heterocyclic compounds [3] etc. Therefore, it is better to look for environmentally safe inhibitors [4-6]. Some researchers investigated the inhibition effect of environment friendly inhibitors like amino acids on metal corrosion [6-13]. This is due to the fact that amino acids are non-toxic, biodegradable, relatively cheap, and completely soluble in aqueous media and produced with high purity at low cost. The environmental friendly arginine is chosen as the corrosion inhibitor for this present work. The literature presents some studies involving amino acids having the ability to prevent the corrosion of iron [14], steel [15-17], aluminum [18,19], nickel [20] and copper [21-25]. The electrochemical studies such as polarization and AC impedance spectra [26-30] and cyclic voltammetry [19] have been studied by using amino acids. The adsorption of amino acids on carbon steel in acidic environment has been investigated by Akiyama *et al.* [31].

Experimental

Preparation of specimens

Carbon steel specimens (0.0267 % S, 0.067 % P, 0.4 % Mn, 0.1 % C and the rest iron) of the dimensions 1.0 cm × 4.0 cm × 0.2 cm were polished to mirror finish and degreased with trichloroethylene and as such used for weight-loss method and surface examination studies.

Weight - loss method

Relevant data on the well water used in this study are given in Table 1. Carbon steel specimens, in triplicate were immersed in 100 ml of well water and various concentrations of arginine in the presence and absence of Zn^{2+} (as $ZnSO_4 \times 7H_2O$) for a period of seven days. The corrosion products were cleaned with Clarke's solution [32]. The weight of the specimens before and after immersion was determined using Shimadzu balance AY62. The corrosion inhibition efficiency (*IE*) was calculated using the equation (1).

$$IE = 100\,[1 - (W_2/W_1)] \tag{1}$$

where W_1 is the corrosion rate in the absence of the inhibitor and W_2 is the corrosion rate in the presence of inhibitor. From the weight loss, the corrosion rate (mm year^{-1}) was calculated using the equation (2):

$$CR = \frac{LW}{A\,t\,0.0365\rho^{-1}} \tag{2}$$

where CR is corrosion rate in mm year^{-1}; LW is the weight loss in mg; A is surface area of the specimen in dm^2; t is period of immersion in days and ρ is density of the metal in g cm^{-2} (7.86).

Potentiodynamic polarization study

Potentiostatic polarization studies were carried out using a CHI electrochemical impedance analyzer, model 660 A. A three-electrode cell assembly was used. The working electrode was a rectangular specimen of carbon steel with one face of the electrode (1 cm^2 area) exposed and the rest shielded with red lacquer. A saturated calomel electrode (SCE) was used as the reference electrode and a rectangular platinum foil was used as the counter electrode. Polarization curves were recorded using *iR* compensation. The results, such as Tafel slopes, and I_{corr}, E_{corr} and linear polarization resistance (LPR) values were calculated. During the polarization study, the scan rate was 0.01 V s^{-1}; hold time at E_f was 0 s and quiet time was 2 s.

AC impedance measurements

A CHI electrochemical impedance analyzer (model 660A) was used for AC impedance measurements. A time interval of 5 to 10 minutes was given for the system to attain its open circuit potential. The real part Z' and the imaginary part Z'' of the cell impedance were measured in ohms at various frequencies. The values of the charge transfer resistance R_t, double layer capacitance C_{dl} and impedance value were calculated.

$$R_t = (R_s + R_t) - R_s \quad (3)$$

where R_s = solution resistance

$$C_{dl} = ½ \pi R_t f_{max} \quad (4)$$

where f_{max} = maximum frequency

AC impedance spectra were recorded with initial E = 0 V; high frequency = 10^5 Hz; low frequency = 10 Hz; amplitude = 0.005 V; and quiet time = 2 s.

FTIR spectra

The structure of arginine is shown in Fig.1. The carbon steel specimens immersed in various test solutions for seven days were taken out and dried. The film formed on the metal surface was carefully removed and thoroughly mixed with KBr, so as to make it uniform throughout. The FTIR spectra were recorded in a Perkin-Elmer 1600 spectrophotometer.

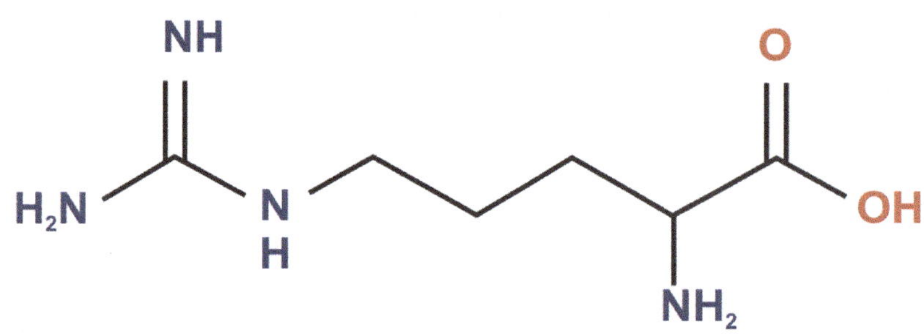

Figure1. *Structure of DL-arginine*

Scanning electron microscopy (SEM analysis)

SEM provides a pictorial representation of the surface. It helps to understand the nature of the surface film in the absence and presence of inhibitors and the extent of corrosion of carbon steel. The scanning electron microscopy photographs were recorded at various magnifications using Hitachi scanning electron microscopy machine S-3000 H.

Atomic force microscopy

Atomic force microscope (AFM) is an exciting new technique that allows surface to be imaged at higher resolutions and accuracies than ever before [33-35]. The microscope used for the present study was VEECO, Lab incorporation. Polished specimens, prior to the initiation of all corrosion experiments, were examined through an optical microscope to find out any surface defects such as pits or noticeable irregularities like cracks, *etc.* Only those specimens, which had a smooth pit free surface, were subjected to AFM examination. The protective films formed on the carbon steel specimens after immersion in the inhibitor systems for different time durations were examined for a scanned area of 05 × 05 µm at a scan rate of 6.68 µm s^{-1}. The two dimensional and three-dimensional topography of surface film gave various roughness parameters of the film.

Results and Discussion

Analysis of the weight loss method

Corrosion rates (*CR*) of carbon steel immersed in well water in the absence and presence of inhibitor (DL-arginine) are given in Tables 1 to 3. The inhibition efficiencies (*IE*) are also given in these Tables. The corrosion rates of the DL-arginine - Zn^{2+} systems as a function of concentrations of DL-arginine are shown in Fig.2.

Table 1. *Corrosion rates (*CR*) of carbon steel immersed in well water in the presence and absence of inhibitor system at various concentrations and the inhibition efficiencies (IE) obtained by weight loss method. Inhibitor system: DL-arginine – Zn^{2+}(amount of Zn^{2+}= 0 ppm); Immersion period: 7 days; pH 8*

Amount of DL-arginine, ppm	*CR* mm year^{-1}	*IE* / %
0	0.0874	--
50	0.0455	48
100	0.0472	46
150	0.0525	40
200	0.0533	39
250	0.0577	34

Table 2. *Corrosion rates (*CR*) of carbon steel immersed in well water in the presence and absence of inhibitor system at various concentrations and the inhibition efficiencies (IE) obtained by weight loss method. Inhibitor system: DL-arginine – Zn^{2+}(amount of Zn^{2+}= 5 ppm); Immersion period: 7 days; pH 8*

Amount of DL-arginine, ppm	*CR* mm year^{-1}	*IE* / %
0	0.0743	15
50	0.0157	82
100	0.0131	85
150	0.0087	90
200	0.0069	92
250	0.001748	98

Table 3 : *Corrosion rates (CR) of carbon steel immersed in well water in the presence and absence of inhibitor system at various concentrations and the inhibition efficiencies (IE) obtained by weight loss method. Inhibitor system: DL-arginine – Zn^{2+}(amount of Zn^{2+}= 10 ppm); Immersion period: 7 days; pH 8*

Amount of DL-arginine, ppm	*CR* mm year^{-1}	*IE* / %
0	0.0699	20
50	0.0280	68
100	0.0218	75
150	0.0192	78
200	0.0175	80
250	0.0122	86

It is observed from Table 1 that DL-arginine shows some inhibition efficiencies. 50 ppm of DL-arginine has 48 percent IE. As the concentration of DL-arginine increases, the IE decreases. This is due to the fact that as the concentration of DL-arginine increases, the protective film (probably iron DL-arginine complex) formed on the metal surface diffuses into solution. That is, the system goes from passive region to active region [36].

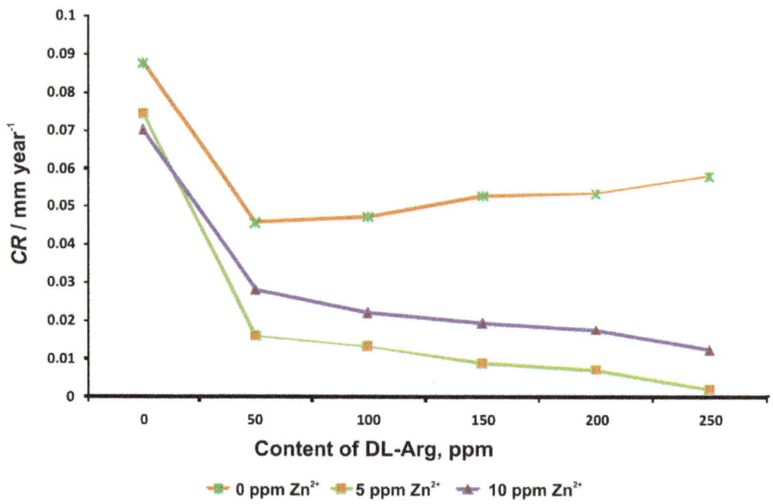

Figure 2. *Corrosion rates of carbon steel immersed in various test solutions*

Influence of Zn^{2+} on the inhibition efficiencies of DL-arginine

The influence of Zn^{2+} on the inhibition efficiencies of DL-arginine is given in Tables 2 and 3. It is observed that as the concentration of DL-arginine increases the IE increases. Similarly, for a given concentration of DL-arginine the *IE* increases as the concentration of Zn^{2+} increases. It is also observed that a synergistic effect exists between DL-arginine and Zn^{2+}. For example, 5 ppm of Zn^{2+} has 15 percent *IE*; 250 ppm of DL-arginine has 34 percent *IE*. Interestingly their combination has a high *IE*, namely, 98 percent.

In the presence of Zn^{2+} more amount of DL-arginine is transported towards the metal surface. This is due to fact that when Zn^{2+} is added to the environment, Zn^{2+}- DL-arginine complex is formed, it diffuses towards metal surface. On the metal surface, iron-amino acid complex is formed and Zn^{2+} is released. This combines with OH^- to give insoluble $Zn(OH)_2$ formed on cathodic sites which is controlled the cathodic reaction. On the metal surface Fe-DL-arginine complex is formed on the anodic sites of the metal surface and anodic reaction is controlled. Thus, the anodic reaction and cathodic reaction are controlled effectively. This accounts for the synergistic effect existing between Zn^{2+} and DL-arginine.

$Fe \rightarrow Fe^{2+} + 2e^-$ (Anodic reaction)

Fe^{2+} + Zn^{2+} - DL-arginine complex $\rightarrow$ Fe^{2+} - DL-arginine complex + Zn^{2+}

$O_2 + 2H_2O + 4e^- \rightarrow 4OH^-$ (Cathodic reaction)

$Zn^{2+} + 2OH^- \rightarrow Zn(OH)_2\downarrow$

Analysis of potentiodynamic polarization study (pH = 8)

Polarization study was used to confirm the formation of protective film formed on the metal surface during corrosion inhibition process [37-42]. If a protective film is formed on the metal surface, the linear polarization resistances value (LPR) increases and the corrosion current value (I_{corr}) decreases.

The potentiodynamic polarization curves of carbon steel immersed in well water in the absence and presence of inhibitors are shown in Fig. 3 and the corrosion parameters are given in Table 4. When carbon steel was immersed in well water the corrosion potential was -668 mV *vs* SCE. When DL-arginine (250 ppm) and Zn^{2+} (5 ppm) were added to the above system, the corrosion potential shifted to the noble side, i.e. to -626 mV *vs* SCE. This indicates that a film is formed on anodic sites of the metal surface. This film controls the anodic reaction of metal dissolution by forming Fe^{2+}- DL-arginine complex on the anodic sites of the metal surface. The formation of protective film on the metal surface is further supported by the fact that the anodic Tafel slope (b_a) increases from 104 to 173 mV.

Table 4 : *Corrosion parameters of carbon steel immersed in well water in the absence and presence of inhibitor system obtained from potentiodynamic polarization study*

System	E_{corr} mV *vs.* SCE	b_c mV decade^{-1}	b_a mV decade^{-1}	I_{corr} A cm^{-2}	LPR Ω cm^2
Well water	-668	268	104	5.775×10^{-7}	5.630×10^{4}
Well water + DL-Arg (250 ppm) + Zn^{2+} (5 ppm)	-626	128	173	5.125×10^{-7}	6.253×10^{4}

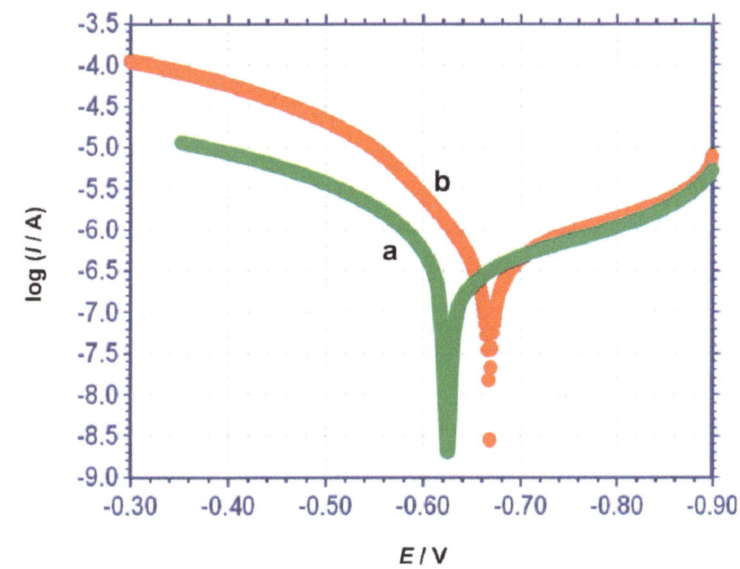

Figure 3. *Polarization curves of mild steel immersed in various test solutions:* ***(a)*** *well water + DL-arginine (250 ppm) + Zn^{2+} (5 ppm);* ***(b)*** *well water (blank)*

Further, the LPR value increases from 5.630×10^4 ohm cm^2 to 6.253×10^4 ohm cm^2; the corrosion current decreases from 5.775×10^{-7} A cm^{-2} to 5.125×10^{-7} A cm^{-2}. Thus, polarization study confirms the formation of a protective film on the metal surface.

Analysis of AC impedance spectra

AC impedance spectra (electrochemical impedance spectra) have been used to confirm the formation of a protective film on the metal surface [43-45]. If a protective film is formed on the metal surface, charge transfer resistance (R_t) increases; double layer capacitance value (C_{dl}) decreases and the impedance log (Z/Ω) value increases. The AC impedance spectra of carbon steel immersed in well water in the absence and presence of inhibitors (DL-arginine-Zn^{2+}) are shown in Figs. 4a and 4b (Nyquist plots) and Figs. 5a and 5b and Figs. 6a and 6b (Bode plots). The AC impedance parameters namely charge transfer resistance (R_t) and double layer capacitance (C_{dl}) derived from Nyquist plots are given in Table 5. The impedance log (Z/Ω) values derived from Bode plots are also given in Table 5.

a **b**

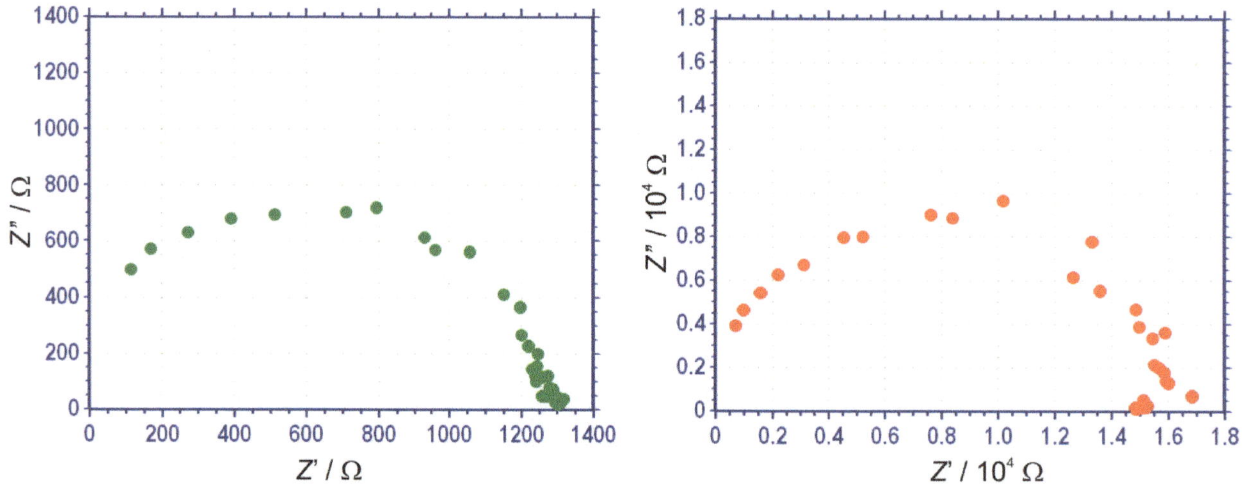

Figure 4. *AC impedance spectra of carbon steel immersed in various test solution (Nyquist plot) (pH 8):* ***a)*** *well water (blank);* ***b)*** *well water + DL-arginine (250 ppm) +* Zn^{2+} *(5 ppm)*

a **b**

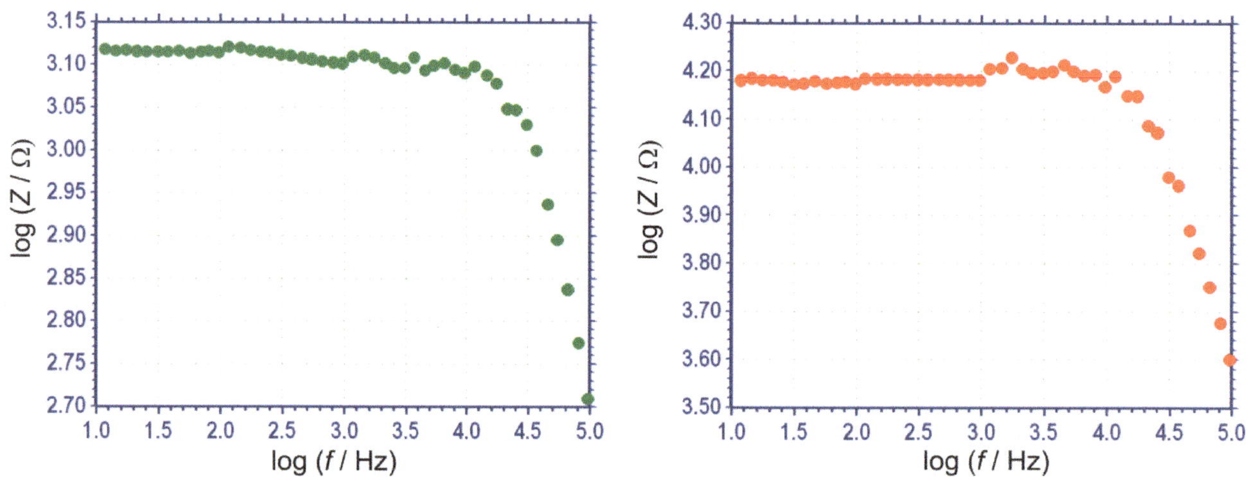

Figure 5. *AC impedance spectra of carbon steel immersed in various test solution (Bode plot) (pH 8):* ***a)*** *well water (blank);* ***b)*** *well water + DL-arginine (250 ppm) +* Zn^{2+} *(5 ppm)*

a **b**

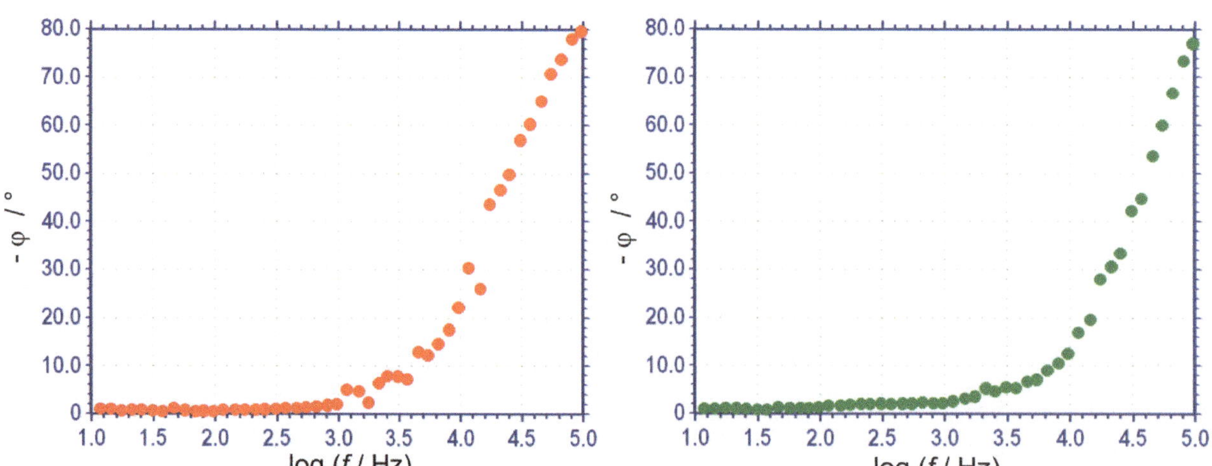

Figure. 6. *AC impedance spectra of carbon steel immersed in various test solution (Phase – Bode plot) (pH 8):* ***a)*** *well water (blank);* ***b)*** *well water + DL-arginine (250 ppm) +* Zn^{2+} *(5 ppm)*

Table 5. *Corrosion parameters of carbon steel immersed in well water in the absence and presence of inhibitor system obtained from AC impedance spectra (pH 8).*

System	Nyquist plot		Bode plot
	R_t / Ω cm^2	C_{dl} / F cm^{-2}	log (*Z* / Ω)
Well water	1212	3.8424×10^{-9}	3.120
Well water + DL-arginine (250 ppm) + Zn^{2+} (5 ppm)	16240	3.1373×10^{-10}	4.181

It is observed that when the inhibitors [DL-arginine (250 ppm) + Zn^{2+} (5 ppm)] are added, the charge transfer resistance (R_t) increase from 1212 Ω cm^2 to 16240 Ω cm^2. The C_{dl} value decreases from 3.8424×10^{-9} to 3.1373×10^{-10} F cm^{-2}. The impedance value log(*Z*/Ω) increases from 3.120 to 4.181. These results lead to the conclusion that a protective film is formed on the metal surface.

It is observed that weight loss study gives very high *IE*. But this is not the case with the electrochemical studies. This may be explained by the fact that weight loss study is the result of 7 days whereas electrochemical studies are an instantaneous study. Similar observations were already reported [37-45].

Analysis of FTIR spectra

FTIR spectra have been used to analyze the protective film formed on the metal surface [39,46-53]. The FTIR spectrum (KBr) of pure DL-arginine is shown in Fig. 7a. The C=O stretching frequency of carboxyl group appears at 1610 cm^{-1}. The CN stretching frequency appears at 1098 cm^{-1}. The NH stretching frequency of the amine group appears at 3182 cm^{-1}.

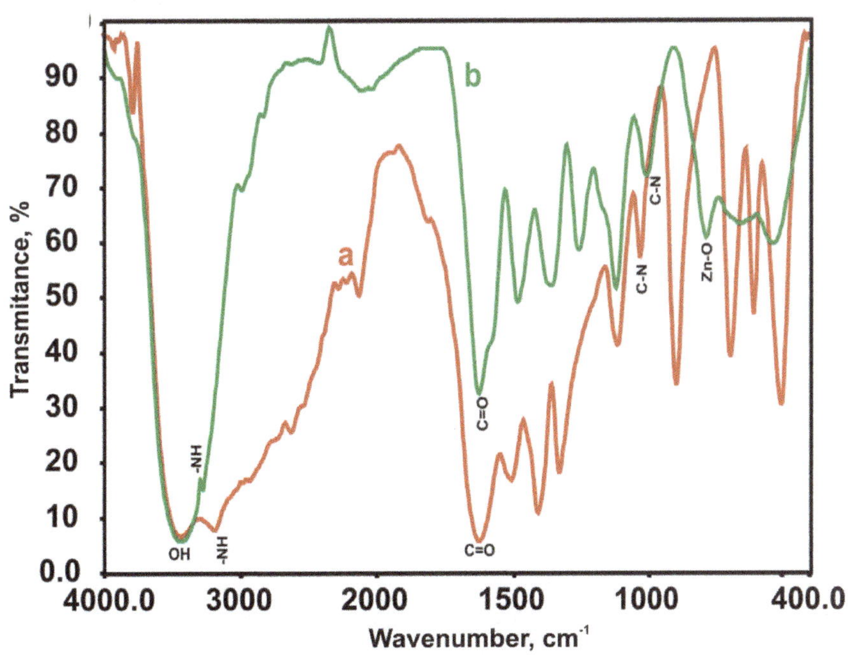

Figure 7. *FTIR spectra of:* ***a)*** *pure DL-arginine;* ***b)*** *film formed on the metal surface after immersion in well water + DL-arginine (250 ppm) +* Zn^{2+} *(5 ppm)*

The FTIR spectrum of the film formed on the metal surface after immersion in the solution containing well water, 250 ppm of DL-arginine and 5 ppm Zn^{2+} is shown in Fig.7b. The C=O stretching frequency has shifted from 1610 to 1619 cm^{-1}. The CN stretching frequency has shifted from 1098 to 1005 cm^{-1}. The NH stretching frequency has shifted from 3182 to 3285 cm^{-1}. This observation suggests that DL-arginine has coordinated with Fe^{2+} through the oxygen atom of the carboxyl group and nitrogen atom of the amine group resulting in the formation of Fe^{2+}-DL-arginine complex on the anodic sites of the metal surface. The peak at 788 cm^{-1} corresponds to Zn-O stretching. The peak at 3442 cm^{-1} is due to OH^- stretching. This confirms that $Zn(OH)_2$ is formed on the cathodic sites of metal surface [46,51-53]. Thus the FTIR spectral study leads to the conclusion that the protective film consists of Fe^{2+}- DL-arginine complex and $Zn(OH)_2$.

SEM Analysis of Metal Surface

SEM provides a pictorial representation of the surface. This helps to understand the nature of the surface film in the absence and presence of inhibitors and the extent of corrosion of carbon steel. The SEM micrographs of the surface are examined [54-56]. The SEM images of different magnification (500 x, 1000 x) of carbon steel specimen immersed in well water for 7 days in the absence and presence of inhibitor system are shown in Figs. 8a and 8b and 8c respectively.

The SEM micrographs of polished carbon steel surface (control) in Figs. 8a and 8b show the smooth surface of the metal. This shows the absence of any corrosion products (or) inhibitor complex formed on the metal surface.

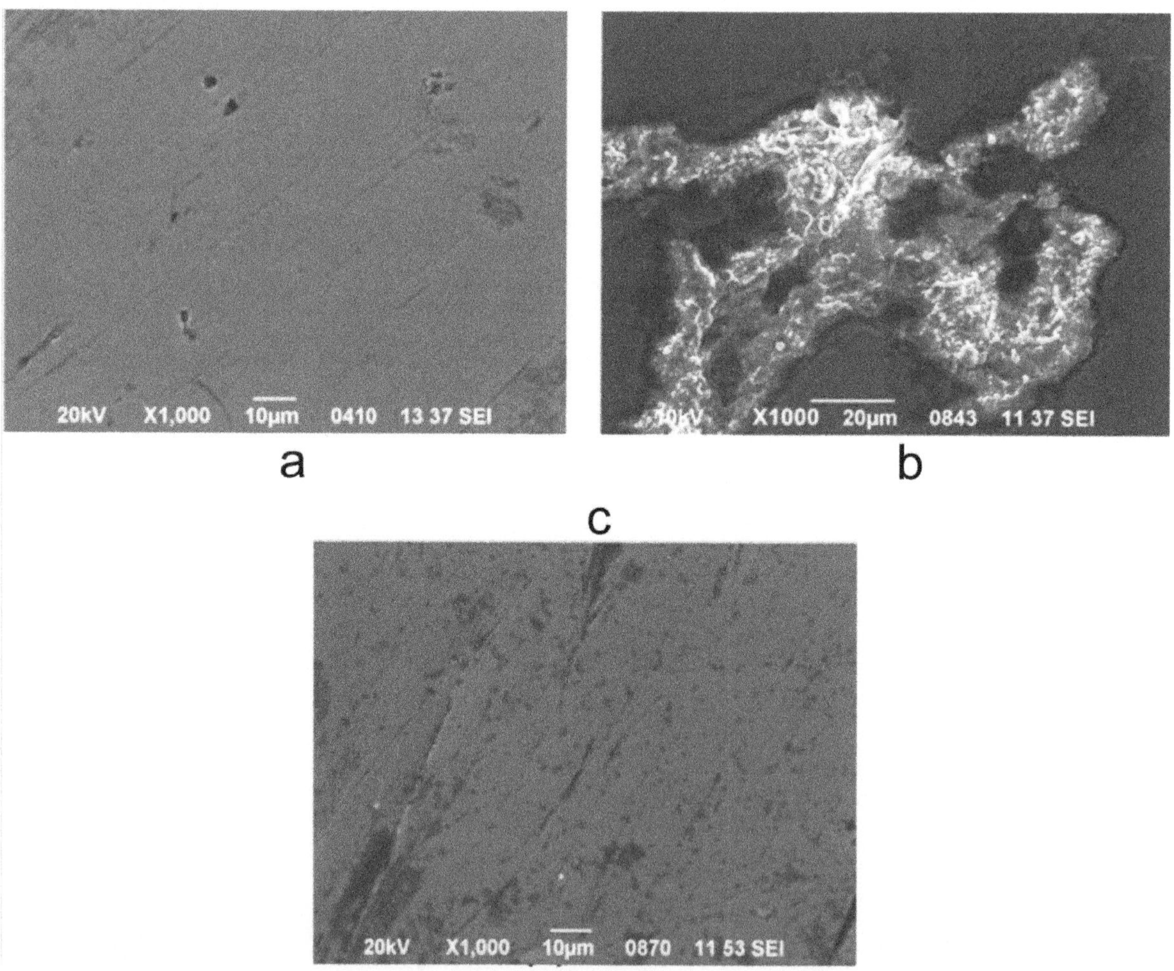

Figure 8. *SEM analysis of:* ***a)*** *Carbon steel (Control);* ***b)*** *Carbon steel immersed in well water (Blank);* ***c)*** *Carbon steel immersed in well water + 250 ppm of DL-arginine + 5 ppm of* Zn^{2+}

The SEM micrographs of carbon steel surface immersed in well water (Figs. 8b) show the roughness of the metal surface which indicates the highly corroded area of carbon steel in well water. However, Figs. 8c indicate that in the presence of inhibitor (250 ppm DL-arginine and 5 ppm Zn^{2+}) the rate of corrosion is suppressed, as can be seen from the decrease of corroded areas. The metal surface is almost free from corrosion due to the formation of an insoluble complex on the surface of the metal [54]. In the presence of DL-arginine and Zn^{2+}, the surface is covered by a thin layer of inhibitors which effectively controls the dissolution of carbon steel.

Atomic Force Microscopy Characterization

Atomic force microscopy is a powerful technique for gathering of roughness statistics from a variety of surfaces [57]. AFM has become an accepted method of roughness investigation [58-64].

All atomic force microscopy images were obtained in a VEECO Lab incorporation AFM instrument operating in contact mode in air. The scan size of all the AFM images are 0.5 µm × 0.5 µm areas at a scan rate of 6.835 µm s^{-1}.

The two dimensional (2D), three dimensional (3D) AFM morphologies and the AFM cross-sectional profile for polished carbon steel surface (reference sample), carbon steel surface immersed in well water (blank sample) and carbon steel surface immersed in well water containing the formulation of DL-arginine 250 ppm and 5 ppm of Zn^{2+} are shown as Figs. 9a, 9d and 9g, Figs. 9b, 9e and 9h and Figs. 9c, 9f and 9i, respectively.

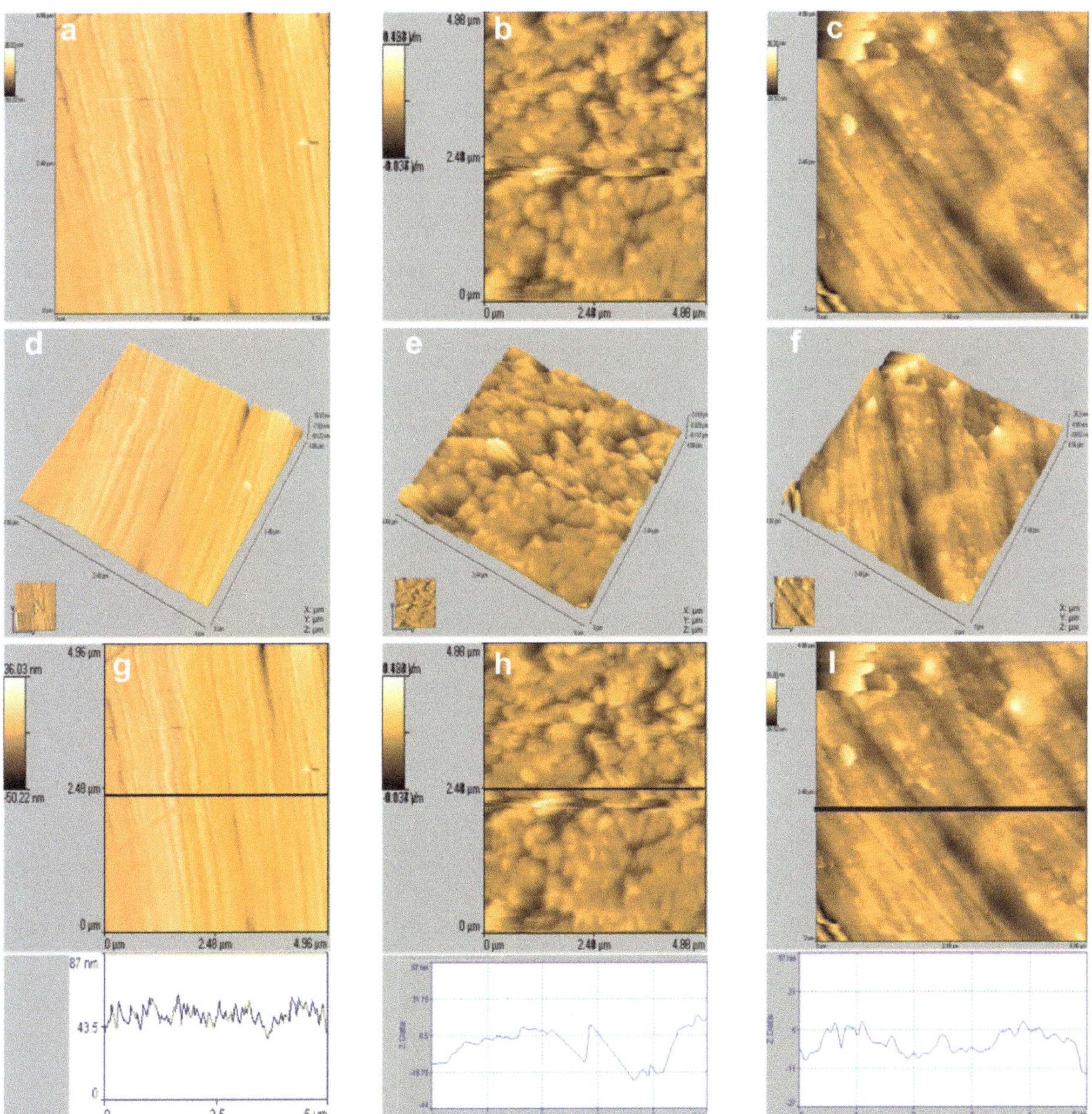

Figure 9. a), b) c) *2D AFM images;* ***d), e), f)*** *3D AFM images and*
g), h), f) *Cross sectional profile which are corresponding to as shown lines on 2D AFM images of*

a), d), g) *Polished carbon steel (Control);* ***b), e), h)*** *Carbon steel immersed in well water (blank) and* ***c), f), i)*** *Carbon steel immersed in well water containing DL-arginine 250 ppm + Zn^{2+} 5 ppm*

Root– mean-square roughness, average roughness and peak-to-valley value

AFM image analysis was performed to obtain the average roughness, R_a (the average deviation of all points roughness profile from a mean line over the evaluation length), root-mean-square roughness, R_q (the average of the measured height deviations taken within the evaluation length and measured from the mean line) and the maximum peak-to-valley (P-V) height values (largest single peak-to-valley height in five adjoining sampling heights) [63]. R_q is much more sensitive than R_a to large and small height deviations from the mean [64].

Table 6 is the summary of the average roughness (R_a), RMS roughness (R_q) maximum peak-to-valley height (P-V) value for carbon steel surface immersed in different environments.

Table 6. *AFM data for carbon steel surface immersed in inhibited and uninhibited environments*

Samples	R_q / nm	R_a /nm	P-V value, nm
Polished carbon steel (control)	5	3	50
Carbon steel immersed in well water	28	21	137
Carbon steel immersed in well water DL-arginine (250 ppm) + Zn^{2+} (5 ppm)	6	4	26

The value of R_{RMS}, R_a and P-V height for the polished carbon steel surface (reference sample) are 5 nm, 3 nm and 50 nm respectively, which show a more homogeneous surface, with some places in which the height is lower than the average depth [57]. Figs. 9a, 9d and 9g display the uncorroded metal surface. The slight roughness observed on the polished carbon steel surface is due to atmospheric corrosion. The RMS roughness, average roughness and P-V height values for the carbon steel surface immersed in well water are 28 nm, 21 nm and 137 nm respectively. These data suggest that carbon steel surface immersed in well water has a greater surface roughness than the polished metal surface. This shows that the unprotected carbon steel surface is rougher and this is due to the corrosion of the carbon steel in well water. Figs. 9b, 9e and 9h display the corroded metal surface with few pits.

The presence of 250 ppm of DL-arginine and 5 ppm of Zn^{2+} in well water reduces the R_q by a factor of 6 nm from 28 nm and the average roughness is significantly reduced to 4 nm when compared with 21 nm of carbon steel surface immersed in well water. The maximum peak-to-valley height also was reduced to 26 nm from 137 nm. These parameters confirm that the surface appears smoother. The smoothness of the surface is due to the formation of a compact protective film of Fe^{2+}- DL-arginine complex and $Zn(OH)_2$ on the metal surface thereby inhibiting the corrosion of carbon steel.

Also the above parameters observed are somewhat greater than the AFM data of polished metal surface which confirms the formation of the film on the metal surface, which is protective in nature.

When carbon steel immersed in well water the R_q, R_a and maximum peak-to-valley (P-V) height are very high. In the presence of inhibitor system, these values are very close to those of polished carbon steel. This indicates that the formation of protective film starts even before the starting of corrosion process. Suppose the film is formed on the iron oxide surface then roughness would be greater than the R_q, R_a, maximum peak-to-valley (P-V) height of the carbon steel surface immersed in well water. But this is not the case. Hence this is concluded that the protective film is formed in the initial stage itself.

Mechanism of Corrosion inhibition

The results of the weight-loss study show that the formulation consisting of 250 ppm DL-arginine and 5 ppm of Zn^{2+} has 98% IE in controlling corrosion of carbon steel in well water. A synergistic effect exists between Zn^{2+} and DL-arginine. Polarization study reveals that this formulation functions as anodic inhibitor. AC impedance spectra reveal that a protective film is formed on the metal surface. FTIR spectra reveal that the protective film consists of Fe^{2+}– DL-arginine complex and $Zn(OH)_2$. In order to explain these facts the following mechanism of corrosion inhibition is proposed [65-71].

- When the solution containing well water, 5 ppm Zn^{2+} and 250 ppm of DL-arginine is prepared, there is formulation of Zn^{2+} – DL-arginine complex in solution.
- When carbon steel is immersed in this solution, the Zn^{2+} - DL-arginine complex diffuses from the bulk of the solution towards metal surface.
- Zn^{2+}- DL-arginine complex diffuses from the bulk solution to the surface of the metal and is converted into a Fe^{2+}- DL-arginine complex, which is more stable than Zn^{2+}- DL-arginine [66].
- On the metal surface Zn^{2+} - DL-arginine complex is converted in to Fe^{2+} - DL-arginine on the anodic sites. Zn^{2+} is released.

  Zn^{2+} - DL-arginine + Fe^{2+} $\rightarrow$ Fe^{2+} - DL-arginine + Zn^{2+}
- The released Zn^{2+} combines with OH^- to form $Zn(OH)_2$ on the cathodic sites [65,66].

  $Zn^{2+} + 2OH^- \rightarrow Zn(OH)_2 \downarrow$
- Thus the protective film consists of Fe^{2+}- DL-arginine complex and $Zn(OH)_2$.
- The SEM micrographs and AFM images confirm the formation of protective layer on the metal surface.

Conclusions

Inhibition of corrosion of carbon steel in well water by arginine-Zn^{2+} system has been evaluated by weight-loss method. A synergistic effect exists between arginine and Zn^{2+} system. The formulation consisting of 250 ppm of arginine and 5 ppm of Zn^{2+} offers good inhibition efficiency of 98%. Polarization study reveals that this formulation functions as an anodic inhibitor. AC impedance spectra reveal that a protective film is formed on the metal surface. The FTIR spectral study leads to the conclusion that the Fe^{2+}- DL-arginine complex formed on anodic sites of the metal surface controls the anodic reaction. $Zn(OH)_2$ formed on the cathodic sites of the metal surface controls the cathodic reaction. The SEM micrographs and AFM images confirm the formation of the protective layer on the metal surface.

Acknowledgements: *The authors are thankful to their respective management and UGC for their encouragement.*

References

[1] A. Baral, R.D. Engelken, *Environ. Sci. Policy* **5** (2002) 121-133
[2] J.M. Gaidis, *Cement Concrete Comp.* **26** (2004) 181-189.
[3] E. Stupnisek-Lisac, A. Loncaric Bozic, I. Cafuk, *Corrosion* **54** (1998) 713-720
[4] A.Y. El-Etre, *Corros. Sci.* **45** (2003) 2485-2495.
[5] A.Y. El-Etre, *Corros. Sci.* **40** (1998) 1845-1850.

[6] H. Ashassi-Sorkhabi, M.R. Majidi and K. Seyyedi, *Appl. Surf. Sci.* **225** (2004) 176-185.
[7] Z. Ghasemi, A. Tizpar, *Appl. Surf. Sci.* **252** (2006) 3667-3672.
[8] D.Q. Zhang, L.X. Gao, G.D. Zhou, *J. Appl. Electrochem.* **35** (2005) 1081-1085.
[9] O. Olivares, N.V. Likhanova, B. Gomez, J. Navarrete, M.E. Llanos-Serrano, E. Arce, J.M. Hallen, *Appl. Surf. Sci.* **252** (2006) 2894-2909.
[10] W.A. Badawy, K.M. Ismail, A.M. Fathi, *Electrochim. Acta* **51** (2006) 4182-4189.
[11] K.M. Ismail, *Electrochim. Acta* **52** (2007) 7811-7819.
[12] H. Ashassi-Sorkhabi, Z. Ghasemi, D. Seifzadeh, *Appl. Surf. Sci.* **249** (2005) 408-418.
[13] E.E. Oguzie, Y. Li, F.H.Wang, *J. Colloid Interf. Sci.* **310** (2007) 90-98.
[14] A.A. Aksut, A.N. Onal, *B. Electrochem.* **11** (1995) 513-521.
[15] C.K. Gomma, *B. Electrochem.* **14** (1998) 291-297.
[16] D.J. Kalota, D.C. Silverman, *Corrosion* **50** (1994) 138-145.
[17] L.H. Madkour, M.M. Ghoneion, *B. Electrochem.* **13** (1997) 1-7.
[18] M.S.S. Morad, A.E.A. Hermas, M.S.A. Aal, *J. Chem. Technol. Biot.* **77** (2002) 486-494.
[19] A.A. El-Shafei, M.N.H. Moussa, A.A. El-Far, *J. Appl. Electrochem.* **27** (1997) 1075-1078.
[20] A. Fouda, M. El-Semongy, *J. Indian Chem. Soc.* **59** (1982) 89-96.
[21] A. Aksut, S.Bilgic, *Corros. Sci.* **33** (1992) 379-387.
[22] G. Gomma, H. Wahdan, *Mater. Chem. Phys.* **39** (1994) 142-148.
[23] G. Moretti, F. Guidi, *Corros. Sci.* **44** (2002) 1995-2011.
[24] M. Rylkina, A. Chikanova, L. Trubacheva, S. Reshetinikov, *Prot. Met.* **35** (1999) 23-26
[25] H. Baba, T.Kodama, *Corros. Sci.* **41** (1980) 1987-2000.
[26] B.I. Ita, *B. Electrochem.* **21** (2005) 319-323.
[27] M.A. Pech Canul, P. Bartolo-Perez, *Surf. Coat. Techol.* **184** (2004) 133-140.
[28] M.A. Pech-Canul, M. Echeverria, *Corr. Eng. Sci. Technol.* **38** (2003) 135-138.
[29] S.K. Rajappa, T.V. Venkatesha, *Turk. J. Chem.* **27** (2003) 189-196.
[30] M.A. Pech-Canul, L.P. Chi-Canul, *Corrosion* **55** (1999) 948-956.
[31] A. Akiyama, N. Nobe, *J. Electrochem. Soc.* **117** (1970) 999-1003.
[32] G. Wranglen, *An introduction to corrosion and protection of metals*, Chapman & Hall, New York, USA (1985).
[33] R.J. Holness, G. Williams, D.A. Worsley and H.N. McMurray, *J. Electrochem. Soc.* **152** (2005) B73-B81.
[34] J. Telegdi, M.M. Shaglouf, A. Shaban, F.H. Karman, I. Betroti, M. Mohai, E. Kalman, *Electrochim. Acta* **46** (2001) 3791-3799.
[35] J. A. Selvi, S. Rajendran and J. Jeyasundari, *Zastita Materijala* **50 (2)** (2009) 91-98.
[36] J. Hilden, J. Virtanen, O. Forsen, J. Aromaa, *Electrochim. Acta* **46** (2001) 3859-3866
[37] J.M. Roque, T. Pandiyan, J. Cruz. E Garcia-Ochoa, *Corros. Sci.* **50** (2008) 614-624.
[38] O. Benali, L. Larabi, M. Traisnel, L. Gengembre, Y. Harek, *Appl. Surf. Sci.* **253** (2007) 6130-6139.
[39] H. Amar, T. Braisaz, D. Villemin, B. Moreau, *Mater. Chem. Phys.* **110** (2008) 1-6
[40] J. A. Selvi, S.Rajendran, V. G. Sri, A. J. Amalraj, B. Narayanasamy, *Port. Electrochim. Acta* **27(1)** (2009) 1-11.
[41] S. Rajendran, J.Paulraj, P. Rengan, J. Jeyasundari and M. Manivannan, *J. Dent. Oral Hyg.* **1** (2009) 1–8.
[42] E. Kalman, I. Felhosi, F.H. Karman, et al., *Corrosion and Environmental Degradation,* M. Schutze, Ed., Weinheim: Wiky-VCH, 1 (2000) 471.
[43] S. Zhang, Z. Tao, W. Li, B. Hou, *Appl. Surf. Sci.* **255** (2009) 6757-6763.
[44] F. El-Taib Heakal, A.S. Fouda, M.S. Radwan, *Mater. Chem. Phy.* **125** (2011) 26-36.
[45] S. Rajendran, M. Kalpana Devi, A.P.P.Regis, A. J. Amalraj, J. Jeyasundari and M. Manivannan, *Zastita Materijala* **50** (2009) 131-140.

[46] R. Kalaivani, B. Narayanasamy, J.A. Selvi, A.J. Amalraj, J. Jeyasundari and S. Rajendran, *Port. Electrochim. Acta* **27** (2009) 177-187.
[47] J. Sathiyabama, S.Rajendran, J. A. Selvi and J. Jeyasundari, *Open Corrosion J.* **2** (2009) 77-82.
[48] R.M. Silverstein, G.C. Bassler, T.Moril, *Spectrometric Identification of Organic Compound*, John Wiley and Sons, New York (1981) 95.
[49] A.D. Cross, *Introduction to practical infrared spectroscopy,* Butterworths, Scientific Publication, London (1990) 73
[50] K. Nakamoto, *Infrared and Raman spectra of inorganic coordination compound,* Willey Interscience, New York (1986).
[51] S. Rajendran, B.V. Apparao and N. Palaniswamy, *B. Electrochem.* 12 (1996) 15-19.
[52] I. Sekine and V. Hirakawa, *Corrosion* **42** (1986) 272-277.
[53] S. Rajendran, B.V. Apparao and N. Palaniswamy, *8th Europ. Symp. Corros. Inhibitors*, Ferrara, Italy, (1995) Proc. Vol 1, p. 465.
[54] D. Gopi, S. Manimozhi, K. M. Govindaraju, P. Manisankar and S. Rajeswari, *J. Appl. Electrochem.* **37** (2007) 439-449.
[55] M.A. Amin, S.S. Abd El-Rehim, E.E.F. Ei-Sherbini, R.S. Bayoumi, *Electrochim. Acta* **52** (2007) 3588-3600.
[56] M.A. Amin, S.S. Abd El-Rehim, H.T.M. Abdel-Fatah, *Corros. Sci.* **51** (2009) 882-894.
[57] R. Vera, R. Schrebler, P. Cury, R. Del Rio, H. Romero, *J. Appl. Electrochem.* **37** (2007) 519-525.
[58] Ph. Dumas, B. Butffakhreddine, C. Amra, O. Vatel, E. Andre, R.Galindo and F. Salvan, *Europhys. Lett.* **22** (1993) 717-722.
[59] J. M. Bennett, J. Jahannir, J.C. Podlesny, T. L. Baiter and D.T. Hobbs, *Appl. Optics* **34** (1995) 213-230.
[60] A. Duparre, N. Kaiser, H. Truckenbrodi, M Berger and A. Kohler, *Int. Symp. on Optics, Imaging and Instrumentation*, 11-16 Jul 1993., San Diego, CA, Proc. SPIE. 1995 (1993) p. 181-192.
[61] A. Duparre, N. Kaiser and S. Jakobs, *Annual Symp. on Optical Materials for High Power Lasers.* 27-29 October 1993, Boulder, CO, Proc. SPIE, 2114 (1993) p. 394.
[62] C. Amra. C. Deumie. D. Torricini. P, Roche, R. Galindo. P. Dumas and F. Salvan, *Int. Symp. on Optical Interference Coatings.* 6-10 June 1994. Grenoble, Proc. SPIE. 2253 (1994) p. 614-630.
[63] T.R. Thomas, *Rough Surfaces*, Longman, New York, 1982.
[64] K.J. Stout, P.J. Sullivan and P.A. McKeown, *CIRP Annals-Manuf. Techn.*, **41** (1992) 621.
[65] D. Gopi, N. Bhuvaneswaran, S. Rajeswari, *B. Electrochem.* **18** (2002) 29-34.
[66] D. Gopi, S. Rajeswari, *J. Solid State Electr..* **6** (2002) 194-202.
[67] D. Gopi, S. Rajeswari, *International Conference on Advances in Surface Science and Engineering (INSURE)*, Chennai, India, 21 – 23 Feb. (2001) Proc. p. 210.
[68] D. Gopi, S. Rajeswari, *NACE International Conf., Corrosion its Mitigation and Preventive Maintenance*, Mumbai, India, 20-23 Nov. (2000) Proc. Vol 1, p. 435.
[69] D. Gopi, S. Rajeswari, *Tenth National Congress on Corrosion Control, Conf.,* Madurai, India, 6-8 Sep. (2000) Proc. p. 353.
[70] A. Veres, G. Reinhard, E. Kalman, *Brit. Corros. J.* **27** (1992) 147-150.
[71] E. Kalman, *Corrosion Inhibitors*, Published for EFC No.11, Institute of Materials London, UK (1994).

5

Characterization of an hrp-aox-polyaniline-graphite composite biosensor

Ana Carolina O. Santana, Erica F. Southgate, João Paulo B. G. Mendes*,
Jo Dweck, Eliana Mosse Alhadeff and Ninoska Isabel Bojorge Ramirez**,✉

Escola de Química, Universidade Federal do Rio de Janeiro, Av. Horácio Macedo, 2.030, Centro de Tecnologia, Bloco E, E-203, Cidade Universitária,CEP 21941-909, Rio de Janeiro, Brasil
**Instituto de Química, Universidade Federal do Rio de Janeiro, Av. Horácio Macedo, 2.030, Centro de Tecnologia, Bloco A, A-302, Cidade Universitária, CEP 21941-909, Rio de Janeiro, Brasil*
***Universidade Federal Fluminense, Dep. Engenharia Química e de Petróleo, R. Passo da Pátria, 156, Bl E-226, São Domingos , Niterói, CEP 24210-240, Rio de Janeiro, Brasil*

✉*Corresponding author*

Abstract

Nowadays there is an increasing demand to develop new and robust biosensors in order to detect low concentrations of different chemicals, in practical and small devices, giving fast and confident responses. The electrode material was a polyaniline-graphite-epoxy composite (PANI/GEC). Alcohol oxidase (AOX) and horseradish peroxidase (HRP) enzymes were immobilized and the responses were tested by cyclic voltammetry. The conductivities for the composites of graphite/polyaniline were determined. The cyclic voltammograms allowed detecting ethanol in pure diluted samples in a range from 0.036 to 2.62 M. Differential scanning calorimetry (DSC) and thermal gravimetry analysis (TGA) were used to verify the thermal characteristics of the composites (0, 10, 20, 30 and 100 % of graphite). The I_{max} value was determined for the dual enzyme biosensor (0.0724 μA), and the K_m^{app} as 1.41 M (with R^2 =0.9912).

Keywords

Cyclic voltammetry; ethanol; immobilized enzymes; PANI/GCE

Introduction

Ethanol is the most frequently analyzed aliphatic alcohol and several methods have been developed for its quantitative determination [1-3]. Measurement of alcohol levels in liquors and alcoholic drinks is a common necessity as is clinical analysis of patient tissue samples. The method approved by the Association of Official Analytical Chemists [4] for quantitative volumetric determination of alcohol in beer, wine and distilled spirits is pycnometry, which is the most common method for determining solution density. This method is considered as reference and has the advantages of accuracy and no need for comparison against a standard solution. The principal disadvantage is that the methodology is laborious, requiring a significant amount of time for its performance. Another disadvantage is that it requires pre-distillation, generally regarded as the first step in which error is introduced during the process of quantitative detection and analysis [5]. Other more accurate analytical methods include spectrophotometry and chromatographic techniques: gas chromatography (GC) or high performance liquid chromatography (HPLC). However, these methodologies are often less favorable due to high equipment prices and the need for well-trained operators. There is currently a movement toward replacing these methods with low-cost, fast and reliable electrodes working in conjunction with immobilized enzymes [6].

There is a growing need for the development of disposable devices for clinical and/or environmental monitoring. This need has stimulated the development of new technologies and methodologies that can efficiently monitor an increasing number of analytes on site in the environmental field or support clinical diagnoses as quickly and as cheaply as possible; offering even the possibility of on-site field monitoring. Besides selectivity, an analytical device must also be sensitive. In this respect, biosensors have shown great potential in recent years and thus appear to be useful components of effective analytical tools [7-8].

Biosensors that link enzyme catalyzed chemical reactions with amperometric detectors are having a great impact on fields such as environmental monitoring [9-10], analysis of the quality of food and beverages [11-12], biomedical monitoring process [13-14] and biomedicine [15]. These analytical tools, prepared by immobilization of enzymes on an electrode surface, are simple, sensitive and offer a fast response. The main problem that appears in the operation of these devices is the transfer of electrons from the active site of the enzyme to the electrode.

Immobilization of the biological material on the electrode surface constitutes a crucial step in development of the biosensor, since the enzyme's structure must be maintained in order to enable its action on the sample of interest [16-18]. Horseradish peroxidase (HRP) is widely used in enzyme-linked biosensors. However, there are at least two main drawbacks shown by this enzyme: (1) It exhibits a very broad specificity to reduce substrates [19-20], which results in low selectivity of the biosensor; (2) although it displays good stability at room temperature, it is unstable at high temperatures [21-22]. The co-immobilization of alcohol oxidase with horseradish peroxidase is expected to increase the selectivity and amplify the sensitivity of the biosensor for the quantitative determination of ethanol [23-24].

Immobilization of dual enzymes provides an excellent basis for increasing the selectivity, sensitivity and the thermal stability of the biosensor, depending on the strategy adopted for immobilizing the enzymes [15]. The immobilized enzymes may be reused several times or employed in an economical continuous flow path. Dual enzyme-linked sensors are amenable to automation for analytical measurements, scale up of enzymatic biotransformation reactors, or to recover a product with greater purity [19, 25]. Xie, *et al.* [26] reported recent advances in enzyme immobilization technologies that enhance enzyme properties such as activity, stability, specificity and reduced inhibition effects. The authors suggest that in the future multi-enzyme sensors based on co-immobilization would be the solution to many of the applications for the biotechnology industry and analytical devices.

The objective of this study was to characterize a composite-based on PANI / epoxy / graphite and evaluate its performance as a substrate for horseradish peroxidase (HRP) and alcohol oxidase (AOX) enzymes immobilized to an electrode creating a biosensor for ethanol detection. The development of such a method for the immobilization of multiple enzymes is highly attractive, especially for economic reasons because as enzymatic activity decays the support can be regenerated and reloaded with fresh enzyme. In fact, the cost of support is often a primary factor in the overall cost of the immobilized catalyst. In order to build the biosensor a composite prepared with graphite and an electron conductor polymer as polyaniline was studied and characterized in terms of its electrochemical conductance capacity and thermal stability [18,27]. We find critical compositions of the material that works with improved sensitivity over a relatively broad range of ethanol concentrations.

Experimental

Materials

Horseradish peroxidase (HRP; EC 1.11.1.7) was purchased from Toyobo, Brazil and alcohol oxidase (AOX, EC 1.1.1.1, specific activity of 200 units/mg of protein), graphite powder and polyaniline (emeraldine salt) were purchased from Sigma-Aldrich. For the incorporation of enzymatic solutions, a 2.5 % (v/v) of glutaraldehyde (Sigma-Aldrich) and 1 mg/mL of protein albumin were used. The ethanol standard solutions were prepared with 0.1 M mono potassium phosphate buffer (pH 7.0). All reagents were of analytical-reagent grade. All solutions were prepared with distilled water.

Apparatus

Amperometric measurements were carried out using an AUTOLAB PGSTAT12 (Ecochemie) connected to a personal computer via a serial RS232 port for data acquisition. The obtained amperometric alcohol dual enzyme sensors were evaluated by means of cyclic voltammetry in a three-electrode configuration with Ag/AgCl/KCl (3M) reference electrode and Pt-wire counter electrode. When not in use, the electrode was stored dry at 4 °C in a refrigerator.

The thermal properties (thermogravimetric analysis (TGA) and differential scanning calorimetry (DSC)) for the composites of graphite : PANI prepared with 0, 10, 20, 30 and 100 % of graphite were performed by TA Instruments SDT Q600. Analyses were conducted in a 30 mL/min flow rate of air atmosphere, with a ramp of 5 °C/min from 30 to 800 °C.

The electrical conductivities of the composites pellets were evaluated using the techniques of two electrodes, between which a pellet of known composition of the composite was fixed with the aid of a sleeve of Teflon. The tests were done using a bench meter ICEL Manaus MD - 6700 coupled to a computer. The disks pellets prepared with 0% and 100% of pure graphite and composites with 1, 3, 5, 10, 20, 30, 50, 70, 100% of graphite mixed with PANI were measured. Pure samples of PANI and graphite were also determined.

Preparation and evaluation of the AOX–HRP-based biosensors

HRP (3.60 g) was dissolved in 30 mL of 50 mM phosphate buffer (PB, pH 7.0). After filtration and dialysis steps, a 0.133 mg/mL of HRP solution was mixed with AOX (47 mg/mL) in buffer pH 7.0. The 10 % (w/v) of bovine serum albumin and 2.5% (v/v) of glutaraldehyde were also prepared in 50 mM PB (pH 7.0) solution. A 10 μL volume of bovine serum albumin and 10 μL of glutaraldehyde were deposited on the electrode surface sequentially. The excess of glutaraldehyde was rinsed off with water. Teflon cylindrical electrodes (5.0 × 0.7 cm and 0.13 cm inside orifice) were used to construct the working dual enzyme biosensor with a 20 mL electrolytic cell and an Ag/AgCl reference electrode and a platinum counter-electrode.

Procedure for immobilization

The methodology used was the ionic immobilization of AOX and HRP enzymes on the electrode surface constructed using a graphite matrix with polyaniline and an epoxy resin. During the immobilization step, a solution containing 2.5 % (v/v) glutaraldehyde, 0.5 % (v/v) BSA and 97 % (v/v) enzyme solution containing 1100 μL HRP and 15 μL of AOX was deposited on the electrode surface. The electrode was left at 4 °C for 24 hours [28].

Measurement procedure

Cyclic voltammetry (CV) measurements on the electrode were performed in a 3-electrode system containing a Ag/AgCl/KCl 3M (Microquímica®) reference electrode, coiled platinum wire (99.99 % pure) mounted at the end of a chemically-resistant epoxy rod as counter electrode in addition to the modified working electrode based on PANI/GCE. The potential was cycled between –400 and 400 mV *vs.* Ag/AgCl.

Determination of ethanol in samples

Ethanol (95 %) samples (0.15 mL) were diluted in a 10 mL flask with 0.1 M mono potassium phosphate buffer solution (pH 7.0). Voltammetric determination was carried out by applying the standard addition method. Diluted sample and standard ethanol solution (0.15 μL) were added to the voltammetric cell containing 10 mL of 0.1 M mono potassium phosphate buffer solution (pH 7.0).

Results and discussion

Study of differential scanning calorimetry and thermal gravimetry analysis

The thermal stability of the graphite composite samples was analyzed by TG, derivatived thermogravimetric analysis (DTG) and DSC. Results of TG and DTG analyses are presented in Figure 1. The curves in Figure 1 follow the mass as a function of temperature of composite samples containing 100 % graphite, 20 % graphite and 0% graphite (100 % polyaniline), respectively. The curve for the composite of Graphite/PANI (red dashed line) shows an intermediate stability between pure samples of Graphite and PANI polymer. The presence of the PANI introduces four decomposition steps. In the first stage, beginning at 150 °C, there is a slow weight loss associated to the release of trapped water or organic solvents in the polymer structure. The second stage of weight loss is observed from 270 °C to 550 °C and is attributed to decomposition of the oligomers. The third decay, from 350 °C to 450 °C, was assigned to the thermal decomposition of the PANI chains. The DTG curves fully support the above mentioned losses. The pure graphite sample decomposes above 600 °C, whereas the 20 % graphite composite presents four degradation steps (three attributed to the pure polyaniline and one to the pure graphite). Similar results were found by Kowner, *et al.* [29], Bourdo, *et al.* [30] and Mo, *et al.* [31].

A linear fitting between the data of polymer content estimated by TG and composition on a dry basis of raw materials in the composites was proposed, showing a good correlation coefficient (R^2 = 0.9811). The difference between the values estimated from the correlation with those of the components in the composite has an average value of -0.1 % with standard deviation of 1.86 %.

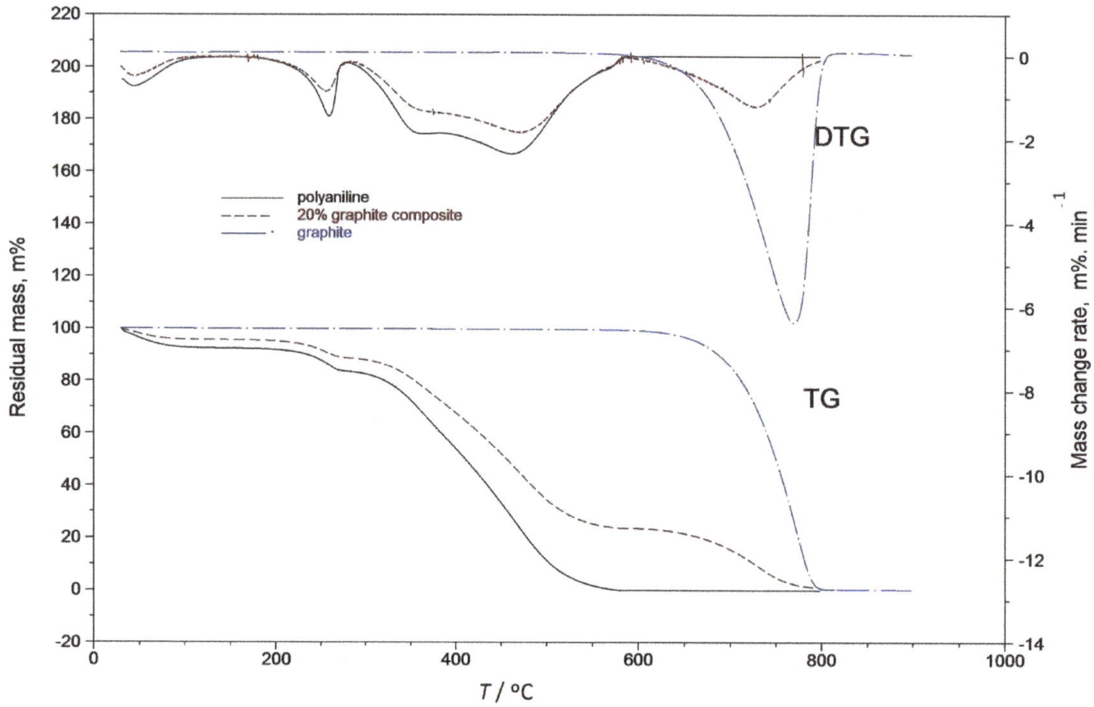

Fig. 1. *TG and DTG curves for the samples prepared with 100, 20 and 0 % of graphite.*

Conductivity of graphite/PANI composite

Song and Choi [32] have reported that the most conductive form of PANI is the fully protonated, half-oxidized emeraldine salt form. A decrease in conductivity was observed when the polymer was deprotonated or either fully oxidized or reduced. This work intends to develop a prototype composite-biosensor based on typical PANI that maintains the conductivity. Conductivity values were determined for different concentrations of graphite : polyaniline composites (1, 3, 5, 10, 20, 30, 50, 70, 100 % of graphite) and for the pure alcohol oxidase. As shown in Figure 2 an increase in the conductivity of the composite samples was observed.

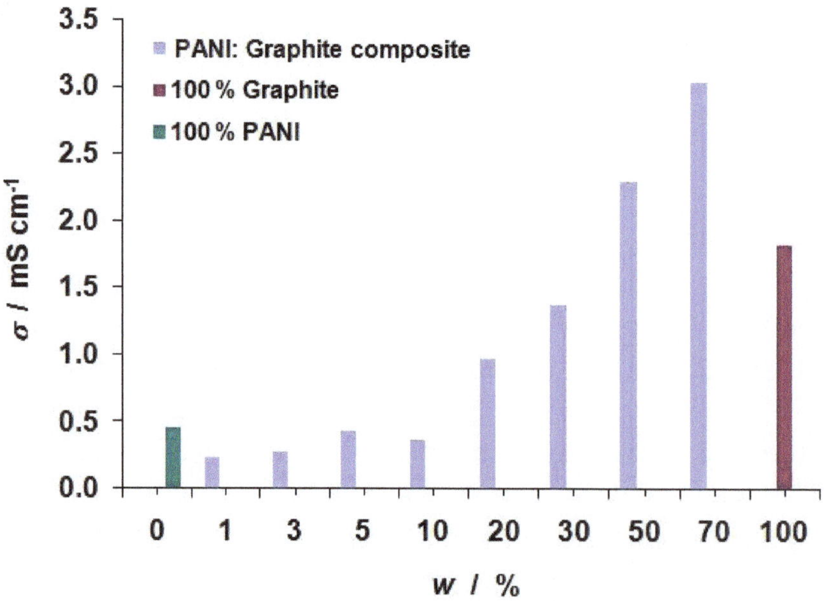

Fig. 2. *Electrical conductivities of PANI/GEC as a function of graphite concentration in the composites. Also shown are samples with 100 % PANI (green bar) and 100 % graphite (red bar).*

This is probably due to synergistic effects of mixing the conductive polymer PANI and graphite powder. The mixture better supports electron transfer and, consequently, displays enhanced electrical conductivity. The conductivity values determined for the pure samples of graphite and PANI were 1.82×10^{-3} S / cm and 4.64×10^{-4} S / cm, respectively. The maximum value was obtained with the 70 : 30 (graphite:PANI) composite; higher than the value measured for the 100 % graphite sample. For the composites prepared with 1 to 10 % of graphite, there were not any significant variations in conductivity. A linear relationship between graphite content and conductivity was observed from 20 to 70 % with the conductivity values for 50 to 70 % of graphite surpassing those for the 100 % graphite sample. Mo *et al.* [31] has detected an increase in the electrical conductivity as a function of graphite nanosheet content in a composite prepared with graphite nanosheets and PANI. Bourdo *et al.* [30] also found similar behavior for pure PANI and graphite samples and for PANI/graphite composites. In the present study, the 30 % PANI composite compound was employed due to the improved performance of its electrical response.

Electrochemical behaviour of the biosensor

The biosensor bi-enzymatic HRP/AOX was characterized using cyclic voltammetry to demonstrate the electrochemical performance of the system. Figure 3 shows the cyclic voltammograms obtained from 5 to 150 mV s^{-1} in a solution of 1mM $K_4Fe(CN)_6$ mixture in 0.1 M KCl and phosphate buffer pH 7.0. The peaks currents of the CVs indicating quasi-reversible processes between $Fe(CN)_6^{4-}/Fe(CN)_6^{3-}$ couple and the electrodes at the faster scan rates Each curve has the same form but it is apparent that the total current increases with increasing scan rate. This again can be rationalized by considering the size of the diffusion layer and the time taken to record the scan. Clearly the voltammogram will be slower to record as the scan rate is decreased. Hence the size of the diffusion layer above the electrode surface will be different depending upon the voltage scan rate used. In spite of that, working with lower scan rates a well--defined cathodic peak and a small anodic could be identified, and the scan rate of 10 mV s^{-1} applied to analyze the ethanol samples. So, the best quality voltammogram was obtained working with a scan rate of 10 mV s^{-1}. Therefore, that was the scan rate applied to analyze the ethanol samples

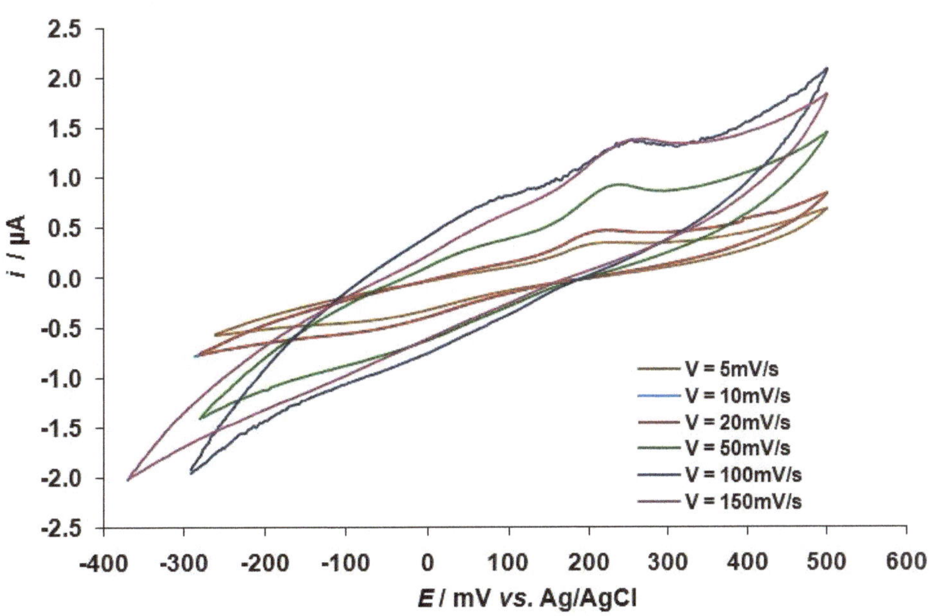

Fig. 3. *Cyclic voltammograms of the AOX/HRP/Graphite/PANI in 0.1 M PBS, pH 7.0 at various scan rates (from inner to outer curves: 5, 10, 20, 50, 100, 150 mV s^{-1}).*

Figure 4 shows shows the current intensity for the calibration curve changed between 0.61 mA (0.316 M) to 0.25 mA (2.62 M) The concentration range of standard ethanol solutions used in the electrochemical measurements was 0.316 - 2.62 M (R^2 = 0.991). This clearly demonstrates that the current density reduces linearly with increased ethanol concentration in the samples. This is attributed to an inhibition of the enzyme. The effect is especially evident at the higher ethanol concentrations, probably due to the reaction end-products (acetaldehyde) or external mass transfer limitations. A similar behavior was reported for AOX and HRP that was covalently immobilized on controlled pore glass [23]. However, that study showed that all the supports exhibited less than 20 % of the specific activity of the free enzyme, as a consequence of conformational changes in the 3-D structure of the protein caused by the covalent binding of AOX to the supports. In this work the enzymes were immobilized by adsorption, which is less aggressive than the covalent immobilization. However, the enzyme may be coupled to the support in a way that hinders the access of substrates to the active center, promoting the mass transfer limitations. The amperometric response exhibited by the different immobilized AOX preparations was also very similar although the highest value was obtained when the support was activated using glutaraldehyde in phosphate buffer pH 7. Sirkar, *et al.* [33] observed an increase in the current density (~60 %) of the electrochemical biosensor response for a multilayer nanocomposite thin film using glutaraldehyde as a crosslinking agent in a trial for stabilizing the structure. The authors proposed that arginine and lysine residues of the enzyme react with amines present on the redox polymer and, as a consequence, the activity was maintained near 100 % for three weeks.

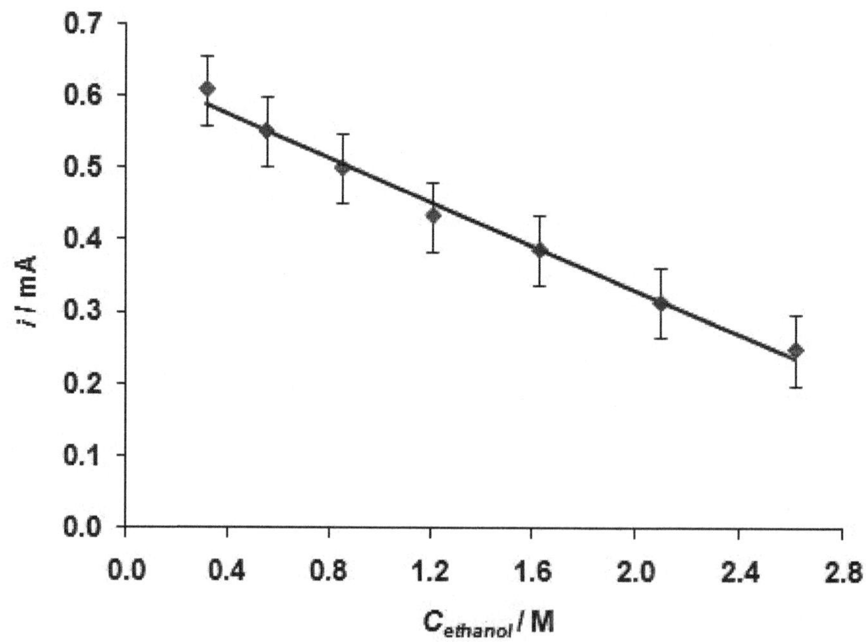

Fig. 4*. Ethanol biosensor calibration curve. Scan rate 10 mV s^{-1} (n = four measurements).*

Wu, *et al.* [34] reported an inverse calibration curve for oxygen consumption by a sensitive ethanol biosensor nanocomposite of carbon nanofiber with immobilized ADH. They observed decreased oxygen consumption with the increase of ethanol concentration in the sample. Chronoamperometric curves showed a decreasing response, upon addition of ethanol aliquots (0 - 112 µM) to static air-saturated pH 7.0 phosphate buffer saline. Wen, *et al.* [35] reported an ethanol biosensor constructed with alcohol oxidase/chitosan immobilized eggshell membrane and a commercial oxygen sensor. Those measurements were based on the depletion of dissolved oxygen upon exposure to ethanol solution (0.15 – 0.75 mM). Al-Mhanna and Hueber [36] reported an economic system that worked with one enzyme in a differential pH measurement device for alcohol oxidase and β-nicotinamide adenine dinucleotide ($NADH^{+}$) reaction and obtained a logarithmic curve for ethanol concentrations against change in pH for standard samples. These authors described an inverse correlation between the signal response and the analyte concentration for the indirect detection measurements working with a wide range of ethanol standard concentration solutions (17.14 μM – 17.14 M).

Mackey, *et al.* [37] optimized the proportion of dual enzyme horseradish peroxidase:glucose oxidase biosensor working with ratios of 1 : 7 to 7 : 1, immobilized on a polyaniline-polyvinylsulphonate modified screen-printed carbon paste electrode and identified the proportion that produced the best response signal was 1 : 1. Rondeau, *et al.* [38] identified the optimal proportion of two enzymes in the biosensor composite by monitoring electrical response signals to establish idealized conditions for glucose oxidase : horseradish peroxidase immobilized with a modified carbon paste for in order to increase the selectivity, sensitivity, accuracy and stability [38]. The intensity of the electrochemical signal response was analyzed by Alpat and Telefoncu [39] who measured the amount of alcohol dehydrogenase immobilized on the electrode surface (47.1 to 200 U cm^{-2}) and found that the linear response was between 0.01 mM and 0.04 mM for 117.6 U cm^{-2}.

Nicell and Wright [40] reported the dependence of horseradish peroxidase activity over a wide range of hydrogen peroxide concentrations. They observed an increase in the inhibitory effect on the enzyme catalytic activity. The static procedures of the electrochemical measurements of this work for the ethanol concentration solutions (0.330 – 2.62 M) probably promotes an increase of the peroxide hydrogen concentration in the electrolytic cell, and hence the inhibition of the HRP.

Yotova and Medhat [41] reported the inhibition effect in a multi-enzyme immobilized biosensor system constructed to analyze residue from pesticides with acetylcholinesterase and choline oxidase. The relative inhibition percentage of each measurement was calculated using the following equation:

$$I,\% = \frac{Ipc_0 - Ipc}{Ipc_0} 100 \qquad (1)$$

where I is the relative inhibition; Ipc_0 is the initial inhibited cathode current intensity measured for the lower ethanol concentration and Ipc the inhibited cathode current intensity determined for each sample. Assuming a possible inhibition effect on the cyclic voltammetric response signal with the increase of the ethanol concentration in the sample, this treatment was adopted for this work. A linear correlation was observed, confirming the inhibitory effect of the ethanol on the enzyme.

Amine, *et al.* [42] published a review that discusses horseradish peroxidase among the enzymes that could be used for inhibition-based biosensors applied for food safety and environmental monitoring. Kuusk and Rinken [43] classified the carbaril inhibition of tyrosinase biosensor by excess substrate and considered the reasons behind their inability to determine low carbaryl concentrations by a classical steady state kinetic approach. The K_m and I_{max} kinetics parameters were calculated from Lineweaver–Burk plots by using the relative inhibition values as described in equation 2:

$$\frac{1}{RI} = \frac{K_m^{app}}{RI_{max}} \frac{1}{C_{ethanol}} + \frac{1}{RI_{max}} \qquad (2)$$

where $1/C_{ethanol}$ is the concentration of the ethanol in the solution sample, RI and RI_{max} represent the initial and the maximum relative inhibition current, respectively, and K_m^{app} is the apparent Michaelis constant.

The Lineweaver-Burk plot for the dual enzyme AOX-HRP biosensor showing $1/I$ versus $1/C_{ethanol}$ is illustrated in Figure 5.

The I_{max} value determined considering the inhibition effect on the dual enzyme biosensor was 0.0724 μA, and the K_m^{app} was 1.41 M (R^2=0.9912).

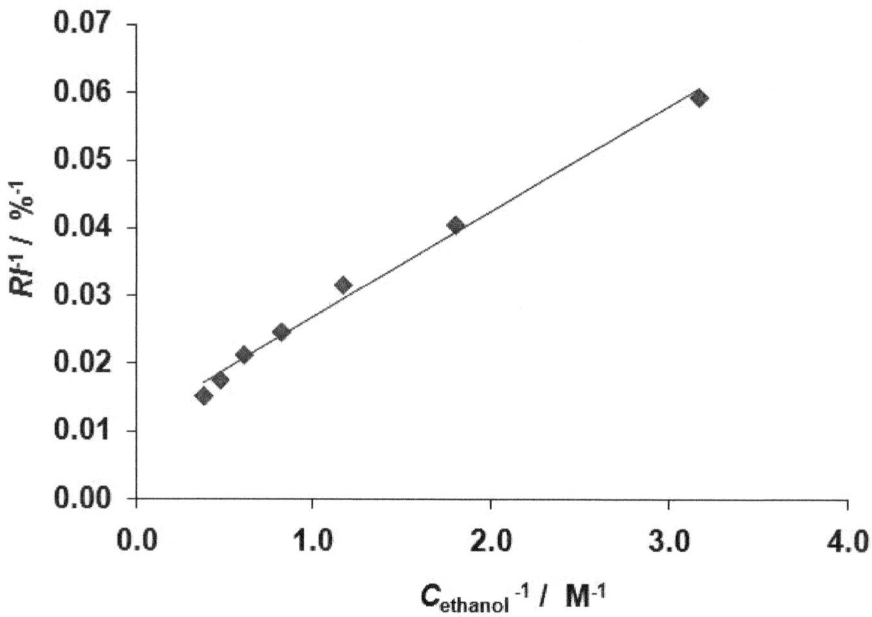

Fig. 5. *Line-weaver-Burk plot for the bienzimatic AOD-HRP biosensor for different ethanol concentration*

Table 1 shows the analytical performance of the proposed ethanol biosensor towards ethanol detection compared with various electrochemical biosensors modified for dual enzymes that also reported K_m^{app} and I_{max}. Despite the low affinity for substrate observed in this work, the sensitivity was higher when compared with those determined for both redox hydrogel dual enzyme films previously reported in the literature [44-45]. This suggests that the linear range and detection limit of the proposed ethanol biosensor mentioned above appear to be beneficial compared to other previously reported modified electrodes.

Table 1. *Comparison of analytical characteristics of ethanol dual enzyme biosensors.*

Film/Composite/Enzymes	*I.D. / cm*	K_m^{app} / mM	I_{max} / nA	*Sensitivity,* nA/μM	Reference
HRP+AOX+PVI-Os	0.305	4.71	813.95	0.17	[44]
HRP/PVI_{10}-Os/PEG-DGE/AOX/CP5	0.305	9.6 ± 0.3	572 ± 7	0.06	[45]
PANI-GEC/HRP/BSA/AOX	0.130	1.410	72.4	51.3	This work

I.D. - internal diameter; PVI - Poly(vinyl-imidazole; PVI_{10}-Os - redox hydrogel synthesized;
PEG-DGE - Poly(ethylene glycol) (400) diglycidyl ether; CP5 - electrodeposition polymer;
Os - complex: redox polymers synthesized (4,4'dimethylbipyridine); PANI-GEC: polyaniline in Graphite epoxy composite;
BSA - Bovine serum albumin. The applied potentials for all configurations are –50 mV vs. *Ag/AgCl.*

Conclusions

The composite material prepared from differing proportions of graphite and PANI displayed enhancement in the conductivity for compositions of less than 20 % graphite and a synergistic effect that increased its response for mixtures with more than 50 % of graphite. The thermal analysis techniques applied to characterize the prepared composites showed a good agreement with the original proposed formula composition. The electrochemical results confirm that it is possible to detect ethanol with this biosensor in the ethanol concentration range of 0.316 to 2.62 mol L^{-1} limited by a significant inhibition effect observed in the enzyme.

Acknowledgements: *Thanks to Toyobo of Brazil (enzyme horseradish peroxidase) and CNPq support from the Announcement Universal - 2008/2010 and PIBIC.*

References

[1] ASTM. *D5501-12* Standard Test Method, 100 Barr Harbor Drive, West Conshohocken, PA, USA, *ASTM International* (2012).
[2] G. Hall, W. M. Reuter, *HPLC Analysis for the Monitoring of Fermentation Broth During Ethanol Production as a Biofuel,* www.perkinelmer.com/pdfs/downloads/abr_ethanol-asbiofuelbyhplcappbrief.pdf (17. 07. 2014).
[3] K. Schugerl, *J. Biotechnol.* **85** (2001)149-173.
[4] W. Horwitz, Official methods of Analysis of AOAC International. Gaithersburg, Maryland 20877-2417, USA, *AOAC International* (2005).
[5] M. L. Wang, J. T. Wang, Y. M. Choong, *Food Chem.* **86** (2004) 609-615.
[6] L. Cao, *Introduction: Immobilized Enzymes: Past, Present and Prospects*. In: Carrier-bound Immobilized Enzymes: Principles, Application and Design. KGaA, Weinheim, WILEY-VCH Verlag GmbH & Co, 2005, p.100.
[7] H. Nakamura, I. Karube, *Anal. Bioanal. Chem*. **377(3)** (2003) 446-468.
[8] A. P. F. Turner, *Chem. Soc. Rev.* **42(8)** (2013) 3184-3196.
[9] R. W. Bogue, *Sensor* **4** (2003) 302-310.
[10] A. K. Wanekaya, W. Chen, A. Mulchandani, *J. Environ. Monit.* **10** (2008) 703-712.
[11] M. C. Blanco-Lopez, M. J. Lobo-Castanon, A. J. Miranda-Ordieres, *J. Chem. Education* **84**(4) (2007) 677-678.
[12] M. F. Barroso, M. F. C. Delerue-Matos, M. B. P. P. Oliveira, *Food Chem.* **132** (2012), 1055-1062.
[13] N. I. Bojorge-Ramírez, A. M. Salgado, B. Valdman, *Assay Drug Dev. Technol.* **5(5)** (2007) 673-682.
[14] M. S. Belluzo, M. E. Ribone, C. M. Lagier, *Sensors* **8** (2008) 1366-1399.
[15] J. Liu, J. Wang, *Biotech. Applied Biochem.* **30** (1999) 177-193.
[16] N. I. Bojorge-Ramirez, A. M. Salgado, B. Valdman, *Brazilian J. Chem. Eng*. **26(**2) (2009) 227-249.
[17] E. M. Alhadeff, A. M. Salgado, O. Cós, N. Pereira Jr., B. Valdman, F. Valero, *Appl. Biochem. Biotech.* **146** (2008) 129-136.
[18] N. I. Bojorge-Ramírez, E. Alhadeff, *Graphite-Composites Alternatives for Electrochemical Biosensor.* In: Metal, Ceramic and Polymeric Composites for Various Uses, Uses. J. Cuppoletti: (2011) p. 684.
[19] N. C. Veitch, *Phytochemistry* **65** (2004) 249-259.
[20] A. M. Azevedo, V. C. Martins, D. M. Prazeres, J. Vojinovic, J. M. S. Cabral, L. P. Fonseca, *Biotechnol. Annu. Rev.* **9** (2003) 199-247.
[21] K. Chattopadhayay, S. Mazumdar, *Biochemistry* **39** (2000) 263-270.

[22] A. S. L. Carvalho, E. Pinto e Melo, B. S. Ferreira, M.T. Neves-Petersen, S. B. Petersen, M.R. Aires-Barros, *Arch. Biochem. Biophys.* **415** (2003) 257 - 267.

[23] A. M. Azevedo, J. M. S. Cabral, T. D. Gibson, L. P. Fonseca, *J. Mol. Catal. B-Enzym.* **28(2-3)** (2004) 45-53.

[24] A. M. Azevedo, D. M. Prazeres, J. M. S. Cabral, L. P. Fonseca, *Biosens. Bioeletron.* **21** (2005) 231-247.

[25] S. Datta, L. R. Christena, Y. R. S. Rajaram, *Biotech.* **3(1)** (2013) 1- 9.

[26] T. Xie, A. Wang, L. Huang, H. Li, Z. Chen, Q. Wang, X. Yin, *Afr. J. Biotechnol.* **8(19)** (2009) 4724-4733.

[27] J. Dweck, B.F. Andrade, E.E.C. Monteiro, R. Fischer, *J. Therm. Anal. Calorim.* **67** (2002) 321-326.

[28] R. S. Lima, G. S. Nunes, T. Noguer, J. Marty, *Quím. Nova* **30(1)** (2007) 9-17.

[29] S. Konwer, J. P. Gogoi, A. Kalita, S.K. Dolui, *J. Mater. Sci.: Mater. Electron.* **22** (2011) 1154 - 1161.

[30] S. Bourdo, B. A. Warford, T. Viswanathan, *J. Solid State Chem*. **196** (2012) 309-313.

[31] Z. Mo, H. Shi, G. Niu, Z. Zhao, Y. Wu, *J. App. Polymer Sci.* **112** (2009) 573-577.

[32] E. Song, J.W. Choi, *Nanomaterials* **3**, (2013) 498-523.

[33] K. Sirkar, A. M. Revzin, V. Pishko, *Anal. Chem*. **72**, (2000) 2930-2936.

[34] L. Wu, J. Lei, X. Zhang, H. Ju, *Biosens. Bioelectron.* **24** (2008) 644-649.

[35] G. Wen, Y. Zhang, S. Shuang, C. Dong, M. M. F. Choi, *Biosens. Bioelectron.* **23** (2007) 121-129.

[36] N. M. M. Al-Mhanna, H. Huebner, *Int. J. Chem.* **3(1)** (2011) 47-56.

[37] D. Mackey, A. Killard, A. Ambrosi, M. Smyth, *Sens. Actuator B-Chem.* **122** (2007) 395-402.

[38] A. Rondeau, N. Larsson, M. Boujtita, L. Gorton, N. El Murr, *Analysis* **27** (1999) 649-656.

[39] S. Alpat, A. Telefoncu, *Sensors* **10**, (2010) 748-764.

[40] J. A. Nicell and H. Wright, *Enzyme Microb. Technol.* **21(4)** (1997) 302-310.

[41] L. Yotova, N. Medhat, *Int. J. Bioautomation* **15(4)** (2011) 267-276.

[42] Amine A., H. Mohammadi, I. Bourais, G Palleschi, *Biosens. Bioelectron.* **21** (2006) 1405 -142.

[43] E. Kuusk, T. Rinken, *Enzyme Microb. Technol.* **34** (2004) 657-661.

[44] J. Castillo, S. Gáspar, I., Sakharov, E. Csöregi, *Biosens. Bioelectron.* **18** (2003) 705-714.

[45] I. S. Alpeeva, A. Vilkanauskyte, B. Ngounou, E. Csöregi, I. Y. Sakharov, M. W. Gonchar, *Microchim. Acta* **152** 21-27 (2005).

6

Electrochemical extraction of oxygen using PEM electrolysis technology

BOULBABA ELADEB, CAROLINE BONNET, ERIC FAVRE and FRANÇOIS LAPICQUE✉

Laboratoire Réactions et Génie des Procédés, CNRS- Université de Lorraine, ENSIC, 1 rue Grandville, F-54001 Nancy, France

✉Corresponding Author: E-mail: francois.lapicque@ensic.inpl-nancy.fr

Abstract

Electrochemical extraction of oxygen from air can be carried out by chemical reduction of oxygen at the cathode and simultaneous oxygen evolution by water anode oxidation. The present investigation deals with the use of an electrolysis cell of PEM technology for this purpose. A dedicated 25 cm^2 cell provided with a commercial water electrolysis MEA and titanium grooved plates has been designed for continuous operation at pressures close to the ambient level. The MEA consisted of a Nafion 117 membrane sandwiched between a Pt/C cathode and a non-supported Pt-Ir anode. Oxygen partial consumption in long-term runs was evaluated by analysis of the outlet air by gas chromatography, depending on the cell voltage - or the current density - and the excess in air oxygen fed to the cathode. Runs over more 50 hours indicated the relative stability of the components used for current densities ranging from 0.1 to 0.2 A cm^{-2} with high efficiency of oxygen reduction. Higher current density could be envisaged with more efficient MEA's, exhibiting lower overpotentials for oxygen evolution to avoid too significant degradation of the anode material and the membrane. Interpretation of the data has been carried out by calculation of the cathode current efficiency.

Keywords

Oxygen extraction; PEM electrolysis; electrode stability; Pt-Ir anode; current efficiency

1. Introduction

Extraction of oxygen from air can be used for various applications, as follows. For instance oxygen impoverished air reduces the oxidation content from the atmosphere over food for the sake of higher preservation in its packing. Lower oxygen contents in air also reduce the corrosion rate at non-noble metals pieces. Another potential application is the preparation of the gas phase to be injected to biological reactors – upon addition of carbon dioxide – for the possible growth of microorganisms such as bacteria, fungi or microalgae or undifferentiated cells of higher-grade vegetal organisms in dedicated bioreactors.

Various techniques for oxygen removal from a gas phase to be used for biological applications or food processing can be listed as follows. The easiest technique, which can be qualified as indirect, simply consists of adding of pure nitrogen into the air stream: the cost of pressured nitrogen renders the technique somewhat expensive for practical applications. Conventional, non electrochemical techniques for oxygen removal rely upon the following various principles: (i) oxygen combustion by a gas microburner, however other gases of significant safety or environment impact such as carbon monoxide or nitrogen oxides can be formed; (ii) selective reaction of oxygen upon addition of reducing agents *e.g.* sulfites that have been used for decades for food preservation: however the sulfate produced by oxygen oxidation of sulfite have to be removed from the medium; (iii) selective adsorption on a suitable compound which has nevertheless to be regenerated for its possible reuse; (iv) membrane separation of oxygen from carbon dioxide and nitrogen at temperatures close to ambient: however, because of the intermediate behaviour of oxygen in the three-species, this technique cannot be used for biological applications.

Besides, various electrochemical techniques can be used for oxygen extraction from an air flow, as reviewed by Winnick [1]. In a pioneering paper, Langer and Haldermann [2] showed that oxygen could be extracted from air and recovered in the form of pure oxygen in a two-compartment electrolytic cell, in the presence of acidic or alkaline solutions and separated by a conventional cationic membrane:

Cathode $O_2 + 4H^+ + 4e \rightarrow 2\,H_2O$ (1)

Anode $2\,H_2O \rightarrow O_2 + 4H^+ + 4e$ (2)

The equilibrium voltage of such a cell is nil, assuming similar conditions of H^+ activity and oxygen pressure in the two compartments; the cell voltage is the sum of the two overpotentials – in absolute value- and the ohmic drop of the cell. The two reactions involved are nevertheless slow, even at platinum surfaces, and the cell voltage attains easily 1 V even at low current densities. Since then, the technique was improved by Wynween and Montgomery [3]. In the eighties, Fujita *et al.* [4] improved the efficiency of the technique by using a Nafion membrane, an air cathode whereas the anode was platinum particles deposited on the membrane surface: this technology derived from PEM fuel cell technology allowed current density up to 200 mA cm^{-2} for cell voltage at 1.4 V for continuous operation: the corresponding energy consumption for the oxygen extraction was near 4.7 kWh kg^{-1} O_2. It can be observed that the process consumes less energy than the conventional electrolysis, with specific energy consumption near 6.7 kWh kg^{-1} O_2 for a cell voltage of 2 V, even though the current densities of the two processes differ quite a lot. More recently, General Electrics [5] developed an oxygen supply system for high-altitude aircraft relying upon similar technology: oxygen at high pressures could be produced with cell voltage claimed to be near 1 V at 125 mA cm^{-2}.

The electrochemical separation of oxygen can be carried out upon specific diffusion in mixed, non stoichiometric oxides exhibiting oxygen vacancies in the lattice, as described in several patents [6-8]. The technology used derived from solid oxide cells: oxygen is reduced to O^{2-} ions at the cathode and migrates to the anode for back oxidation to oxygen; temperatures higher than 500 °C are required for sufficient diffusivity of O^{2-} ions in the non-stoichiometric oxide mixture. However the current density is usually below 100 mA cm^{-2}.

Oxygen extraction by its reduction to hydrogen peroxide on a graphite-based cathode through exchange of two electrons only was imagined by Tseung and Jasem [9]

Cathode $O_2 + H_2O + 2e \rightarrow HO_2^- + OH^-$ (3)

As expected form the electron number, the energy consumption could be reduced to 2.7 kWh kg^{-1} O_2 in a concentrated KOH solution. Peroxide decomposes chemically to oxygen and peroxide on the surface of $NiCo_2O_4$ mesh, whereas oxygen is formed at the $NiCo_2O_4$ anode after

Anode $2\ OH^- \rightarrow ½\ O2 + H_2O + 2e$ (4)

Brillas *et al.* [10] developed the above technique by using carbon-polytetrafluorethylene air-fed cathode. Peroxide anions HO_2^- are formed at the carbon-based cathode, whereas oxygen evolution occurs at the anode by oxidation of both HO_2^- and OH^- anions. Pure oxygen is produced with a voltage lower than in conventional water electrolyzers, because of the lower standard potential of $OH^-/\ H_2O_2$, which further reduces the energy consumption. However, the current density is limited below 0.2 A cm^{-2} by the finite concentration of peroxide in the solution.

Recently, we demonstrated the use of a PEMFC for oxygen extraction, the anode reaction being hydrogen oxidation [11]. The technique was shown to be of relatively high efficiency, with current density up to 0.6 A cm^{-2}, *i.e.* 0.015 mol O_2 s^{-1} per m^2 membrane. Long-term tests confirmed the validity of the method, which however requires the presence of hydrogen to feed the fuel cell. To avoid this drawback, it was preferred to replace hydrogen oxidation at the anode by oxygen evolution. The present investigation deals with the use of an electrolysis cell of PEM technology involving reactions (1) and (2) at the electrodes. A dedicated 25 cm^2 cell provided with a commercial water electrolysis MEA has been designed for operation at pressures close to the ambient level. Oxygen partial consumption in long term tests was evaluated by analysis of the outlet air, depending on the cell voltage - or the current density - and the excess in air oxygen fed to the cathode. Interpretation of the data has been conducted by calculating the cathode current efficiency.

2. Experimental section

2.1. Electrochemical cell and measurement devices

The 25 cm^2 cell has been designed and built up from regular PEM technology. The cell consisted of two Ti bipolar plates being 10 mm thick: the flow pattern of serpentine type with five 1×1 mm^2 parallel channels with 1mm broad edge was designed and machined by the mechanical workshop of the lab. The grooved part of the plate was covered by a 0.5 µm gold layer by ion sputtering. Gas diffusion layer at the cathode was of carbon material (SGL 30 BC, UBZM, Germany) including a macroporous carbon fiber paper and a carbon black microporous layer (MPL), whereas a one mm thick high porosity Ti fleece was used at the anode. The MEA (Fumatech) designed for PEM water electrolysis, consisted in a 117 Nafion membrane, a Pt/C (0.4 mg cm^{-2}) cathode and a non-supported Pt/Ir (0.45 mg cm^{-2}) anode (Figure 1).

All experiments were conducted at temperature levels ranging from 50 to 80°C. Liquid water was driven to the anode compartment by a peristaltic pump at 5 cm^3 min^{-1} and preheated before entering the cell. Purified compressed dry air at flow rate ranging from 20 to 200 STP cm^3 min^{-1} was humidified in lab-made trickled bed at RH=62 %. The cell was operated either at fixed voltage or at controlled current density using a PGSTAT 30 Autolab potensiostat connected to a 20 Amp. Autolab booster.

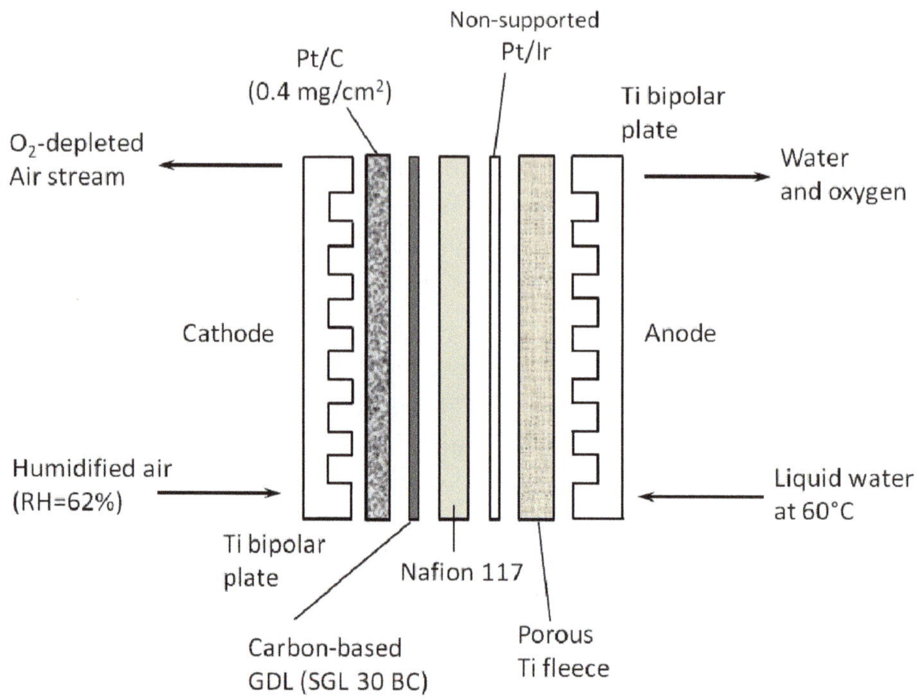

Figure 1. *Schematic view of the 25 cm^2 cell for oxygen extraction*

Prior to its use for oxygen extraction, the MEA was run up by carrying out water electrolysis at 80°C: the anode was fed with hot water and nitrogen at 40 STP cm^3 min^{-1} was fed to the cathode. As recommended by the supplier the MEA was conditioned by 24 hour long operation at 2.0 V for maturation of the anode Ir/Pt catalyst.

2.2. Current versus voltage measurements

Current density *vs.* voltage curve was established either in potensiostatic or in galvanostatic modes, so that the cell voltage was below 1.4 V for long term runs. Because the reactions of interest are oxygen reduction and evolution, the two electrode potentials are distributed around the equilibrium potential of H_2O/O_2 couple. Moreover the two reactions are relatively slow processes and exhibit absolute overpotentials in the same order of magnitude, it can therefore be estimated that the maximal cell voltage corresponds to anode potentials in the order of 1.8 V *vs.* NHE *i.e.* in a domain where corrosion phenomena at the anode can become significant and affect the stability of the anode catalyst. Most measurements were carried out with air but oxygen was also used for comparison.

2.3. Oxygen extraction runs and gas analysis

Oxygen extraction was achieved through numerous runs conducted at fixed current density for which the cell voltage was below 1.4 V as justified above. Taking into account Faraday's law and the oxygen content in air ($y_{O_2}^{in}$ =0.21), the inlet oxygen flow rate was large enough for all runs. The cell voltage at fixed current and air flow rate was monitored for lapses of time ranging from 4 to 24 hours: in most cases, the voltage attained a steady value within ten to thirty minutes.

The gas leaving the cathode was stored in a 400 cm^2 polymeric sampling bag. This bag has been thoroughly emptied first, the outlet was then sampled for a couple of minutes and the bag was purged again. The second, longer sampling procedure for analysis was then achieved. After sufficient filling, the sampling gas was shut. The decrease in temperature to the ambient level was the cause of water condensation and the bag slightly shrunk. The ambient temperature, T_{amb}, recorded for all runs ranged from 19 to 27°C. Analysis of the gas containing in the bag, *i.e.* nitrogen, oxygen and slight amount of water vapor was carried out off line by gas chromatography using a molecular sieve microcolumn with a TCD detector in the µGC equipment (Varian 490 GC). Three replicate analyses were conducted for each sample. The TCD detector was calibrated by analysis of six synthetic air-nitrogen mixtures of oxygen concentration perfectly known and ranging from 5 to 21 vol. %. The vapor amount in the gas mixture at T_{amb} was estimated using Antoine's law:

$$\log P_{H_2O} = 5.20389 - \frac{1737.93}{T_{amb} - 39.485} \tag{5}$$

where the partial pressure of water is expressed in Atm and T_{amb} is in K. The composition of the dry outlet could then be deduced from y_{H_2O}.

3. Experimental results and interpretation

3.1. Electrochemical behavior of the cell

Steady values of the current or the voltage – depending on the electrical mode - were usually obtained within 15 minutes or so and for voltammetric measurements the voltage or current level was changed every 30-60 minutes. The MEA response could slightly depend on the long-term run carried out immediately before the measurement. Temperature was shown to exert a moderate effect (data not shown) and the results presented here were obtained at 60°C.

At a given voltage gas flow rate exerted also a moderate influence on the cell current at a given cell voltage, as exemplified by Figure 2. Over a given flow rate depending on the current density, the effect is however of reduced importance. The observed phenomenon probably expresses the partial control by oxygen diffusion from the air fed to the cathode.

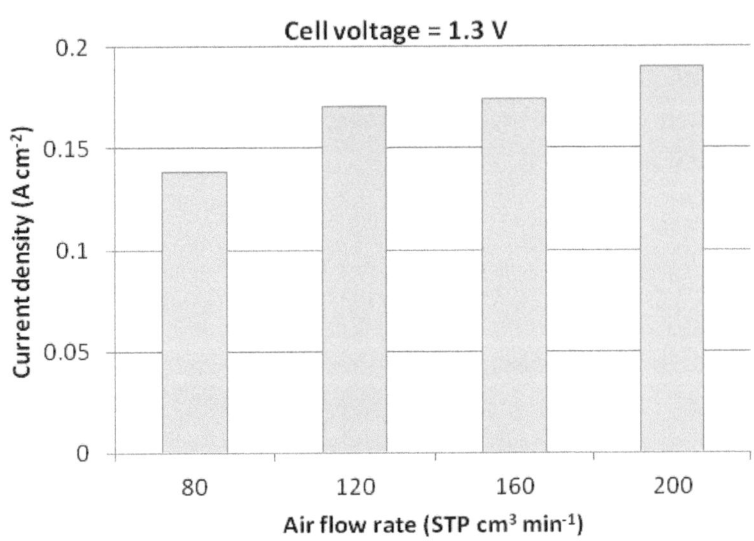

Figure 2. *Effect of the air flow rate on the current density recorded at fixed cell voltage.*

As expected, current density is an increasing function of the cell voltage (Figure 3). Feeding the cell with pure oxygen largely increases the performance of the cell, with for instance a current density twofold larger with pure oxygen than upon air feed at 1V. In comparison to what occurs with air pure oxygen accelerates charge transfer rates in both oxygen reduction and evolution, while suppressing the gas-side mass transfer occurring in the presence of air nitrogen. Current density up to 0.34 A cm^{-2} could be measured at 1.2 V with pure oxygen. The voltammetric data were compared to those obtained by Langer [2] or by Fujita [4] in Figure 3. In spite of the different *i-V* profiles reported, the available current densities are in the same order of magnitude over 1 V, also with better performance with pure oxygen. The differences between the three sources of data can be explained by the different features of the electrochemical cells: in addition to the likely different ohmic resistances, the electrodes differed a lot, in particular the anode, consisting of deposited Pt in the quoted papers and a Pt-Ir mixture in the present investigation.

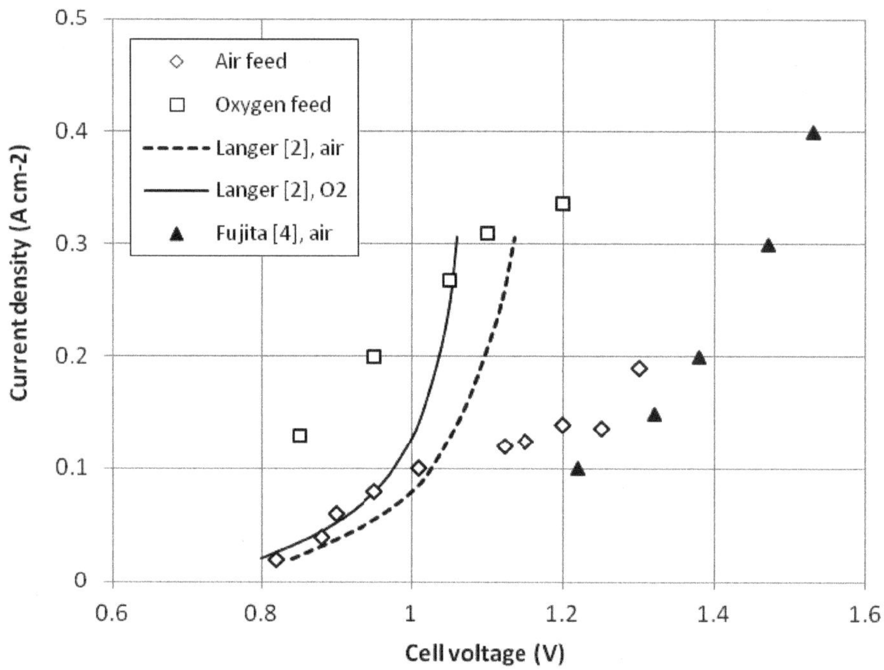

Figure 3. *Current density versus cell voltage at steady state. Present work with air at 200 STP cm^3 min^{-1} and oxygen at 120 STP cm^3 min^{-1}; comparison with literature data.*

Besides, the cell behaviour in long term runs was examined during oxygen extraction runs. As shown by Figure 4 for fixed values of the cell voltage, the current monitored remained at a stable level for periods up to 50 hours.

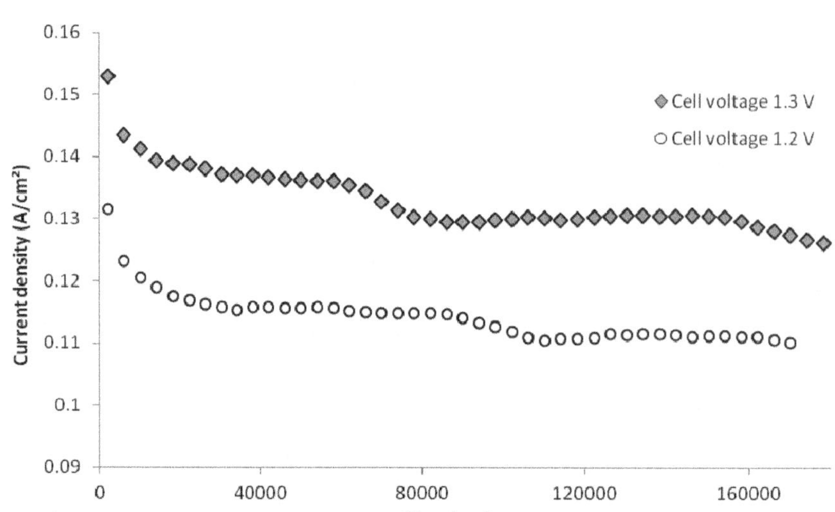

Figure 4. Current density in the oxygen extraction cell operated at fixed voltage.

3.2. Model of the cell oxygen extraction

A simple zero-D model has been written for the cathode compartment of the cell. The inlet variables are the molar flow rate of dry air F_{air}^{in} and the cell current *I*, whereas outlet variables are the molar flow rate of oxygen $F_{O_2}^{out}$ and the temperature of the sampling gas, T_{amb}: water was assumed to be at vapor-liquid equilibrium at T_{amb}. Faraday's law is written for oxygen consumption at the cathode according to reaction (1):

$$F_{O_2}^{out} = F_{O_2}^{in} - \frac{\Phi I}{4\Im} \quad (6)$$

where Φ is the current efficiency of oxygen reduction and $\Im$ is the Faraday's constant. The inlet oxygen flow rate is related to the inlet air flow as follows:

$$F_{O_2}^{in} = F_{air}^{in} - F_{N_2}^{in} = F_{air}^{in} y_{O_2}^{in} \quad (7)$$

where fraction $y_{O_2}^{in}$ refers to the oxygen content in air. The oxygen molar fraction in the sampling bag is defined as:

$$y_{O_2}^{out} = \frac{F_{O_2}^{out}}{F_t^{out}} \quad (8)$$

where subscript *t* refers to the total number of moles in the gas, covering nitrogen, oxygen and water vapor. Finally the total molar flow rate of gas is related to the flow of dry gas and the vapor molar fraction in the bag:

$$F_t^{out} = \frac{F_{air}^{out}}{1 - y_{H_2O} T_{amb}} = \frac{F_{air}^{in} - \frac{\Phi I}{4\Im}}{1 - y_{H_2O} T_{amb}} \quad (9)$$

For the sake of simplicity, the ambient temperature is no more mentioned in the following. The fraction of vapor in the sampled gas was calculated by Antoine's law. Finally, the stoichiometric factor λ of fed air oxygen related to the flow of oxygen consumed by reaction (1) with a current efficiency at unity is defined by the relation:

$$F_{O_2}^{in} = y_{O_2}^{in} F_{air}^{in} = \lambda \frac{I}{4\Im} \quad (9)$$

From relations (6) to (9) the current efficiency of oxygen reduction Φ is expressed as follows:

$$\Phi = \lambda \frac{1 - y_{H_2O} - \dfrac{y_{O_2}^{out}}{y_{O_2}^{in}}}{1 - y_{O_2}^{out} - y_{H_2O}} \quad (10)$$

For current efficiency equal to unity, the molar fraction of the oxygen in the sampling bag can be directly expressed as a function of operating parameter λ as:

$$y_{O_2}^{out} = \frac{\lambda - 1}{\dfrac{\lambda}{y_{O_2}^{in}} - 1} \left(1 - y_{H_2O}\right) \quad (11)$$

3.3. Experimental data and comparison with model predictions

The oxygen molar fractions determined by gas chromatography are compared to the values predicted by eq. (11) in Figure 5, considering the actual T_{amb} value. Good agreement between theory and practice is generally observed, even though the predicted line is slightly below the experimental data, indicating that the current efficiency could be somewhat lower than unity. For very large λ values, corresponding to large excess in oxygen, air is only slightly depleted in oxygen and the outlet gas fraction of oxygen is little different from the inlet value. Conversely, with low excess of oxygen, high abatement of oxygen from the fed air can be obtained, as expected.

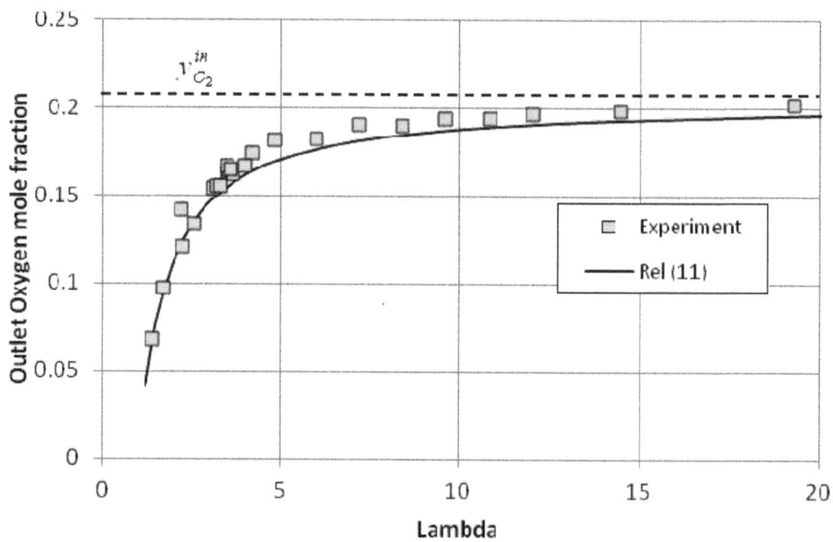

Figure 5. Oxygen molar fraction at the outlet of the cell versus the stoichiometric factor of air oxygen λ

Finally, because the outlet molar fractions of oxygen are in the same order of magnitude as the inlet fraction, in particular for large excess in fed air, the uncertainty in the determination of the current efficiency had to be estimated. The related calculations reported in the appendix show that high values of the current and moderate values for stoichiometric factor are favorable for higher accuracy in the determination of Φ: although the inlet flow rate of air could be moderate, the outlet molar fraction of oxygen is noticeably different from the inlet fraction, which prevails on the overall estimate for the uncertainty. On the contrary, for large λ values, the outlet molar fraction of oxygen is little below the inlet fraction and, from the expressions of terms A and B, the determination of the current efficiency is little accurate, with uncertainty exceeding 30 %.

The variations of the current efficiency with the stoichiometric factor of air oxygen and current density are shown in Figure 6 and 7 respectively, in the form of the maximum and the minimum estimates for each experiment. In spite of a significant dispersion, the data clearly show that the current efficiency is somewhat lower than unity.

The separate analysis of the effect of operating conditions on the current efficiency is rendered uneasy because of the large dispersion of the data in the representations (Φ *vs.* λ) or (Φ *vs.* i) in Figure 6 and 7. This apparent dispersion is mainly caused by the uncertainty in the determination of the current efficiency but also by the slight interaction between the two parameters. Nevertheless the stoichiometric factor of air oxygen exerts a negative influence on the current efficiency (Figure 6) whatever the current density applied. The positive effect of current density on Φ appears clearly with very little effect from λ (Figure 7): for current density over 100 mA cm^{-2}, the current efficiency is near or larger than 0.9, which is promising for the investigated process.

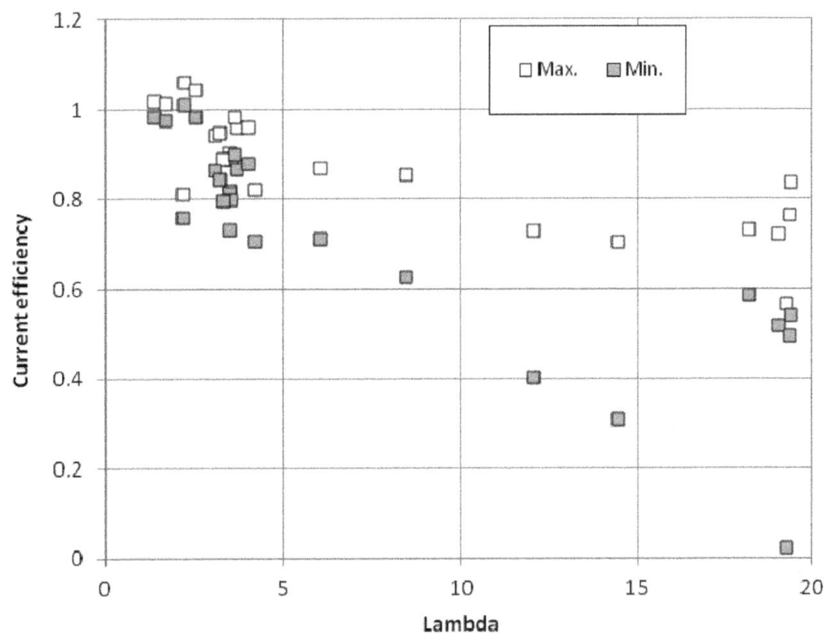

Figure 6. Current efficiency for oxygen reduction in the cell versus stoichiometric factor λ: maximum and minimum estimates

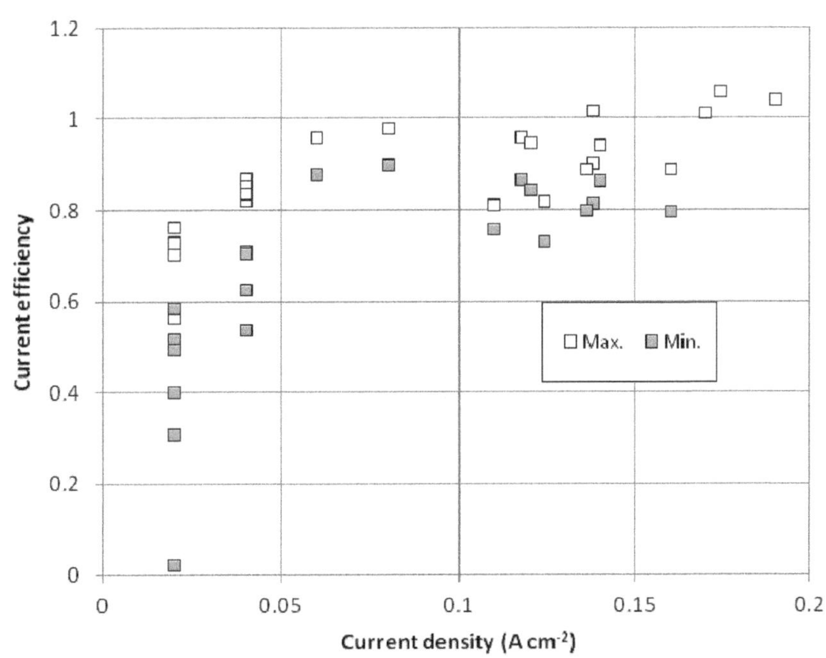

Figure 7. *Current efficiency for oxygen reduction in the cell versus the current density: maximum and minimum estimates*

4. Conclusions and significance

The validity of the oxygen extraction using PEM electrolysis technology has been validated in long term runs for current density up to 200 mA cm^{-2}, with a cell voltage near 1.3 V. Assuming that the reductive extraction is carried out with a current efficiency of 90 %, the corresponding energy consumption can be estimated to be near 4.4 kWh kg^{-1} O_2. Although the state of health of the PEM electrodes could not be evaluated by measurements of the electrode active surface [12], as done for PEMFC, no real decay in activity could be observed after more than 400 hour tests with the same MEA.

Considering the current progresses in water electrolysis technology, the current efficiency can be expected to attain 100 %, which would result in energy consumption near 4 kWh kg^{-1} O_2. Besides, the current improvement in the electrodes of PEM electrolysers is to allow substantial reduction in cell voltage. However, to our point of view, the main advantage offered by PEM electrodes is that the cathode reaction involving air oxygen should be carried out with current density to 1 A cm^{-2}, provided that the cell voltage does not exceed 1.4 V to limit degradation issues at an acceptable level: progress in electrolysis MEA has to be achieved for this purpose. In contrast, the reduction from a liquid phase as done through the peroxide route cannot be conducted for current densities over 100 or 200 mA cm^{-2}. Although the PEM technique is not to allow lower energy consumption, higher production rates can be expected.

Long-term tests will be carried out at larger current densities, *i.e.* higher cell voltages, together with investigation of ageing phenomena, to evaluate the highest current density offered by the technique. Further work has to be done by improving developing electrode materials exhibiting faster kinetics for both oxygen reduction and evolution.

Acknowledgements: *The authors are indebted from Region Lorraine and Institut Carnot ICEEL which funded the investigation cost and B. Eladeb's PhD grant.*

Appendix: Uncertainty in determination of the current efficiency

Relation (10) shows that efficiency Φ is a function of parameters or variables λ, $y_{O_2}^{out}$ and y_{H_2O}. The differential of the current efficiency was expressed as a function of partial derivatives with respect to the three above variables:

$$d\Phi = \left(\frac{\partial \Phi}{\partial \lambda}\right) d\lambda + \left(\frac{\partial \Phi}{\partial y_{O_2}^{out}}\right) dy_{O_2}^{out} + \left(\frac{\partial \Phi}{\partial y_{H_2O}}\right) dy_{H_2O} \tag{A1}$$

Derivation of the current efficiency with respect to λ, $y_{O_2}^{out}$ and y_{H_2O} followed by algebraic rearrangement led to:

$$\frac{d\Phi}{\Phi} = \frac{d\lambda}{\lambda} + \frac{dy_{O_2}^{out}}{y_{O_2}^{out}}\left[y_{O_2}^{out}\left(\frac{1}{B} - \frac{1}{A y_{O_2}^{in}}\right)\right] + \frac{dy_{H_2O}}{y_{H_2O}}\left[y_{H_2O}\left(\frac{1}{B} - \frac{1}{A}\right)\right] \tag{A2}$$

with

$$A = 1 - y_{H_2O} - \frac{y_{O_2}^{out}}{y_{O_2}^{in}} \tag{A3}$$

and

$$B = 1 - y_{O_2}^{out} - y_{H_2O} \tag{A4}$$

Factor λ is linked to the molar flow rate of fed oxygen, which is proportional to the volume flow rate: the relative uncertainty of l can therefore be estimated by that for the inlet air flow rate. The flow meters employed for the measurements had a 1 % relative uncertainty in a large range of flow rates, *i.e.* for flow rates larger than 2 % of the full scale flow rate Q_{air}^{max}. Below this threshold value, larger uncertainty can be expected. The following expressions were then considered:

$$\frac{\Delta\lambda}{\lambda}=\frac{\Delta Q_{air}^{in}}{Q_{air}^{in}}=1\%\frac{0.02\,Q_{air}^{max}}{Q_{air}^{in}}\quad \text{for } Q_{air}^{in}<0.02\,Q_{air}^{max} \tag{A5}$$

$$\frac{\Delta\lambda}{\lambda}=\frac{\Delta Q_{air}^{in}}{Q_{air}^{in}}=1\%\quad \text{for } Q_{air}^{in}>0.02Q_{air}^{max}$$

In practice, because of the flow rate scale considered, the relative uncertainty varied from 3 to 10 %.

Moreover, in spite of the replicate analysis in the gas chromatograph and the thorough calibration procedure, it can be estimated that the relative error in the analytical determination is near 1 %:

$$\frac{\Delta y_{O_2}^{out}}{y_{O_2}^{out}}=1\% \tag{A6}$$

Finally, the uncertainty in y_{H_2O} is due to uncertainty in the ambient temperature. Considering that this temperature was estimated within 1 °C and using Antoine's law (5), the relative uncertainty in the molar fraction of vapour could be estimated in the temperature range considered:

$$\frac{\Delta y_{H_2O}}{y_{H_2O}}=4.5\% \tag{A7}$$

Applications of relations (A2)-(A7) yielded estimates for the relative uncertainty on the current efficiency, in the range 2-30 % for most cases.

References

[1] J. Winnick, "Electrochemical separation of Gases", in *"Advances in Electrochemical Science and Engineering"*, edited by H. Gerischer and C.W. Tobias, **Vol. 1**, VCH, New-York (1990), 210-248.

[2] S.H. Langer and R.G. Haldemann, *J. Phys. Chem.* **68** (1964) 962-963.

[3] R.A. Wynveen and K.M. Montgomery, *J. Electrochem. Soc.* **114** (1967) 589-592.

[4] Y. Fujita, H. Nakamura, T. Muto, *J. Appl. Electrochem.* **16** (1986) 935-940.

[5] J.W. Harrison, ASME 75-ENAs-51 (1975).

[6] W.N. Lawless, US Patent N° 4,462,891 (July 31, 1984); 4,547,277 (Oct. 15, 1985).

[7] L.G. Marianowski and R.J. Remick, US Patent N° 4,589,296 (1989).

[8] M. Riecke, US Patent N° 6,632,400 B2 (2003).

[9] A.C. Tseung, S.M. Jasem, *J. Appl. Electrochem.* **11** (1981) 209-215.

[10] E. Brillas, A. Maestro, M. Moratalla, *J. Appl. Electrochem.* **27** (1997) 83-92.

[11] S. Altmeyer, E. Favre, C. Bonnet, P. Carré, F. Lapicque « Etude de procédés pour application au flux gazeux d'un photobioréacteur de culture d'algues », Report 08 GPS 155, Nancy (Jan. 2009) [in French].

[12] A. Pozio, M. De Franceco, A. Cemmi,F. Cardellini and L. Giorgi, *J. Power Sources* **105** (2002) 13-19.

7

Electrochemical degradation of Reactive Brilliant Red K-2BP on Ti/RuTiIrSnMn oxide anode in a batch cell

LUO JIANCHENG*, YANG JIE*, LI WEISHAN*,**,***,✉ HUANG QIMING*,**,*** and XU HONGKANG****

*School of Chemistry and Environment, South China Normal University, Guangzhou 510006, China

**Key Laboratory of Electrochemical Technology on Energy Storage and Power Generation of Guangdong Higher Education Institutes, South China Normal University, Guangzhou 510006, China

***Engineering Research Center of Materials and Technology for Electrochemical Energy Storage (MOE), South China Normal University, Guangzhou 510006, China

****Dongguan Hongjie Environmental Technologies Ltd, Dongguan 523039, China

✉Corresponding Author: E-mail: liwsh@scnu.edu.cn

Abstract

Electrochemical degradation of Reactive Brilliant Red K-2BP on Ti/RuTiIrSnMn oxide anode in chloride containing solution was investigated by voltammetry and electrolysis in a batch cell. It is found that the degradation mechanism of K-2BP on Ti/RuTiIrSnMn oxide anode involves an indirect electrocatalytic oxidation, in which K-2BP is oxidized by the electrochemically generated active chlorine. This degradation reaction follows pseudo-first order reaction kinetics. Ti/RuTiIrSnMn oxide exhibits excellent electrocatalytic activity toward the generation of active chlorine from chloride. Hence, K-2BP can be electrochemically degraded effectively in chloride containing solution. The decolorization efficiency was found to increase with the decrease in pH and with the increase in current density, NaCl concentration, temperature, and flow rate of the solution.

Keywords

Decolorization; Electrooxidation; Active chlorine

Introduction

Dye effluent in textile industry is an environmental concern due to its huge quantity, dark color, low biodegradability, and potential toxicity to aquatic life. Various treatment methods have been proposed and/or tested to effectively degrade the effluent, including biological treatment [1-7], photocatalysis [8-10], ozonation [11,12], wet oxidation [13,14], Fenton reaction [15,16], electrochemical coagulation [17-19], and electrochemical oxidation [20-31].

Compared with other methods for the treatment of the textile effluent, electrochemical technologies are easy to operate, highly efficient and environmentally friendly. Electrooxidation has been widely used to treat various industrial wastewaters, including the wastewaters containing dye, urea, phenol, molasses, microcystins, tannery, olive oil, pesticide, *etc.* [20,26-32].

There are two mechanisms for the electrooxidation treatment of wastewater, direct and indirect oxidation. In the direct oxidation mechanism, pollutants are oxidized directly on the anode surface. In that case, there is no any other substance involved, except the electrons in the electrooxidation of pollutants. However, pollutants and their degraded products are usually adsorbed on the anode, leading to the low degradation efficiency and the complicated treatment technology for the anode cleaning. In the indirect mechanism, pollutants are oxidized in a bulk solution by a medium, such as active chlorine and hydroxyl radicals electrogenerated in-situ on the anode surface, and thus the adsorption of pollutants and their degradation products can be avoided. The medium is utilized immediately after its electrogeneration. Therefore, indirect electrooxidation can lead to a high efficiency of degradation.

Dimensionally stable anode (DSA) was originally designed for chlor-alkali industry [33]. DSA has long lifetime and it is much cheaper than noble metal anodes such as platinum. It has been widely used in a wastewater treatment research and development. Ruthenium- and iridium- oxides are the main composition of DSAs. They exhibit good electrocatalytic activity for active chlorine generation and oxygen evolution. Based on these two oxides, many composite oxides have been developed for DSA use. For example, ruthenium-based oxides, especially $Ti/Ti_{0.7}Ru_{0.3}O_2$, have been reported to treat the refractory effluent [26,28,30,31].

Reactive Brilliant Red K-2BP (K-2BP) is an azo-dye which is extensively used in the textile industry. Its molecule structure is shown in Fig. 1. K-2BP is resistant to either biological or photodegradation. The solution containing K-2BP is brilliant red and can be distinguished even in trace concentration by naked eye.

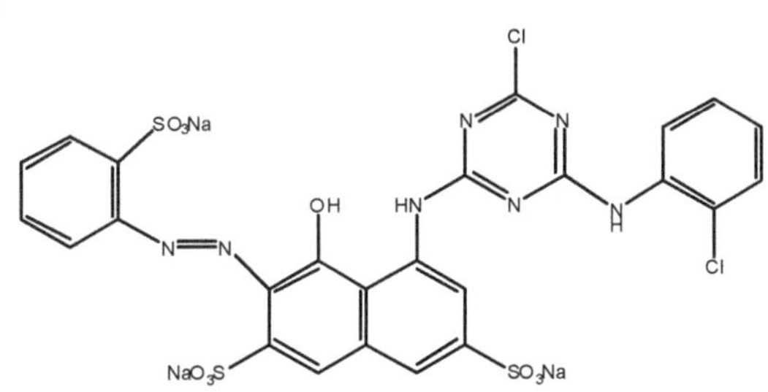

Figure 1. *Molecule structure of Reactive Brilliant Red K-2BP*

In this work, a coating on titanium substrate of composite oxides based mainly on Ru and Ir, Ti/RuTiIrSnMn oxide, was used as an anode to degrade K-2BP by the indirect electrooxidation with electrochemically generated active chlorine as the medium. The parameter effects of electrolytic conditions on the degradation efficiency were investigated and the degradation mechanism was discussed.

Experimental

Reagents and solutions

Deionized water was used for the preparation of all solutions. Sodium hydroxide and sulfuric acid were used to adjust pH of the solution. Sodium chloride was used for the generation of active chlorine. Sodium sulfate and sodium nitrate were also used as electrolytes for comparison. All the reagents were analytical reagents and purchased from Sinopharm Chemical Reagent Co., Ltd, China. The commercially used textile azo dye, mainly containing Reactive Brilliant Red K-2BP, was purchased from the local market. All the chemicals were used as received without further purification.

Electrolysis cell

The experiments were carried out in an un-divided filter-press cell (Fig. 2) with constant current at fixed temperature of 30 °C except otherwise stated. The dye-containing solution was supplied from the feed tank (250 mL) to the electrolysis cell by a peristaltic pump. The treated water can be recycled back to the solution tank if necessary. Constant current was provided by a D. C. power supply (363, EG&G). A magnetic stirrer was used to keep the solution mixed and a thermostatic water bath was employed to control the temperature. The solution volume was 200 mL with 100 mg/L K-2BP. The anode (Ti/RuTiIrSnMn oxide) was supplied by Guangzhou Etsing Plating Research Institute, which was prepared by coating the composites (the compounds containing Ru, Ti, Ir, Sn, and Mn) on Ti plate and treated under high temperature. The cathode was made from stainless steel (SUS304). Both electrodes, anode and cathode were of the same size of geometric area having 21 cm^2 exposed to the solution. The gap between the two electrodes was 10 mm. Before each set of experiments, the anode was immersed in absolute alcohol for 5 min and then rinsed with deionized water. The cathode was polished with sandpaper (2000), then immersed in 1.0 M HCl solution for 5 min, and finally rinsed with deionized water.

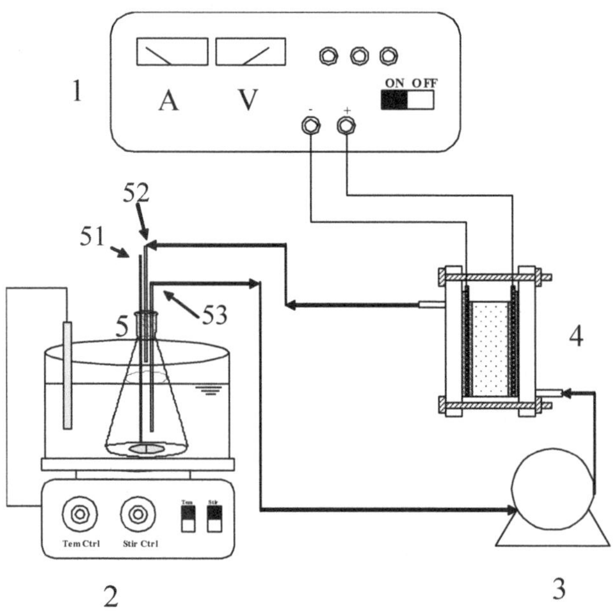

1. D. C. supply
2. magnetic stirrer
3. peristaltic pump
4. un-divided filter-press cell
5. feed tank
51.sampling
52. inlet
53. Outlet

Figure 2. *Schematic diagram of experimental setup for the degradation of K-2BP*

Analytical procedure

At desired time interval, the dye solution was sampled and filtered through a 0.45 μm millipore membrane. Several drops of saturated sodium sulfite solution were added into the sample to prevent further oxidation of the dyestuffs and their intermediates by residual chlorine. The samples were analyzed immediately after their collection using a Shimadzu UV-1700 double-beam spectrophotometer.

The concentration of K-2BP in the samples was determined from the UV-Vis absorbance recorded at 520 nm (max) and the degradation efficiency (A / %) was calculated according to [28]:

$$A\,/\,\% = \frac{(A_0 - A_t)}{A_0}100$$

where A_0 and A_t are the absorbance of the solution before and after electrolysis, respectively.

Voltammetric measurements

In order to understand the degradation mechanism of K-2BP on Ti/RuTiIrSnMn oxide anode, linear and cyclic voltammograms were obtained by using a three-electrode cell (Solartron 1470E, Solartron Co., Ltd.). A Pt plate was used as the counter electrode, a saturated Hg/Hg_2Cl_2 electrode (SCE) was used as the reference electrode, and Ti/RuTiIrSnMn oxide disk with a diameter of 1 cm was used as the working electrode. The potentials reported are with respect to the SCE. All the electrochemical measurements were carried out at room temperature. Linear voltammetry was performed in 0.5 M H_2SO_4 solutions with and without 1.0 M NaCl from 0.5 V to 1.35 V at a can rate of 1 mV s^{-1}. Cyclic voltammetry was performed in 0.5 M H_2SO_4 + 1.0 M NaCl solutions with and without 200 mg L^{-1} K-2BP between 0.5 V and 1.15 V at 1 mV s^{-1}. The cyclic voltammograms reported in this paper are the results of the 5th cycle. Before the experiments, the working electrode was immersed in absolute alcohol for 5 min, rinsed by deionized water and kept in the experimental solutions for 30 min.

Results and Discussion

Effect of salts

In order to enhance the diffusion and adsorption of dyestuffs onto the textile, abundant inorganic salts, such as NaCl, Na_2SO_4 and $NaNO_3$, are added into the dyestuff slurries. This technology results in the real textile effluent containing various electrolytes. On the other hand, the existence of salts improves the conductivity of the effluent and reduces the electric energy needed for the electrochemical treatment of the effluent. Therefore, in this paper the effect of various salts on degradation efficiency of K-2BP on Ti/RuIrSnMnTi anode was investigated.

As shown in Fig. 3, the decolorization efficiency of K-2BP in both Na_2SO_4 and $NaNO_3$ solutions increases slowly and linearly with treatment time. The decolorization efficiency of K-2BP is only 50.9 % after one hour electrolysis in both Na_2SO_4 and $NaNO_3$ solutions. However, the decolorization efficiency of K-2BP in NaCl solution increases drastically in the first 20 min, then increases slowly in the later stage and reaches almost 100 % after 40 min. This indicates that NaCl favors the degradation of K-2BP. The mechanism can be explained as follows.

In the NaCl-free solution, water is decomposed on the surface of metal oxide anode, producing hydroxyl radicals, and the dye is degraded into CO_2 by hydroxyl radicals subsequently (Eqs. 1 and 2):

$$H_2O - e^- \rightarrow \cdot OH + H^+ \tag{1}$$

$$\cdot OH + dye - e^- \rightarrow H_2O + H^+ + CO_2\uparrow \tag{2}$$

The reaction of water decomposition on anode to form hydroxyl radicals is usually accompanied by the oxygen evolution reaction (Eq. 3), which reduces formation efficiency of the hydroxyl radicals and thus leads to the low degradation efficiency of K-2BP in both Na_2SO_4 and $NaNO_3$ solutions.

$$2H_2O - 4e^- \rightarrow 4H^+ + O_2\uparrow \tag{3}$$

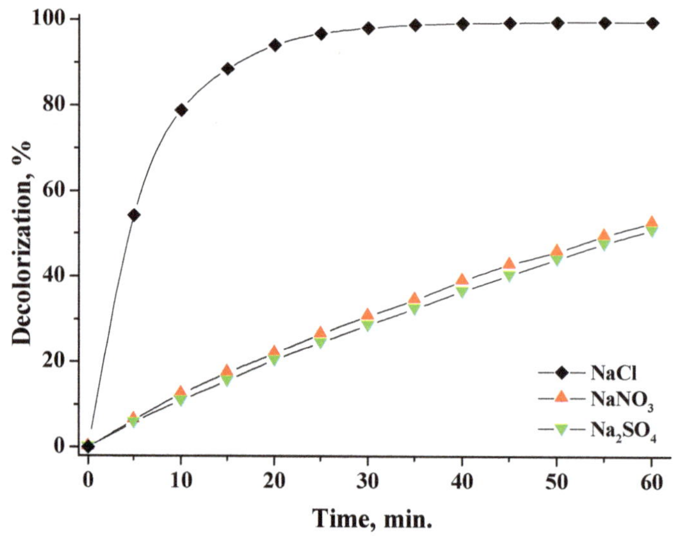

Figure 3. *Effect of supporting electrolyte (SE) on the decolorization efficiency of K-2BP.*
(j = 30 mA cm^{-2}, c_{SE} = 4.0 g L^{-1}, flux = 50 mL min^{-1}, initial pH = 7.0)

The composite of Ru, Ir, Sn, Mn, and Ti oxides exhibits high activity for the oxidation of chloride ions to form active chlorine. In the NaCl-containing solution, chloride ions are oxidized on this kind of composite oxide anode, forming active chlorine [22,27,29] (Eq. 4).

$$2Cl^- - 2e^- \rightarrow Cl_2\uparrow \quad (4)$$

The active chlorine, electrogenerated *in-situ* on an anode surface, transforms into a strong oxidant, such as HClO, which oxidizes dyes into small molecules or even carbon dioxide (Eq. 5). Therefore, higher degradation efficiency of the dye on the anode in the NaCl-containing solution can be expected than that in NaCl-free solution.

$$HClO + \text{dye (and its oxidation intermediates)} \rightarrow H_2O + Cl^- + CO_2\uparrow \quad (5)$$

Fig. 4 presents the linear voltammograms of Ti/RuIrSnMnTi oxide electrode in 0.5 M H_2SO_4 solution with and without NaCl.

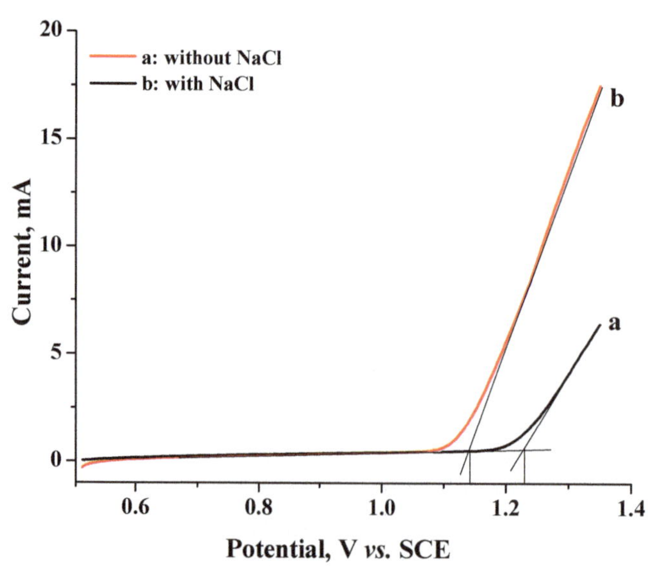

Figure 4. *Linear voltammograms of Ti/RuIrSnMnTi oxide electrode in the solutions with and without NaCl*

In the NaCl-free solution, the decomposition potential of water is 1.231 V. However, when chloride ions are added, the decomposition potential shifts negatively to 1.142 V and the current increases significantly. This indicates that chloride ions can be oxidized more easily on Ti/RuIrSnMnTi oxide electrode than the water decomposition, thus the Ti/RuIrSnMnTi oxide shows a good electrocatalytic activity toward the oxidation of chloride ions.

Fig. 5 presents the voltammograms of Ti/RuIrSnMnTi oxide electrode in 0.5 M H_2SO_4 + 1.0 M NaCl solutions with and without 200 mg/L K-2BP. For the electrode in the solution without K-2BP, the current increases sharply at the potential higher than 1.1 V during the forward scanning. Based on the observation of Fig. 4, it follows that the oxidation current is corresponding to the formation of active chlorine from oxidation of chloride ions. Active chlorine can be detected during the backward scanning, as shown by the current peak at about 1.09 V. For the electrode in the solution with K-2BP, the current for the oxidation of chloride ions is similar to that in the solution without K-2BP, but the peak current for the reduction of active chlorine is reduced. This suggests that K-2BP is not oxidized directly on the Ti/RuIrSnMnTi anode, but indirectly by active chlorine.

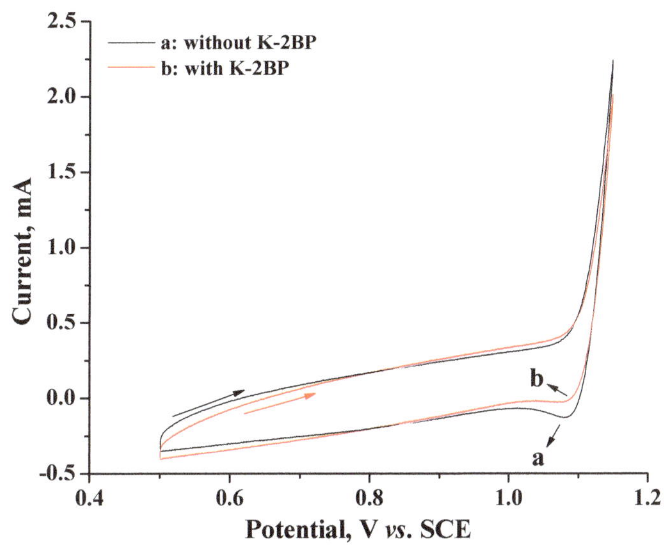

Figure 5. *Cyclic voltammograms of Ti/RuIrSnMnTi oxide electrode in 0.5 M H_2SO_4 + 1.0 M NaCl solutions with and without 200 mg L^{-1} K-2BP*

Effect of current density

Charge transfer step is a critical process in electrocatalysis, where the current density at the anode surface determines the rate of the electrode reaction. The effect of current density on the decolorization efficiency was investigated. Fig. 6 presents the obtained results. It can be seen from Fig. 6 that in the first 15 min, when the applied current density rises from 5 to 20 mA cm^{-2}, the decolorization efficiency increases drastically from 50 to 85 %. There is an optimal decolorization efficiency at 20 mA cm^{-2}. A further increase of current density makes no obvious improvement in the decolorization efficiency. This phenomenon can be explained as follows. When the applied current density is low, the production rate of active chlorine is less than its consumption rate for the oxidation of K-2BP, resulting in the low decolorization efficiency. When the applied current density is high, the production rate of active chlorine increases and the decolorization efficiency is improved. As the applied current density increases further, the concentration of chloride ions on anode surface is deficient and active chlorine is no more available for oxidizing the K-2BP.

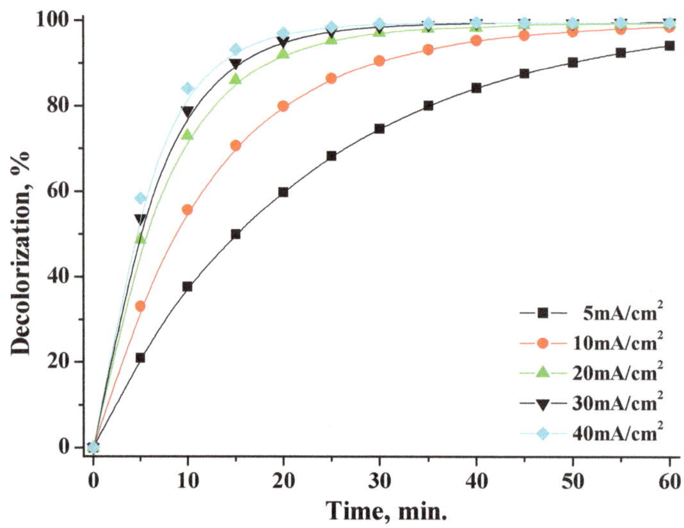

Figure 6. *Effect of current density on the decolorization efficiency of K-2BP (CNaCl=4.0g/L, flux=50mL/min, initial pH=7.0)*

Effect of NaCl concentration

Fig. 7 presents the effect of NaCl concentration on the decolorization of K-2BP. It can be seen from Fig. 7 that in the broad range of NaCl concentration, from 0.5 to 5.0 g L^{-1}, K-2BP can be removed completely, even at low NaCl concentration. For instance, when the dose of NaCl is 0.5 g L^{-1}, the decolorization efficiency reaches 90 % in the first 25 min, and in the later 15 min the decolorization efficiency reaches 100 %. By increasing the dose of NaCl from 0.5 to 2.0 g L^{-1} the decolorization rate is improved. However, a further increase of the NaCl dose does not improve the decolorization rate any further.

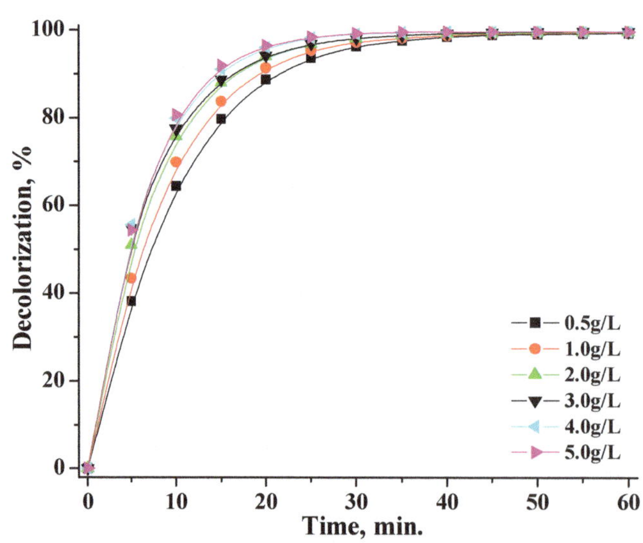

Figure 7. *Effect of NaCl concentration on the decolorization efficiency of K-2BP(j=30 mA/cm², flux=50 mL/min, initial pH=7.0)*

The electrooxidaton process of K-2BP on Ti/RuIrSnMnTi anode is determined not only by the charge transfer step, but also by the mass transfer step. At low NaCl concentration, the production rate of active chlorine is less than its consumption rate with K-2BP, leading to low decolorization rate. The increase of NaCl dose can reduce the effect of the mass transfer process and increases the formation rate of active chlorine. When the dose of NaCl is 2.0 g L^{-1}, the formation rate of active chlorine matches the reaction rate between active chlorine and K-2BP. Further increasing of the dose of NaCl does not increase the decolorization rate, because the formation of active chlorine is controlled by the charge transfer step.

Effect of initial pH

Fig. 8 presents the effect of the initial pH of the solution on the decolorization efficiency. As shown in Fig. 8, the initial pH of the solution affects the decolorization efficiency of K-2BP significantly. In the pH range between 1.0 and 11.0, the decolorization efficiency of K-2BP increases with the decrease of the initial pH of the solution.

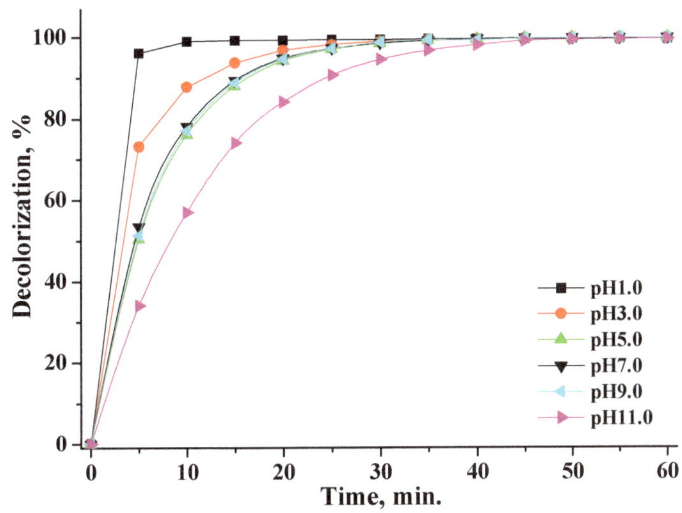

Figure 8. *Effect of initial pH on the decolorization efficiency of K-2BP(j=30 mA/cm^2, c_{NaCl}=4.0 g/L, flux=50 mL/min)*

In aqueous solution, the active chlorine in-situ electrogenerated is in the form of hypochlorous acid and hypochlorite ion, which reach an equilibrium (Eq. 6):

$$HClO \rightleftarrows ClO^- + H^+ \tag{6}$$

According to the Le Chatelier's principle, the increase of the hydrogen ion concentration shifts the equilibrium to the left in favor of forming hypochlorous acid. The standard formation potential of hypochlorous acid and hypochlorite ion is 1.49 V and 0.81 V, respectively, and the potential is a function of hydrogen ion concentration in solution (Eqs. 7 and 8):

$$\varphi(HClO/Cl^-) = \varphi^\theta(HClO/Cl^-) + RT/2F\ \ln([H^+][HClO]/[Cl^-]) \tag{7}$$

$$\varphi(ClO^-/Cl^-) = \varphi^\theta(ClO^-/Cl^-) + RT/2F\ \ln([ClO^-]/[OH^-]^2[Cl^-]) \tag{8}$$

Therefore, oxidative ability of hypochlorous acid is much stronger than of hypochlorite ion, and the smaller the value of pH is, the stronger is the oxidative ability of hypochlorous acid [30,34-37].

Effect of flow rate

The effect of solution flow rate on decolorization efficiency of K-2BP was considered in the range from 10 to 70 ml min^{-1}. Fig. 9 presents the obtained results. It can be seen from Fig. 9 that decolorization efficiency increases as the flow rate increases from 10 to 50 ml min^{-1}. Further increase in the flow rate does not affect the decolorization efficiency significantly. Apparently, the formation process of active chlorine is also controlled by the diffusion of chloride ions. The diffusion layer thickness is proportional to the flow rate, *i.e.* the faster flow rate favors the diffusion of chloride ions to the electrode surface. However, at high flow rate, turbulent flow appears and the effect of flow rate cannot be observed.

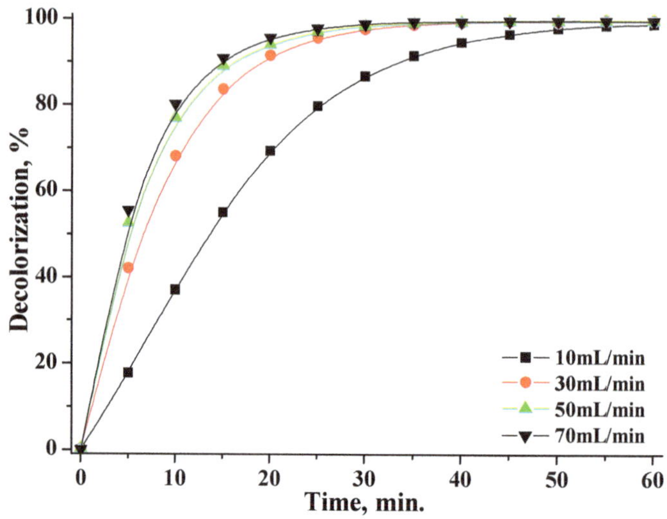

Figure 9. *Effect of flow rate on the decolorization efficiency of K-2BP* *(j = 30 mA cm^{-2}, c_{NaCl}= 4.0 g L^{-1}, initial pH = 7.0)*

Effect of temperature

Fig. 10 presents the effect of temperature on the decolorization efficiency of K-2BP. As shown in Fig. 10 the increasing temperature leads to the enhancement of the decolorization efficiency. The time of complete decolorization is 35 min at 303 K, while it reduces to 20 min at 343 K. The enhancement of decolorization efficiency of K-2BP can be ascribed to the improvement of the charge transfer rate and the reaction rate of the active chlorine with K-2BP in the solution.

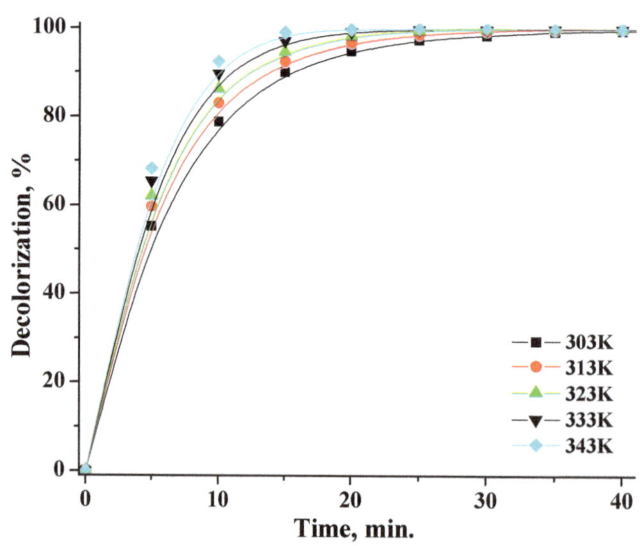

Figure 10. *Effect of temperature on the decolorization efficiency of K-2BP* *(j= 30 mA cm^{-2}, c_{NaCl}= 4.0 g L^{-1}, flux = 50 mL min^{-1}, pH = 7.0)*

Reaction kinetics

The variation of K-2BP concentration (absorbance) with time at different temperatures was modeled by the zero order, the first order and the second order reaction. The obtained results are shown in Table 1. It can be seen from Table 1 that the correlation coefficient (R2) of the first order kinetics model is about 0.999, far higher than those of zero and second order models. This indicates that the decolorization process of K-2BP follows the pseudo-first-order kinetics. Therefore, the decolorization rate equation can be expressed by (Eq. 9):

$$\ln\frac{c_t}{c_0}=\ln\frac{A_t}{A_0}=-K_{obs}t \tag{9}$$

where t is the reaction time (min), c_0 and c_t are the concentrations (mg L^{-1}) of K-2BP at the initial and at t, A_0 and A_t are the absorbance of K-2BP at the initial and at t, and K_{obs} is the pseudo-first-order kinetic constant (min^{-1}).

Table 1. *Results of fitting of the kinetics equation for zero-order, first-order and second-order of Reactive Brilliant Red K-2BP at various temperatures*

Temperature, K	zero order	second order	first order	
	R^2	R^2	R^2	K_{obs}/min^{-1}
303	0.84361	0.91002	0.99933	0.15119
313	0.79883	0.87378	0.99946	0.16941
323	0.76557	0.79324	0.99987	0.19533
333	0.71530	0.71310	0.99903	0.23650
343	0.67154	0.71861	0.99574	0.28096

Fig. 11 shows the linear relationship between the decolorization time t and ln (A_t/A_0). Fig. 12 presents the relationship between rate constant K_{obs} and temperature. It can be seen that the rate constant increases with temperature.

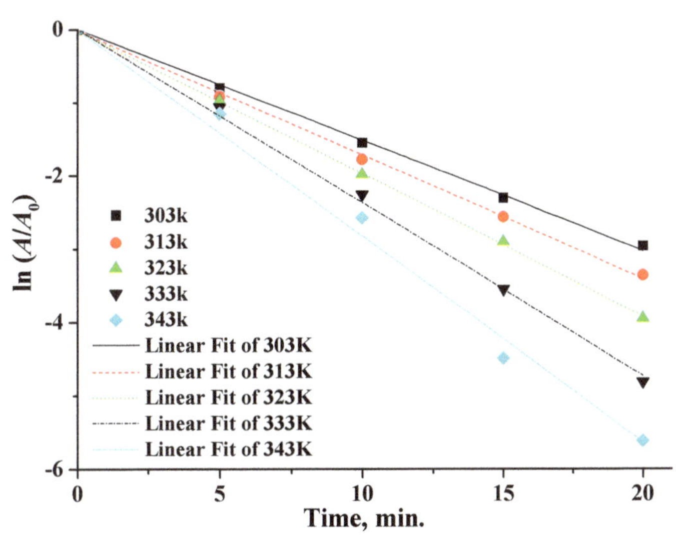

Figure 11. *First-order kinetic fitting results of decolorization rate of K-2BP at various temperatures*

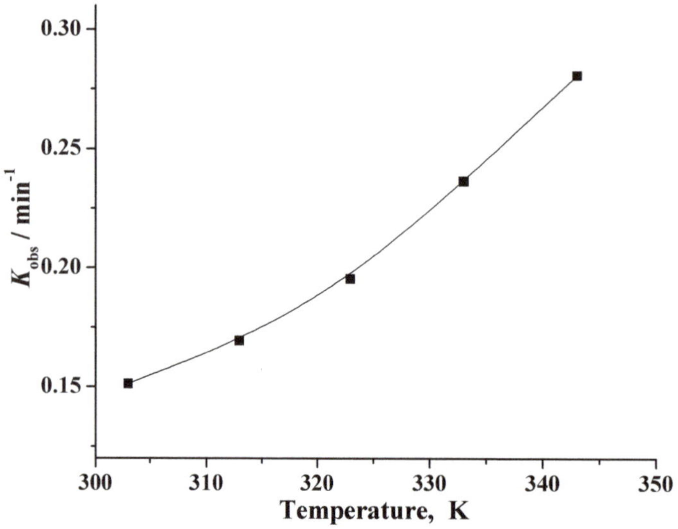

Figure 12. *Relationship between rate constant* K_{obs} *and temperature*

UV-visible spectroscopy

In K-2BP there are aromatic rings, naphthalene rings, triazine rings and -N=N- groups. -N=N- is the chromophore group and the bridge between the aromatic ring and naphthalene to form a big bond conjugate system. Fig. 13 presents the variation of K-2BP UV-visible spectrum with time. K-2BP is characterized by three absorbance peaks of 235 nm, 285 nm and 370 nm in ultraviolet region and only one peak of 520 nm in visible region. The absorbance in the visible region should be ascribed to the chromophore group of -N=N-, and the absorbance in ultraviolet region should be ascribed to the aromatic ring, triazine ring and naphthalene ring, respectively [38-40]. It can be seen from Fig. 13 that the intensity of all the absorbance peaks decreases rapidly with time and these peaks disappear completely in 30 min. This indicates that in the NaCl-containing solution, Ti/RuIrSnMnTi anode can remove the color of K-2BP completely, and the triazine ring, aromatic ring and naphthalene ring can be cleaved into aliphatic acid with low molecular weight or even completely mineralized.

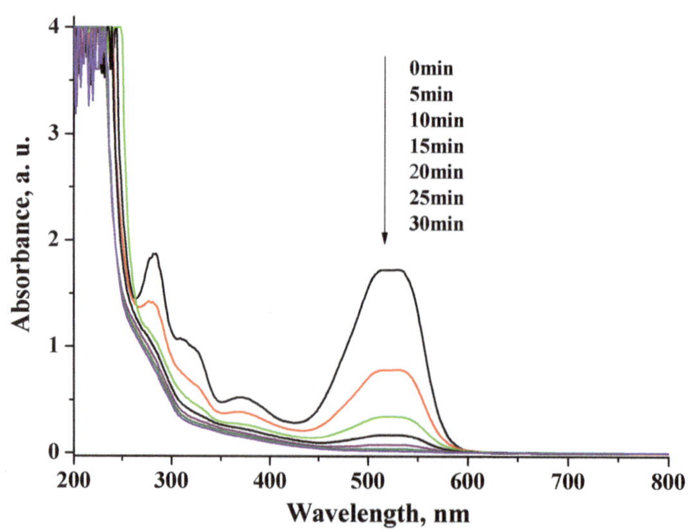

Figure 13. *UV-Vis spectrum variation of K-2BP during electrolysis in NaCl solution*

Conclusions

Ti/RuIrSnMnTi oxide anode can be used to degrade Reactive Brilliant Red K-2BP effectively in the solution containing chloride ions. The degradation mechanism of K-2BP is an indirect electrocatalytic oxidation, in which K-2BP is oxidized by the electrogenerated active chlorine. This degradation reaction follows pseudo-fist-order-reaction kinetics and can be improved by increasing current density and NaCl concentration, enhancing the temperature and flow rate, and lowering the pH of the solution.

Acknowledgements: *This work was financially supported by Natural Science Foundation of Guangdong Province (Grant No. 10351063101000001) and Combination Project of Enterprise, University and Scientific Research of Guangdong Province (Grant No. 2010B090400019).*

References

[1] H. Salvado, M.P. Gracia, J.M. Amigó, *Water Res.* **29** (1995) 1041-1050.
[2] W. Fuchs, C. Resch, M. Kernstock, M. Mayer, P. Schoeberl, R. Braun, *Water Res.* **39** (2005) 803-810.
[3] K.C.A. Bromley-Challenor, J.S. Knapp, Z. Zhang, N.C.C. Gray, M.J. Hetheridge, M.R. Evans, *Water Res.* **34** (2000) 4410-4418.
[4] S. Chinwetkitvanich, M. Tuntoolvest, T. Panswad, *Water Res.* **34** (2000) 2223-2232.
[5] A.B. Dos-Santos, F.J. Cervantes, R.E. Yaya-Beas, J.B. van Lier, *Enzyme Microb. Tech.* **33** (2003) 942-951.
[6] T. Panswad, K. Iamsamer, J. Anotai, *Bioresource Technol.* **76** (2001) 151-159.
[7] C.T.M.J. Frijters, R.H. Vos, G. Scheffer, R. Mulder, *Water Res.* **40** (2006) 1249-1257.
[8] J.F. Lei, W.S. Li, *Acta Phys. Chim. Sin.* **25** (2009) 1173-1178.
[9] I. Peternel, N. Koprivanac, H. Kušić, *Water Res.* **40** (2006) 525-532.
[10] A. Vijayabalan, K. Selvam, R. Velmurugan, M. Swaminathan, *J. Hazard. Mater.* **172** (2009) 914-921.
[11] J.N. Wu, T.W. Wang, *Water Res.* **35** (2001) 1093-1099.
[12] T.Y. Chen, C.M. Kao, A. Hong, C.E. Lin, S.H. Liang, *Desalination* **249** (2009) 1238-1242.
[13] G.H. Chen, L.C. Lei, P. Yue, *Ind. Eng. Chem. Res.* **38** (1999) 1837-1843.
[14] A. Santos, P. Yustos, S. Rodriguez, F. Garcia-Ochoa, M. de Garcia, *Ind. Eng. Chem. Res.* **46** (2007) 2423-2427.
[15] I. Gulkaya, G.A. Surucu, F.B. Dilek, *J. Hazard. Mater.* **136** (2006) 763-769.
[16] C.T. Wang, J.L. Hu, W.L. Chou, Y.M. Kuo, *J. Hazard. Mater.* **152** (2008) 601-606.
[17] A.H. Essadki, M. Bennajah, B. Gourich, C. Vial, M. Azzi, H. Delmas, *Chem. Eng. Process.* **47** (2008) 1211-1223.
[18] X.D. Zhang, J.D. Hao, W.S. Li, H.J. Jin, J. Yang, Y.M. Huang, D.S. Lu, H.K. Xu, *J. Hazard Mater.* **170** (2009) 883-887.
[19] W. Balla, A.H. Essadki, B. Gourich, A.Dassaa, H. Chenik, M. Azzi, *J. Hazard. Mater.* **184** (2010) 710-716.
[20] M. Panizza, G. Cerisola, *Water Res.* **40** (2006) 1179-1184.
[21] M. Panizza, G. Cerisola, *Environ.* Sci. *Technol.* **38** (2004) 5470-5475.
[22] Z.C. Wu, M.H. Zhou, *Environ. Sci.Technol.* **35** (2001) 2698-2703.
[23] M.H. Zhou, Q.Z. Dai, L.C. Lei, C.Ma, D. Wang, *Environ. Sci.Technol.* **39** (2005) 363-370.
[24] M. Li, C.P. Feng, W.W. Hu, Z. Zhang, N. Sugiura, *J. Hazard. Mater.* **162** (2009) 455-462.
[25] J.B. Parsa, M. Rezaei, A.R. Soleymani, *J. Hazard. Mater.* **168** (2009) 997-1003.
[26] S.S. Vaghela, A.D. Jethva, B.B. Mehta, S.P. Dave, S. Adimurthy, G. Ramachandraiah, *Environ. Sci. Technol.* **39** (2005) 2848-2855.

[27] S. Song, J.Q. Fan, Z.Q. He, L. Chan, Z. Liu, J. Chen, X. Xu, *Electrochim. Acta* **55** (2010) 3606-3613.

[28] B.J. Hernlem, *Water Res.* **39** (2005) 2245-2252.

[29] P. Canizares, J.A. Dominguez, M.A. Rodrigo, J. Villasenor, J. Rodriguez, *Ind. Eng. Chem. Res.* **38** (1999) 3779-3785.

[30] H.X. Shi, J.H. Qu, A.M. Wang, J.T. Ge, *Chemosphere* **60** (2005) 326-333.

[31] C.R. Costa, P. Olivi, *Electrochim. Acta* **54** (2009) 2046-2052.

[32] C.A. Martinez-Huitle, A. Battisti, S. Ferro, S. Reyna, M. Cerro-Lopez, M.A. Quiro, *Environ. Sci. Technol.* **42** (2008) 6929-6935.

[33] S. Trasatti, *Electrochim. Acta* **45** (2000) 2377-2385.

[34] A. Maljaei, M. Arami, N.M. Mahmoodi, *Desalination* **249** (2009) 1074-1078.

[35] O. Scialdone, S. Randazzo, A. Galia, G. Silvestri, *Water Res.* **43** (2009) 2260-2272.

[36] S. Raghu, C.A. Basha, *Ind. Eng. Chem. Res.* **47** (2008) 5277-5283.

[37] L. Szpyrkowicz, C. Juzzolino, S.N. Kaul, S. Daniele, M.D. De Faveri *Ind. Eng. Chem. Res.* **39** (2000) 3241-3248.

[38] X.N. Wang, J.P. Jia, Y.L. Wang, *Ultrason. Sonochem.* **17** (2010) 515-520.

[39] J. Yang, J. Wang, J.P. Jia, *Environ. Sci. Technol.* **43** (2009) 3796-3802.

[40] X.J. Lu, B. Yang, J.H. Chen, R.Sun, *J. Hazard. Mater.* **161** (2009) 241-245.

8

Platinized titanium dioxide electrodes for methanol oxidation and photo-oxidation[§]

SVETLOZAR IVANOV, IOANNA MINTSOULI*, JENIA GEORGIEVA, STEPHAN ARMYANOV, EUGENIA VALOVA, GEORGIOS KOKKINIDIS*, IOANNIS POULIOS*, SOTIRIS SOTIROPOULOS*✉

Rostislaw Kaischew Institute of Physical Chemistry, Bulgarian Academy of Sciences, Sofia 1113, Bulgaria

**Department of Chemistry, Aristotle University of Thessaloniki, Thessaloniki 54124, Greece*

✉Corresponding Author: E-mail: eczss@chem.auth.gr

[§] Dedicated to 75th birthday of Professor Sergio Trasatti

Abstract

Platinized deposits have been formed on TiO_2 particulate films supported on Ti substrates, by means of galvanic replacement of pre-deposited metallic Cu and subsequent immersion of the Cu/TiO_2 coatings into a chloroplatinic acid solution. The spontaneous replacement of Cu by Pt results in $Pt(Cu)/TiO_2/Ti$ electrodes. Both the platinized and the precursor TiO_2/Ti electrodes have been characterized by SEM microscopy/EDS spectroscopy, their surface electrochemistry has been assessed by cyclic voltammetry in the dark and their photoelectrochemical properties by photovoltammetry under UV illumination. It has been found that, although platinized rutile-rich electrodes exhibit typical Pt surface electrochemistry, the anatase-rich electrodes show only traces of oxide formation and stripping. The latter has been translated to a suppression of methanol oxidation at anatase-rich electrodes. On the contrary, methanol oxidation at platinized rutile-rich electrodes occurs at significant rates and can be further enhanced upon UV illumination, as a result of Pt and TiO_2 synergism in the photoelectrochemical oxidation of methanol.

Keywords

Titanium dioxide; platinum; galvanic replacement; electrocatalysis; photoelectrocatalysis

Introduction

Platinized Ti or TiO_2 finds a number of practical applications that include industrial anodes, photocatalysis/photoelectrocatalysis and fuel cell catalysts. In more detail, Ti (with native, thermally or electrochemically grown TiO_2 on it) is the basis of dimensionally stable anodes (DSAs) which are coated with precious metal and other oxide catalysts and are used in many industrial processes (e.g. chlorine production, water electrolysis, electrowinning etc) [1-4]. Also, modifying the semiconductor oxides of TiO_2 by precious metals (especially Pt) is a well-established practice to improve their photocatalytic efficiency by reducing photogenerated electron-hole recombination rates [5-7]. Finally, Pt/TiO_2 bi-component powder catalysts mixed with high surface area carbons have recently been tested as alternative anode or cathode catalysts in fuel cells [8-10].

Methods for Pt deposition/incorporation onto/into TiO_2 supported layers include electrodeposition [11-14], vacuum deposition [15], thermal decomposition followed by reduction [16] and photodeposition [17,18]. Platinization of TiO_2 powders is almost exclusively carried out by chemical reduction of Pt complexes in the presence of TiO_2 particles or simultaneously with TiO_2 sol-gel preparation [19]. Electrodeposition allows for accurate control of precious metal loading but requires specialized equipment, uses concentrated precious metal solutions and it is difficult to apply to powder substrates; chemical methods usually involve reducing agents and almost always involve a high temperature annealing in a reducing atmosphere; photodeposition is a lengthy process and can also lead to precious metal losses due to uncontrolled deposition at reactor components other than the substrate [17].

During the last decade an alternative method for the introduction of the noble metal onto the electrode support has been developed. The method is based on the spontaneous galvanic replacement of surface layers of a non-noble metal M (M: Pb, Cu, Fe, Co, Ni) by a noble metal (*e.g.* Pt, Au, Pd, Ir) upon immersion of the former in a solution of metal ions of the latter (the resulting bimetallic system is denoted as Pt(M), Au(M) *etc.*). For example, in the case of Cu and Pt:

$$2\ Cu + PtCl_6^{2-} \rightarrow Pt + 6\ Cl^- + 2\ Cu^{2+} \qquad (1).$$

The method (also known as transmetallation) was first applied by Adžić and co-workers to underpotentially deposited (UPD) Cu monolayers [20-22] and to Cu and Pb bulk deposits by Kokkinidis and co-workers [23-24] on flat electrode substrates. It has since been extended to other transition metals (Fe, Co, Ni) (see for example the works of Sotiropoulos and co-workers [25-30]) as well as to carbon powder supports (see for example [31,32]). Advantages of the new technique include the fact that it is a fast and room temperature processes, it employs low concentration solutions of the precious metal and can lead to the formation of thin precious metal deposits that may decrease its loading.

Despite the rapid evolution of the application of the technique on conducting substrates, the formation of metal/semiconductor (*e.g.* Pt/TiO_2) catalysts by means of transmetallation has yet to be established. These systems should offer the possibility of enhancing anodic reactions occurring at Pt by means of simultaneous photooxidations at n-type semiconducting supports and to that direction, methanol oxidation at UV-illuminated Pt/TiO_2 electrodes has recently been reported [33].

The aim of this work has been to investigate the feasibility of applying the galvanic replacement technique to the metallization of TiO_2 powders. Specific objectives have been:

i. The preparation and microscopic as well as compositional characterization of Pt(Cu)/TiO_2 deposits on Ti substrates ;

ii. The electrochemical characterization of the deposits by means of Pt surface electrochemistry and methanol oxidation;

iii. The study of the effect of UV illumination on the electrochemical behavior of these deposits.

Experimental

Preparation of TiO_2/Ti, Cu/TiO_2/Ti and Pt(Cu)/TiO_2/Ti electrodes

Ti rectangular substrates (0.5 mm thick) had typical dimensions of 1 cm × 0.5 cm and were etched in a HF/HNO_3 3:1 mixture and then washed with doubly distilled water. Known volumes of a 1 g L^{-1} methanol dispersion of Degussa P-25 TiO_2 were pipetted onto one side of the Ti substrate and the coating was dried at 50°C for 10 min. TiO_2 loadings were in the 2-3 mg cm^{-2} range. These coated substrates were sintered in an air atmosphere at 700°C (for 5 h) or 500°C (for 1.5 h) using a Carbolite CWF 1100 oven, resulting in particulate TiO_2 layers. The back side of the samples had been etched again before it was insulated with epoxy resin glue (RS).

Electrodeposition of Cu on the TiO_2/Ti electrode was performed at constant potential (in the mixed control potential regime and more specifically at the half-wave potential of exploratory deposition voltammetry) from 0.1 M $HClO_4$ + 0.01 M $CuSO_4$ deaerated solutions. The total charge density passed was 400 mC cm^{-2} for rutile-rich electrodes and 400 or 1200 mC cm^{-2} for anatase-rich electrodes.

The Cu/TiO_2/Ti electrodes were immersed in 2 ml of a 0.1 M HCl + 10^{-3} M K_2PtCl_6 solution for 30 min so that replacement of Cu by Pt could take place:

$$2\ Cu/TiO_2/Ti + PtCl_6^{2-} \rightarrow Pt\ (Cu)/\ TiO_2/Ti + 2\ Cu^{2+} + 6\ Cl^- \quad (2)$$

Microscopic and spectroscopic characterisation of coatings

Scanning Electron Microscopy (SEM) was carried out using a JEOL JSM-5510 microscope and elemental analysis of the coatings was performed by the accompanying EDS (EDAX) system. X-Ray Diffraction (XRD) deposit characterisation was performed with the help of a Rigakou Miniflex diffractometer.

Electrochemical and photoelectrochemical characterisation of coatings

Cyclic voltammetric and constant potential experiments on TiO_2 electrodes in the dark and under UV illumination were carried out with the Autolab 30 (EcoChimie) system in a three-electrode cell equipped with a flat quartz window opposite the working electrode. A saturated calomel electrode (SCE) was used as the reference electrode and a Pt foil as the counter electrode. Voltammograms were run for at least three consecutive full cycles since preliminary experiments showed that a near-steady state response was observed only after the second run; all results reported correspond to the stabilized voltammetric picture.

A Radium Ralutec 9W/78 UVA lamp (λ=350-400 nm, λ_{max}=366 nm), placed at a distance of 2.5 cm from the sample, was used for front face electrode illumination. The power density on the sample surface position was measured as 3 mW cm^{-2} with a photometer.

Electrode materials and chemicals

Ti plates 0.5 mm thick were from Alfa Aesar (99.5 %, metals basis). TiO_2 powder was P-25 Degussa® from Degussa (now Evonic). $HClO_4$ from Riedel, (puriss *p.a.*, ACS reagent, ≥ 70 %) and $CuSO_4.6H_2O$ from Sigma-Aldrich (ACS reagent) were used in the preparation of Cu deposition solutions. H_2PtCl_6 hexahydrate from Sigma-Aldrich (ACS reagent, ≥37.50 % as Pt) was employed for the Pt exchange solution. MeOH was from Riedel (Chromasolv®, for HPLC, gradient grade, ≥ 99.9 %).

Results and Discussion

TiO_2 particulate films on Ti electrode substrates: microscopic, electrochemical and photoelectrochemical characterization

Figure 1 shows the SEM micrograph of the surface of a typical Ti/TiO_2 (700 °C) electrode. This is characterized by a reticulated network of micrometer-sized aggregates into which the much smaller anatase and rutile particles (22 and 31 nm, as determined by XRD) of the original P-25 TiO_2 material have been merged after annealing. The picture of Ti/TiO_2 (500 °C) electrodes was very similar, indicating that sintering at 500° (for 1.5 h) or 700 °C (for 5h) has the same effect on the overall morphology of the particulate films. However, as confirmed by XRD [34], the 500 °C-treated samples retain the 75÷25 % w/w anatase-to-rutile composition (and are termed anatase-rich or TiO_2-500 samples hereafter) while the 700°C-treated samples are transformed to rutile-rich samples (3÷97 % w/w anatase-to-rutile; TiO_2-700).

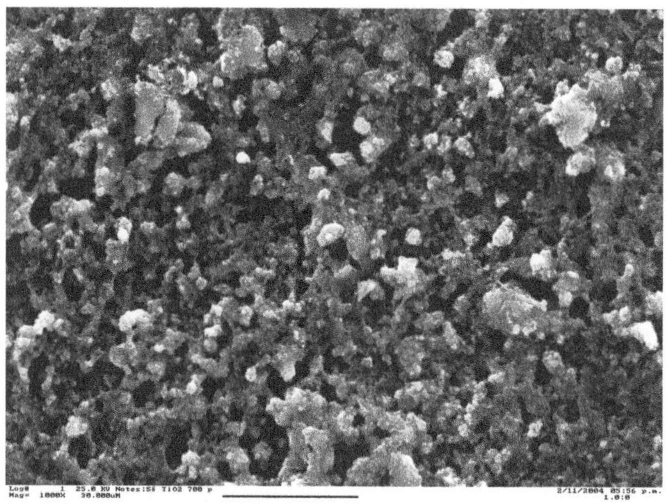

Figure 1. *SEM micrograph of a TiO_2 particulate film prepared on a Ti substrate by annealing solution-cast Degussa P-25® TiO_2 particles at 700 °C for 5h. (Scale bar length: 30 μm.)*

Electrochemical characterisation of TiO_2 coatings in the dark

Figure 2 presents fast (100 mV s^{-1}) cyclic voltammograms in the dark of an anatase-rich/500 °C electrode (with a 4.3 mg cm^{-2} TiO_2 loading) and a rutile-rich/700 °C electrode (with a 2.4 mg cm^{-2} TiO_2 loading), in a 1 M $HClO_4$ solution. Prior to hydrogen evolution, a couple of cathodic and anodic main peaks are observed (together with traces of another couple at more positive potentials), attributed to the surface transformation of the Ti(IV)/Ti(III) couple at potentials negative to the flat band potential of TiO_2 and just prior to hydrogen evolution [17,35-37].

The charge density (per geometric substrate area) under the cathodic peaks was found to be 7.5 mC cm^{-2} for the anatase-rich/500°C electrode and 2.8 mC cm^{-2} for the rutile-rich/700°C one. Since the value of Ti surface density in a TiO_2 crystal has been reported as approximately 9 Ti^{4+} ions /nm^2 [38], the Ti(IV)/Ti(III) transformation translates to an expected charge density of 144 μC cm^{-2}. Then the sintered electrodes presented in Figure 2 have roughness factors of 52 and 19 electroactive TiO_2 cm^2 per substrate cm^2. If we further take into account the TiO_2 loading of these particular electrodes (4.3 mg cm^{-2} and 2.4 mg cm^{-2} respectively) then their mass-specific TiO_2

electroactive area can be calculated as 1.2 $m^2 g^{-1}$ and 0.8 $m^2 g^{-1}$ for the anatase-rich and rutile-rich electrodes, respectively. The similar values of electroactive areas are in line with the similarities observed in the morphology by SEM analysis.

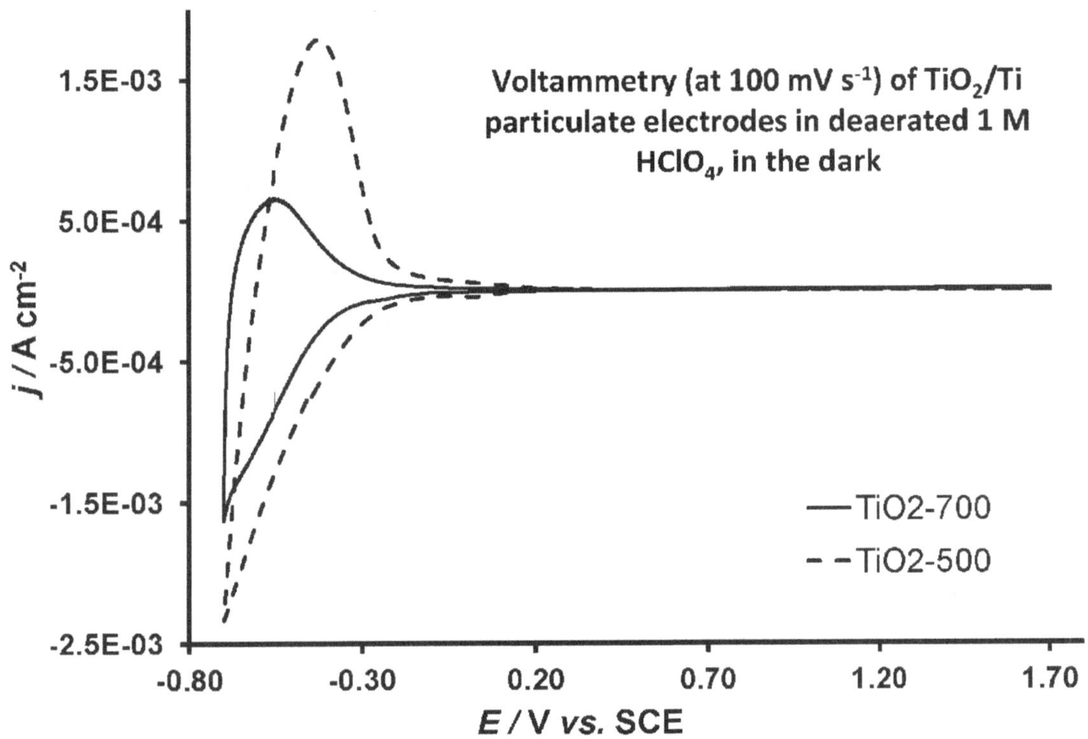

Figure 2. *Cyclic voltammograms of TiO_2/Ti electrodes (anatase-rich/annealed at 500 °C for 1.5 h and rutile-rich/ annealed at 700 °C for 5h) in 1 M $HClO_4$, in the dark.*

Figure 3(A) shows photovoltammograms (at 10 mV s^{-1}) of two TiO_2/Ti electrodes with similar loadings (2.2-2.4 mg cm^{-2}) at anatase- and rutile-rich TiO_2 electrodes under UV illumination (360 nm, 3 mW cm^{-2}). If there are no other oxidizable species in solution, the observed photocurrent is due to water oxidation by photogenerated holes to $OH^{\bullet}$ or other primary oxidation products such as $O_2^{\bullet-}$and H_2O_2. The limiting photocurrent observed at high positive potentials is controlled by the diffusion of photogenerated electrons through the particles of the particulate semiconductor layer [39-41]. The onset potential of the photocurrent (which can be approximated to the flat band potential) for the anatase-rich electrode is ca. -0.150 V vs. SCE, lower by some 200 mV from that of the rutile-rich electrode, in line with anatase reported to have a more negative flatband potential than that of rutile [42]. Also, the limiting photocurrent in the plain electrolyte solution is found to be higher at the rutile-rich electrode, indicating its higher photoelectrocatalytic activity for water oxidation. Figure 3(B) shows the effect of the addition of 0.5 M MeOH to the photovoltammetry. It is seen that, although the shift of the flatband potential remains the same, the limiting photocurrent of the anatase-rich electrode is now higher, most likely due a higher adsorption affinity for methanol on the anatase surface where it is oxidized by the photogenerated holes.

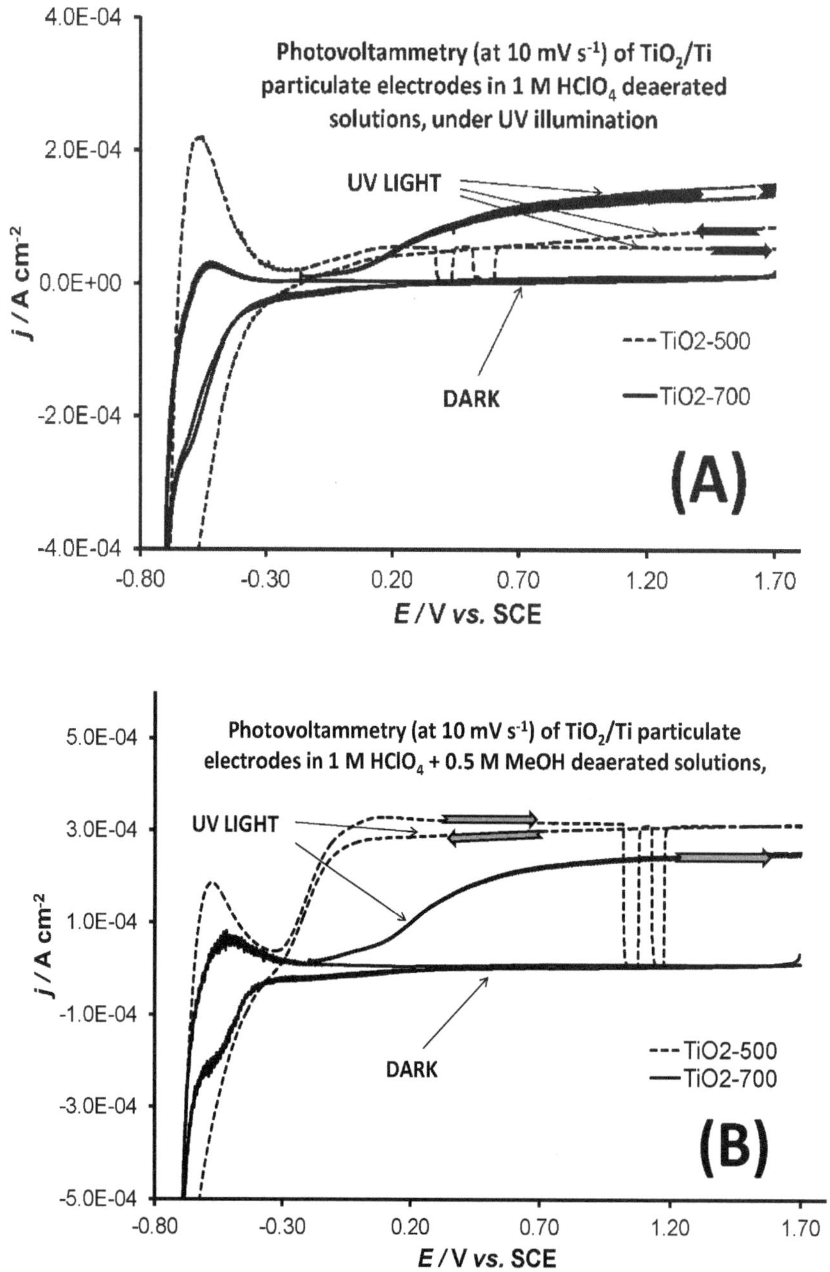

Figure 3. *Cyclic voltammograms of TiO_2/Ti electrodes (anatase-rich/annealed at 500 °C for 1.5 h and rutile-rich/ annealed at 700 °C for 5h) in the dark and under continuous (or occasionally shut) UV illumination, in* ***(A)*** *1 M $HClO_4$ and in* ***(B)*** *0.5 M MeOH + 1 M $HClO_4$ solutions. (Arrows indicate potential scan direction; the photocurrent fall for the TiO_2-500 electrode corresponds to a temporary illumination shut down.)*

Cu electrodeposition on TiO_2/Ti electrodes: deposition voltammetry, microscopic (SEM) and spectroscopic (EDS) characterisation of Cu/TiO_2 precursors

Figure 4 presents the cyclic voltammograms of the TiO_2/Ti electrodes in a Cu^{++} containing solution to obtain an exploratory electrodeposition picture. The cathodic electrodeposition peaks are observed for both types of electrodes (at potentials corresponding to the onset of the Ti(IV)/Ti(III) transformation shown in Figure 2) but interestingly no stripping peaks are recorded throughout the positive potential range. This means that the Cu/TiO_2 interface behaves as a Schottky diode with dark current (including that of electrodeposition) starting to flow only at potentials negative enough for TiO_2 metallization and/or the onset of Ti surface reactions to occur. The fact that the onset of Cu electrodeposition should be related to the Ti(IV)/Ti(III) transformation (possibly via a catalytic cycle) is also supported by both phenomena exhibiting the same trend when one passes from an anatase-rich to a rutile-rich electrode.

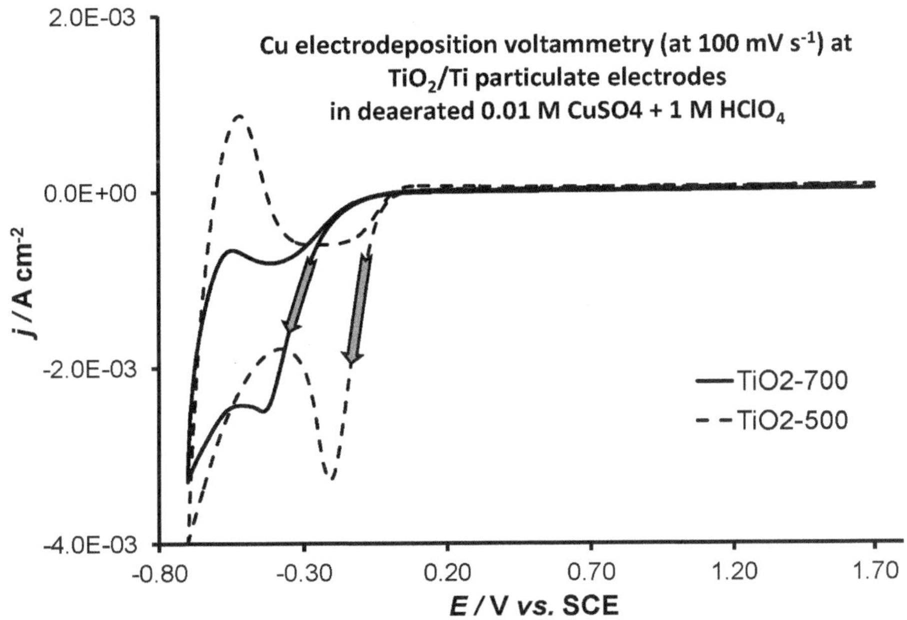

Figure 4. *Cyclic voltammograms of TiO_2/Ti electrodes (anatase-rich/annealed at 500 °C for 1.5 h and rutile-rich/ annealed at 700 °C for 5h) in the dark, in the electrodeposition solution of 0.01 M $CuSO_4$ + 1 M $HClO_4$ solution. (Arrows indicate potential scan direction.)*

Constant potential Cu electrodeposition was carried out at the half-wave cathodic peak potential for each type of electrode and (in initial experiments) the electrodeposited Cu charge density was 400 mC cm^{-2} for both electrodes. Figure 5 presents the SEM images (at different resolution) of such a Cu/TiO_2/Ti rutile-rich electrode. Numerous Cu particles of sub-micron dimensions can be seen to have been deposited onto/into the reticulated matrix of TiO_2. A similar picture has been obtained for anatase-rich electrodes and, for typical TiO_2-loadings of 2.2-2.4 mg cm^{-2} This procedure led to Cu/TiO_2 coating having a 17-18 % w/w Cu content (w.r.t. TiO_2, as confirmed by EDS).

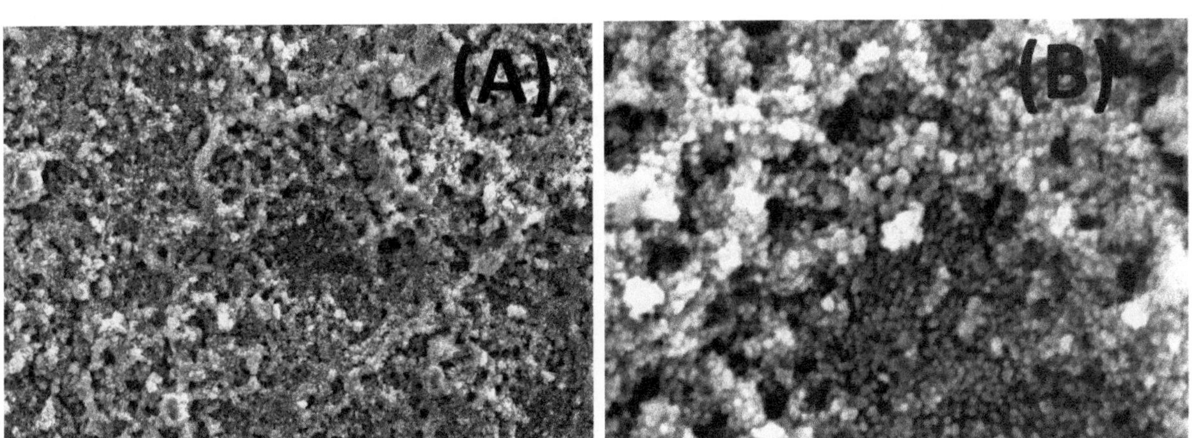

Figure 5. *SEM micrographs of a Cu/TiO_2 particulate film prepared by electrodeposition of 400 mC cm^{-2} on a TiO_2/Ti electrode (rutile-rich/ annealed at 700 °C for 5h). (Scale bar lengths:* ***(A)*** *30 μm and* ***(B)*** *7.5 μm.)*

Galvanic replacement of Cu in Cu/TiO₂/Ti electrodes by Pt: microscopic, spectroscopic and electrochemical/photoelectrochemical characterization of Pt(Cu)/TiO₂/Ti electrodes in acid

Upon immersion of the Cu/TiO_2 electrodes prepared as described in the previous section in a 10^{-3} M chloroplatinate solution for 30 min, Cu was replaced according to (2) by Pt. Figures 6 show SEM images of the thus produced $Pt(Cu)/TiO_2$ coatings, revealing the existence of much smaller Pt nanoparticles (compared to those of the Cu precusrsor shown in Figures 5) and an increased porosity (intermediate of that of the Cu/TiO_2 precursor and of the TiO_2 starting material), due to significant etching of Cu into the solution.

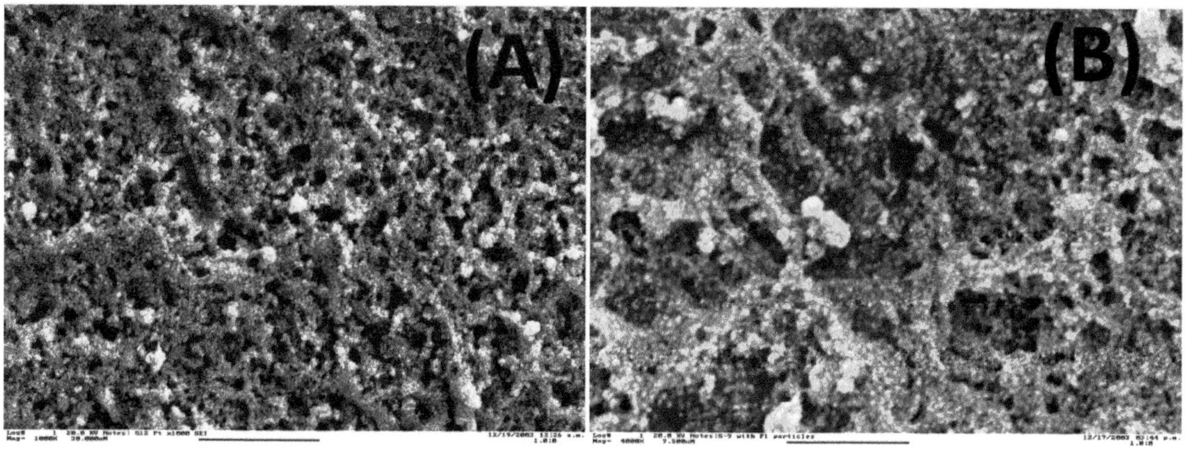

Figure 6. *SEM micrographs of a $Pt(Cu)/TiO_2$ particulate film prepared from the Cu/TiO_2 film shown in Figure 4 above by immersion in a 10^{-3} M chloroplatinic acid + 0.1 M HCl solution for 30 min. (Scale bar lengths:* ***(A)*** *30 μm and* ***(B)*** *7.5 μm.)*

The EDS analysis of the $Pt(Cu)/TiO_2$ deposits indicates that almost all Cu has been replaced or etched (only traces of Cu were measured) and that their Pt content is 6% w/w (w.r.t. TiO_2) for rutile-rich electrodes and 2% w/w for anatase-rich electrodes. An insight into the mechanism of Pt deposition and the differences observed for the two substrates can be acquired by the following reasoning. The absence of significant quantities of Cu suggests that only Pt deposited directly onto TiO_2 locations adjacent to Cu deposits survives (whereas Pt deposited on Cu locations collapses, as layers beneath it corrode from uncovered locations). Since it has already been reported that Pt nuclei are preferably formed on rutile (which has a larger concentration of oxygen vacancies) [43], it follows that larger quantities of Pt are deposited directly onto rutile-rich substrates than on anatase-rich ones. To be able to produce similar Pt loadings on both types of substrates, larger quantities of Cu had to be deposited on anatase-rich substrates. A 3-fold increase in pre-deposited Cu (plating density of 1200 mC cm^{-2}) resulted in a 5 % w/w Pt content. The rutile-rich and anatase-rich electrodes for which results are shown hereafter had Pt loadings of 0.15 mg cm^{-2} and 0.12 mg cm^{-2}, respectively.

Figures 7 present the surface electrochemistry of the $Pt(Cu)/TiO_2/Ti$ electrodes in deaerated acid.

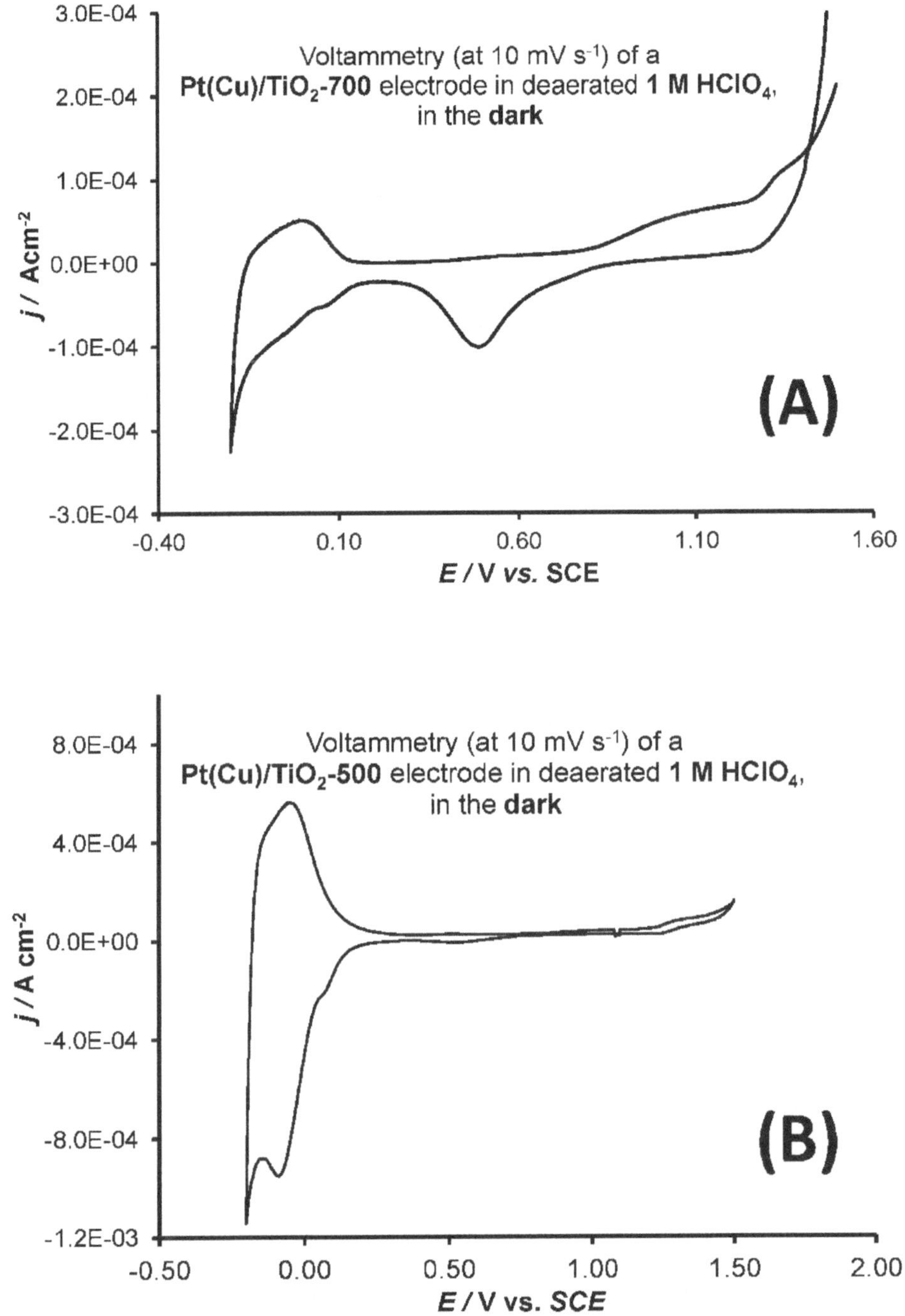

Figure 7. *Cyclic voltammograms of Pt(Cu)/TiO_2/Ti electrodes in the dark, in a deaerated 1 M $HClO_4$ solution.* ***(A)*** *Rutile-rich electrode/annealed at 700 °C for 5 h;* ***(B)*** *Anatase-rich electrode/annealed at 500 °C for 1.5 h.*

Although the rutile-rich electrode shows the typical polycrystalline Pt picture (with the hydrogen adsorption/desorption and oxide formation/stripping peaks), there are only traces of oxide formation and stripping for the anatase-rich electrode. In this case, the formation of a Schottky diode between the metal (Pt) and the n-type semiconductor (TiO_2) which would allow for currents at potentials negative to the flat band potential and cut them off at more positive potentials, is not a plausible scenario. This is because, as shown in Figures 8, hydrogen evolution occurs at Pt(Cu)/TiO_2 at potentials far more positive than the TiO_2 metallization region;

furthermore some anodic reactions (ferrocyanide oxidation and hydrazine-not shown here) did take place on the anatase-rich electrodes in the potential range of Pt oxide formation and stripping.

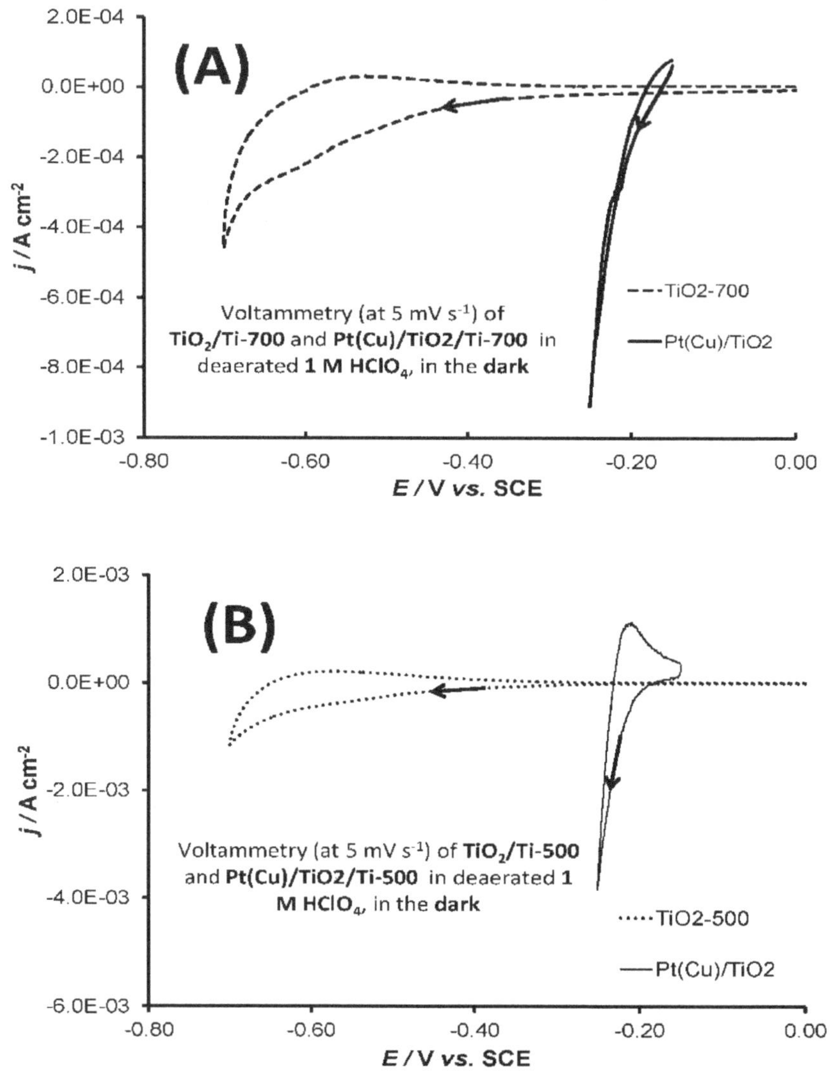

Figure 8. *Slow potential scan voltammograms of Pt(Cu)/TiO_2/Ti and TiO_2/Ti electrodes in the dark, in a deaerated 1 M $HClO_4$ solution.* ***(A)*** *Rutile-rich electrode/annealed at 700 °C for 5 h;* ***(B)*** *Anatase-rich electrode/annealed at 500 °C for 1.5h.*

All this means that electron transfer reactions are possible at the Pt overlayer via resonance tunneling from Pt to the Ti base of the electrode through the TiO_2 semiconductor, with the help of an impurity band introduced by Pt [44]. Therefore, the supression of Pt oxide/formation is to be attributed to strong Pt-anatase interactions. It should be noted that the extraordinary picture of Figure 7(B) has also been reported for platinized reduced anatase [45], monocrystalline rutile [46] or low Pt coverage/small particle size anatase-rich samples [47], and can been interpreted via the formation of a strong Pt-Ti bond [48], inhibiting the formation of PtO. The reason why these interactions are higher for anatase substrates is due to the fact that smaller nuclei (thus more affected by the TiO_2 in contact) and stronger Pt adsorption are observed at anatase [43] (despite the formation of larger Pt clusters at rutile locations). From the charge under the cathodic hydrogen adsorption peak the Pt electroactive area per electrode substrate geometric area (roughness factor) can be calculated as 10 for rutile-rich and 81 for anatase-rich electrodes, close or within the lower part of the 50-1000 range of Pt black or carbon-supported Pt fuel cell catalysts [17]. Taking into account the loading of each electrode, these values of the roughness factor are translated to ca 6.5 and 67.5 m^2 g^{-1} of electroactive Pt (note that a typical fuel cell catalyst has a value of 65 m^2 g^{-1} [49]). The higher surface area/dispersion of Pt in anatase-rich electrodes is in line with the above-mentioned reasoning for stronger Pt-TiO_2 interactions at these substrates.

Figures 9 (A)-(B) show the effect of UV illumination on the cyclic voltammetry of the Pt(Cu)/TiO_2/Ti electrodes in acid. A comparison with the results of the corresponding TiO_2/Ti electrodes of Figure 3(A), reveals that there is approximately a 3-fold decrease of the net photocurrent at both electrodes (from 116 to 40 μA cm^{-2} for the rutile-rich electrode and from 61 to 22 μA cm^{-2} for the anatase-rich electrode, at +0.80 V).

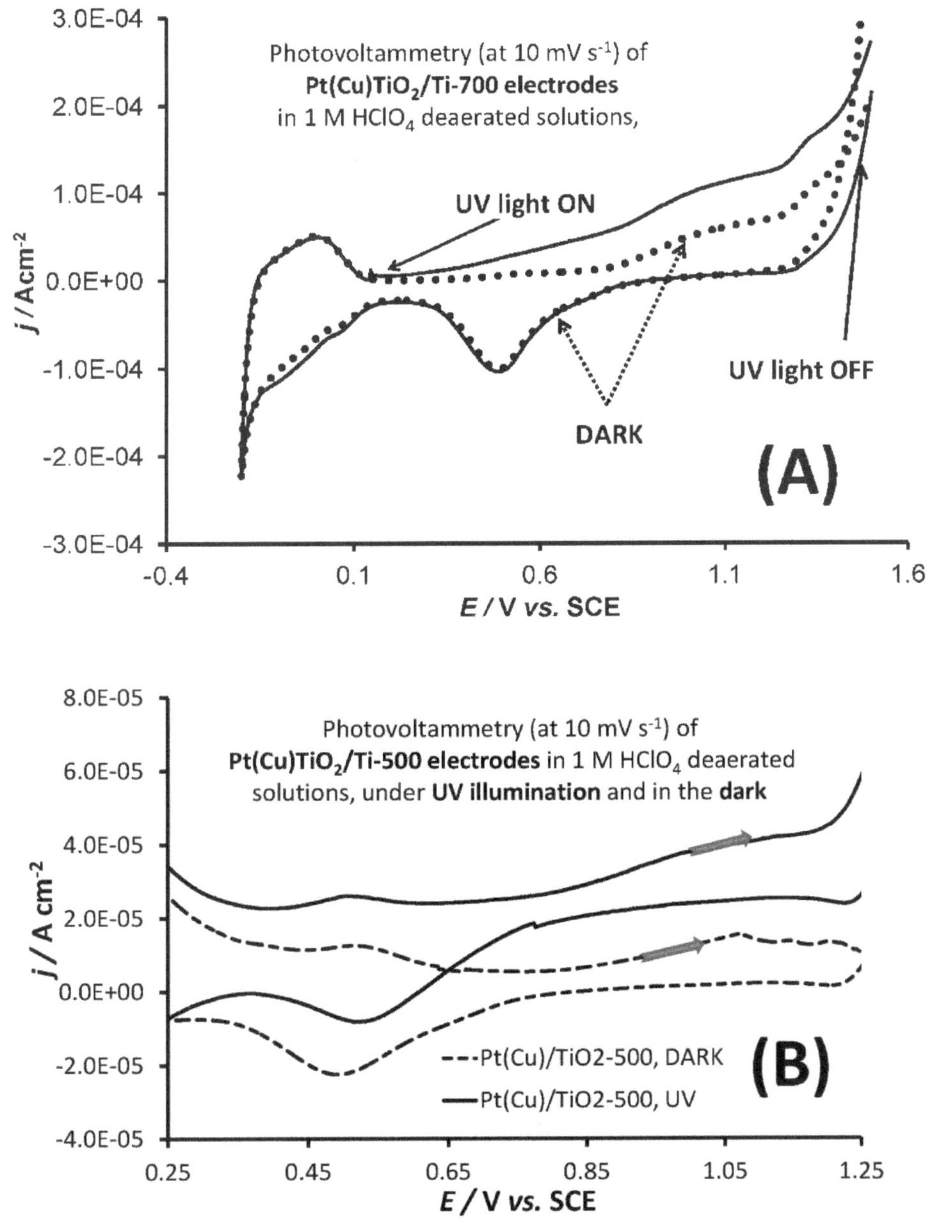

Figure 9. *Cyclic voltammograms of TiO_2/Ti electrodes in the dark and under UV illumination, in 1 M $HClO_4$. **(A)** Rutile-rich electrode/annealed at 700 °C for 5 h; **(B)** Anatase-rich electrode/annealed at 500 °C for 1.5h.*

This is opposite to the beneficial effect that Pt has on plain photocatalysis but it is to be expected since first, Pt loading of the photoelectrocatalysts prepared in this work (8-10 % w/w) is much higher than that of Pt-doped TiO_2 photocatalysts (0.1-1 % w/w) resulting to partial shielding of TiO_2 from light and second, the addition of a metal should not have a key role in decreasing charge recombination during photoelectrocatalysis since this is mainly achieved by the electrical bias.

Methanol oxidation at Pt(Cu)/TiO_2/Ti electrodes in the dark and under UV illumination

Figure 10 presents slow potential sweep cyclic voltammograms (at 10 mV s^{-1}; third cycle) of the Pt(Cu)/TiO_2/Ti electrodes, in a 0.5 M MeOH + 0.1 M $HClO_4$ solution. Although the typical voltammetric picture for methanol oxidation on Pt is observed (an oxidation peak showing distinct hysterisis), the current densities are rather low and the anatase-rich electrode is almost ten times inferior to the rutile-rich electrode. The latter trend can be interpreted by the absence of significant Pt oxide formation at the anatase-rich electrode (see Figure 7(B)) and the crucial role of these oxides in oxidizing/removing the carbonaceous intermediates/poisons of methanol oxidation. Note that a similar correlation has been found for methanol oxidation at Pt/TiO_2 electrodes in [47].

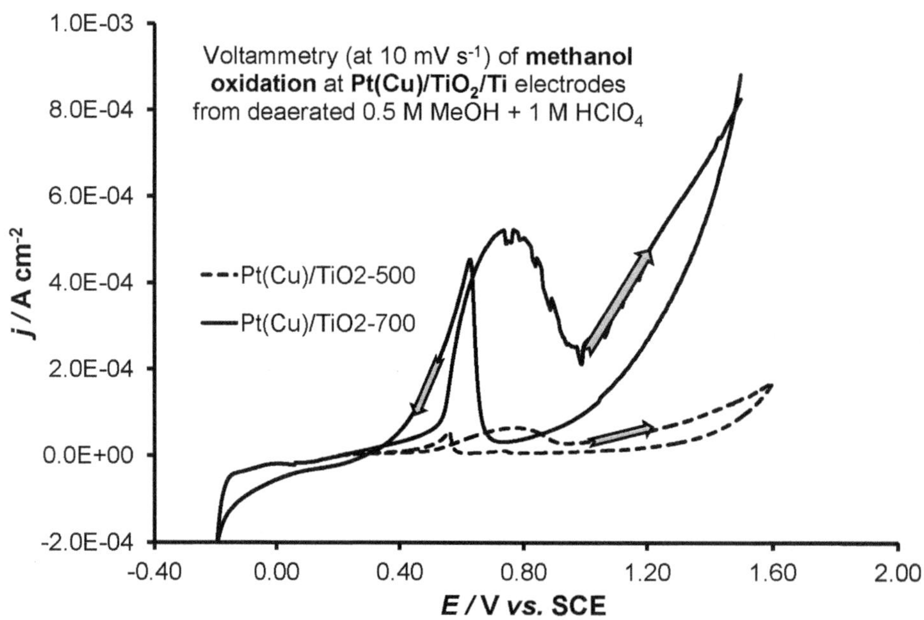

Figure 10. *Cyclic voltammograms of Pt(Cu)/TiO_2/Ti electrodes (anatase-rich/annealed at 500°C for 1.5 h and rutile-rich/ annealed at 700 °C for 5h) in the dark, in a 0.5 M MeOH + 0.1 M $HClO_4$ solution. (Arrows indicate potential scan direction.)*

The Pt mass-specific current density of the rutile-rich Pt(Cu)/TiO_2 catalyst can be estimated as 3.6 mA mg^{-1} (at the peak potential) which is lower than commonly observed values at Pt/C catalysts (see for example the value of ca 15 mA mg^{-1} of a commercial Pt/C(E-Tek) catalyst under same methanol concentration and similar scan rate conditions [50]). This is to be expected for the rutile-rich electrode which (based on Pt surface electrochemistry) was found to have a small Pt electroactive surface area (see discussion above). However, when compared to the performance of the Pt/TiO_2 catalyst prepared by photodeposition in [17] and resulting in a massive 10 mg cm^{-2} Pt loading and/or losses (depending on the amounts deposited onto the substrate or the reactor walls), our catalyst (with a 0.15 mg cm^{-2} Pt loading) is superior. This is because the photodeposited catalyst of [17] gives methanol oxidation currents in the 1.5-3 mA mg^{-1} range (after correction for different concentrations and depending on scan rate).

Figure 11 shows the effect of UV illumination on methanol oxidation at the rutile-rich electrode. The enhancement factor of the peak current is calculated as 1.66 and the net increase is some 344 μA cm^{-2}. The latter is much higher than the net photocurrent of 40 μA cm^{-2} recorded at the same Pt(Cu)/TiO_2/Ti electrode in the acid supporting electrolyte (Figure 9(A)) and even higher than the limiting photocurrent of 260 μA cm^{-2} observed on the plain TiO_2/Ti electrode in the presence of methanol (Figure 3(B)). This in turn indicates that the oxidation current enhancement observed at the Pt(Cu)/TiO_2/Ti electrode upon illumination in the presence of methanol is not merely due to the superposition of methanol oxidation on Pt and photooxidation on TiO_2 but due to a synergistic effect, most likely the photooxidative removal of poisonous intermediates from Pt sites at neighbouring TiO_2 sites.

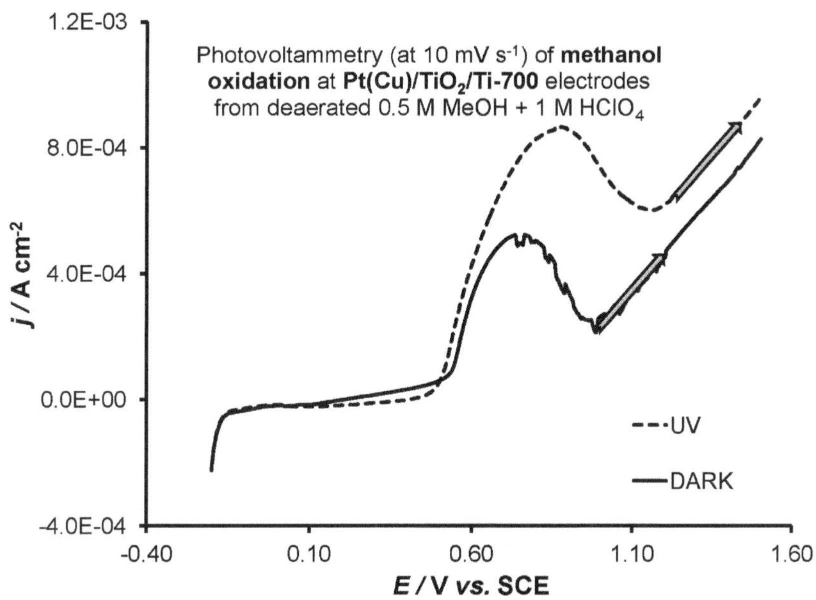

Figure 11. *Voltammograms of a Pt(Cu)/TiO_2/Ti electrode (rutile-rich/ annealed at 700°C for 5h) in the dark and under UV illumination, in a 0.5 M MeOH + 1 M $HClO_4$ solution. (Arrows indicate potential scan direction.)*

Conclusions

i. It has been demonstrated that the two-step electrodeposition-galvanic replacement technique (previously applied to metallic substrates) can be successfully applied to semiconductor oxides too. Hence, following Cu electrodeposition on particulate TiO_2 electrode coatings and its subsequent replacement by Pt, stable Pt(Cu)/TiO_2 catalytic layers can be formed.

ii. Depending on whether the TiO_2 substrate was rutile-rich or anatase-rich, the formation and stripping of Pt surface oxides could or could not be seen in cyclic voltammetry experiments. This indicates stronger interactions between Pt and anatase than between Pt and rutile.

iii. As a direct consequence of the above point, methanol oxidation in the dark is largely suppressed at platinized anatase-rich electrodes. (Since the suppression of Pt oxide formation is known to have a beneficial effect on oxygen reduction, such electrodes are worth testing in that reaction.)

iv. The significant methanol oxidation current enhancement observed upon the UV illumination of the platinized rutile-rich electrodes (due to Pt and TiO_2 synergism) makes more research into their optimisation worth pursuing.

Acknowledgements: *This research has been co-financed by the European Union (European Social Fund – ESF) and Greek national funds through the Operational Program "Education and Lifelong Learning" of the National Strategic Reference Framework (NSRF) - Research Funding Program: Heracleitus II. Investing in knowledge society through the European Social Fund.*

References

[1] S.Trasatti (Ed.), *Electrodes of Conductive Metallic Oxides Parts A and B*, Elsevier, Amsterdam , 1980-1981

[2] A.C.C.Tseung,, S.Yeh, X.Liu, G.H.Kelsall and P.Dykstra, *Novel Acid Resistant Oxygen Evolution Anodes*, Commission of the European Communities, Brussels, (1990) p.9

[3] Ulman's Encyclopedia of Industrial Electrochemistry, vol. **A6**, VCH, Weinheim(1986)

[4] H.I.Warren, *Anodes for Electrowinning*, D.J.Robinson (Ed.), The Metallurgical Society of AIME, (1984) p.79

[5] V. Subramanian, E. Wolf, P.V. Kamat, *J. Phys. Chem. B* **105** (2001) 11439-11446

[6] T. Ohsawa, Y. Yamamoto, M. Sumiya, Y. Matsumoto, H. Koinuma, *Langmuir* **20**(2004) 3018

[7] B. Wen, C. Liu, Y. Liu, J. *Phys. Chem. B* **109** (2005) 12372-12375

[8] J.-M. Lee, S.-B. Han, J.-Y. Kim, Y.-W. Lee, A.-R. Ko, B. Roh, I. Hwang, K.-W. Park, *Carbon* **48(8)** (2010) 2290-2296

[9] H. Song, P. Xiao, X. Qiu, W. Zhu, *J. Power Sources* **195(6)** (2010) 1610-1614

[10] G. Selvarani, S. Maheswari, P. Sridhar, S. Pitchumani, A.K. Shukla, *J. Electrochem. Soc.* **156(11)** (2009) B1354-B1360

[11] C.S. Chen, F.-M. Pan, *Appl. Catal. B-Environ.* **91(3-4)** (2009) 663-669

[12] E. Mahe, D. Devilliers, *Electrochim. Acta* **46** (2001) 629-636

[13] J.V. Macpherson, J.-P. Gueneau de Mussy, *Electrochem. Solid State Let.* **4(9)** (2001) E33-E36

[14] J.V. Macpherson, J.-P. Gueneau de Mussy, *J. Electrochem. Soc.* **149(7)** (2002) B306-313

[15] G.A. Hope, A.J. Bard, *J. Phys .Chem.* **87** (1983) 1979-1984

[16] K. Tammeveski, T. Tenno, A. Rosental, P. Talonen, L.-S. Johansson, L. Niinisto, *J. Electrochem. Soc.* **146(2)** (1999) 669-676

[17] B.E. Hayden, D.V. Malevich, D. Pletcher, *Electrochem. Commun.* **3** (2001) 395-399

[18] X.-Z. Li, F.-B. Li, *J. Appl. Electrochem.* **32** (2002) 203-210

[19] L. Xiong, A. Manthiram, *Electrochim. Acta* **49** (2004) 4163-4170

[20] S.R. Branković, J.X. Wang, R.R. Adžić, *Surf. Sci.* **474** (2001) L173-L179

[21] S.R. Branković, J. McBreen, R.R. Adžić, *J. Electroanal. Chem.* **503** (2001) 99-104

[22] S.R. Branković, J. McBreen, R.R. Adžić, *Surf. Sci.* **479** (2001) L363-368

[23] M. Van Brussel, G. Kokkinidis, I. Vandendael, C. Buess-Herman, *Electrochem. Commun.* **4(10)** (2002) 808-813

[24] M. Van Brussel, G. Kokkinidis, A. Hubin, Cl. Buess-Herman, *Electrochim. Acta* **48** (2003) 3909-3919

[25] S. Papadimitriou, A. Tegou, E. Pavlidou, G. Kokkinidis, S. Sotiropoulos, *Electrochim. Acta* **52** (2007) 6254-6260

[26] A. Tegou, S. Papadimitriou, E. Pavlidou, G. Kokkinidis, S. Sotiropoulos, *J .Electroanal. Chem.* **608** (2007) 67-77

[27] S. Papadimitriou, A. Tegou, E. Pavlidou, G. Kokkinidis, S. Sotiropoulos, *Electrochim. Acta* **53** (2008) 6559-6567

[28] A. Tegou, S. Papadimitriou, S. Armyanov, E. Valova, G. Kokkinidis, S. Sotiropoulos, *J. Electroanal. Chem.* **623(2)** (2008) 187-196
[29] A. Tegou, S. Papadimitriou, G. Kokkinidis, S. Sotiropoulos, *J. Solid State Electrochem.* **14(2)** (2010) 175-184
[30] S. Papadimitriou, S. Armyanov, E. Valova, A. Hubin, O. Steenhaut, E. Pavlidou, G. Kokkinidis, S. Sotiropoulos, *J. Phys. Chem. C* **114(11)** (2010) 5217-5223
[31] Y. Ando, K. Sakaki, R. Adzic, *Electrochem. Commun.* **11** (2009) 1135-1138
[32] B.I. Podlovchenko, T.D. Gladysheva, A.Yu. Filatov, L.V. Yashina, *Russ. J. Electrochem.* **46(10)** (2010) 1189-1197
[33] K.-W. Park, S.-B. Han, J.-M. Lee *Electrochem. Commun.* **9** (2007) 1578–1581
[34] S. Bakardjieva, J. Subrt, V. Stengl, M.J. Dianez. J. Sayagues, *Appl. Catal. B-Environ.* **58** (2005) 193-202
[35] V.B. Baez, J.E. Graves, D. Pletcher, *J. Electroanal. Chem.* **340** (1992) 273-286
[36] V.B. Baez and D. Pletcher, *J. Electroanal. Chem.* **377** (1994) 231-240
[37] V.B.Baez and D.Pletcher, *J. Electroanal. Chem.* **382** (1995) 59-64
[38] D.N. Furlong, D.E. Yates, T.W. Healy, in S. Trasatti (Ed.), *Electrodes of Conductive Metallic Oxides*, Elsevier, Amsterdam, 1981, p.375
[39] B. O'Regan, J. Moser, M. Gratzel, *J. Phys. Chem.* **94** (1990) 8720-8726
[40] K. Vinodgopal, S. Hotchandani, P.V. Kamat, *J. Phys. Chem.* **97** (1993) 9040-9044
[41] S. Sodergren, A. Hagfeldt, J. Olsson, S.-E. Lindquist, *J. Phys. Chem.* **98** (1994) 5552-5556
[42] J. Augustynski, in A.Wieckowski (Ed.), *Interfacial Electrochemistry: Theory, Experiment and Applications*, Marcel Dekker Inc., NY, 1999, p.707
[43] H.Iddir, V. Skavysh, S. Agut, N.D. Browning, M.M. Disko, *Physical Review B - Condensed Matter and Materials Physics*, **73(4)** (2006) art. no. 041403
[44] L. Avalle, E. Santos, V.A. Macagno, *Electrochim. Acta* **39** (1994) 1291-1295
[45] J. Sanchez, M. Koudelka, J. Augustynski, *J. Electroanal. Chem.* **140** (1982) 161-166
[46] K. Uosaki, R. Yoneda, H. Kita, *J. Phys. Chem.* **89** (1985) 4042-4046
[47] B.E. Hayden, D. Pletcher, J.-P. Suchsland, L.J. Williams *Phys. Chem. Chem. Phys.* **11** (2009) 1564-1570
[48] J.A. Horsley, *J. Am. Chem. Soc.* **101(11)** (1979) 2870-2874
[49] M. Hampden-Smith, P. Atanassova, P. Atanassov, T. Kodas, in W. Vieltich, A. Lamm, H.A. Gasteiger (Eds.), Handbook *of Fuel Cells: Fundamentals Technology and Applications*, Vol.**3**, Wiley, Chichester, England, 2003, p.506
[50] M. Chen, Z.-B. Wang, Y. Ding, G.-P. Yin, *Electrochem. Commun.* **10** (2008) 443-446

9

Electrochemical treatment of Acid Red 1 by electro-Fenton and photoelectro-Fenton processes

Camilo González-Vargas, Ricardo Salazar✉ and Ignasi Sirés*,✉

Laboratorio de Electroquímica Medio Ambiental, LEQMA, Departamento de Ciencias del Ambiente, Facultad de Química y Biología, Universidad de Santiago de Chile, USACh, Casilla 40, Correo 33, Santiago, Chile

**Laboratori d'Electroquímica dels Materials i del Medi Ambient, Departament de Química Física, Facultat de Química, Universitat de Barcelona, Martí i Franquès 1-11, 08028 Barcelona, Spain*

✉Corresponding Author
✉Corresponding Author

Abstract

Small volumes (100 mL) of acidic aqueous solutions with 30-200 mg L^{-1} TOC of the toxic azo dye Acid Red 1 (AR1) have been comparatively treated by various electrochemical advanced oxidation processes (EAOPs). The electrolytic system consisted of a BDD anode able to produce $^{\bullet}OH$ and an air-diffusion cathode that generated H_2O_2, which subsequently reacted with added Fe^{2+} to yield additional $^{\bullet}OH$ from Fenton's reaction. Under optimized conditions (i.e., 1.0 mM Fe^{2+}, 60 mA cm^{-2}, pH 3.0, 35 °C), the analysis of the initial rates for decolourization and AR1 decay assuming a pseudo-first-order kinetics revealed a much higher rate constant for photoelectro-Fenton (PEF, ~ 2.7×10^{-3} s^{-1}) compared to electro-Fenton (EF, ~ 0.6×10^{-3} s^{-1}). Mineralization after 180 min was also greater in the former treatment (90 % vs 63 %). The use of UV radiation in PEF contributed to Fe(III) photoreduction as well as to photodecarboxylation of refractory intermediates, yielding a mineralization current efficiency as high as 85% during the treatment of solutions of 200 mg L^{-1} TOC. Primary reaction intermediates included three aromatic derivatives with the initial naphthalenic structure and four molecules only featuring benzenic rings, which were totally mineralized in PEF.

Keywords

Air-diffusion cathode; Azophloxine; boron-doped diamond (BDD); E128; EAOPs; decolourization; food azo dye; mineralization; Red 2G

Introduction

In recent years, great attention has been paid to the electrochemical advanced oxidation processes (EAOPs) based on Fenton's reaction chemistry for water decontamination [1]. Among them, electro-Fenton (EF) and photoelectro-Fenton (PEF) have been the most studied methods for the treatment of waters polluted by organic pollutants such as pharmaceuticals [2-4], pesticides [5-7] and synthetic dyes [8-11] due to their outstanding performance. In EF and PEF, H_2O_2 is continuously electrogenerated in the reaction cell from the two-electron reduction of injected or flushed gaseous O_2 as follows [1]:

$$O_2 + 2\,H^+ + 2\,e^- \rightarrow H_2O_2 \qquad (1)$$

The use of carbonaceous cathodes, particularly in the form of gas-diffusion electrodes (GDEs), promotes the production of a high yield of H_2O_2 at high rate. Therefore, the current efficiency for reaction (1) on GDEs turns out to be the highest among all tested materials. The activation of H_2O_2, which is a weak oxidant, is achieved in the presence of a small amount of metal catalyst, especially Fe^{2+}, at pH around 3. Under such conditions, $^{\bullet}OH$ is produced in the bulk from Fenton's reaction (2) [1]:

$$Fe^{2+} + H_2O_2 \rightarrow Fe^{3+} + {}^{\bullet}OH + OH^- \qquad (2)$$

In PEF, the contaminated acidic solution is irradiated with artificial UV light, thus causing: (i) the photoreduction of $Fe(OH)^{2+}$, which is the main Fe^{3+} species at pH 3, via photo-Fenton's reaction (3), and (ii) the photodecarboxylation of some Fe(III)-carboxylate complexes, which are quite refractory to oxidation by $^{\bullet}OH$ formed from Fenton's reaction, via reaction (4) [1]. While reaction (3) contributes to form additional $^{\bullet}OH$ as well as to maintain the Fenton's cycle by continuously regenerating Fe^{2+}, reaction (4) makes it possible to reach great mineralization degrees.

$$Fe(OH)^{2+} + h\nu \rightarrow Fe^{2+} + {}^{\bullet}OH \qquad (3)$$

$$[Fe(OOCR)^{2+}] + h\nu \rightarrow Fe^{2+} + CO_2 + R^{\bullet} \qquad (4)$$

For some pollutants, the performance of EF and PEF can be further enhanced by using a large O_2-overpotential anode such as boron-doped diamond (BDD). This material favours the electro-oxidation (EO) of the organic molecules by action of $^{\bullet}OH$ adsorbed on the anode surface via reaction (5) [12]:

$$BDD + H_2O \rightarrow BDD({}^{\bullet}OH) + H^+ + e^- \qquad (5)$$

Acid Red 1 (AR1, also called Amido Naphtol Red G, Red 2G, Food Red 10 or Azophloxine, see chemical structure in Table 1) is a synthetic monoazo dye. Azo dyes constitute the most comprehensive and varied family among all synthetic organic dyes available in the industry for dyeing all kinds of fabrics [13]. Until 2007, AR1 was the preferred dye for baby food colouring due to the resulting intense red colour, but the Scientific Panel of the European Food Safety Authority (EFSA) on food additives, flavourings, processing aids and materials in contact with food was then asked to re-evaluate this dye due to food safety concerns related to its high toxicity and carcinogenic effects [14]. Based on this report, the European Union has agreed with its suspension as food colouring, as published in the Official Journal [15]. Not surprisingly, it is also banned in other countries outside Europe.

At present, it is known that AR1 is extensively metabolized to aniline, it may interfere with blood haemoglobin and it is suspected of being a carcinogen [16]. It is therefore one of the dyes that the hyperactive children's support group recommends to be eliminated from the diet of children. In contrast, AR1 has become one the most used dyes in Chile and over the world for dyeing polyester-, nylon-, cellulose- and acrylic-based fibers. Consequently, if such industrial wastewaters are not conveniently treated before discharge, their impact on the surrounding ecosystems may be dramatic.

Some few studies have focused on the fate of AR1 upon application of non-electrochemical advanced oxidation processes (AOPs) [17-19]. Regarding the electrochemical technology, only its electrochemical reduction on an activated carbon fiber cathode has been reported [20], but as far as we are concerned its removal by EF and PEF with BDD or other anodes has not been reported yet. These latter processes are of great interest for the removal of food azo dyes, as some of us recently demonstrated for the treatment of Tartrazine solutions [21].

In this work, the decolourization, AR1 decay and mineralization profiles resulting from the treatment of synthetic aqueous solutions of 100 mL of AR1 by several EAOPs have been investigated. BDD and GDE, both of 2.5 cm^2, have been used at constant current upon addition of 50 mM Na_2SO_4 as supporting electrolyte at pH 3. The optimization of iron catalyst concentration and applied current has preceded the treatment of solutions with up to 200 mg L^{-1} TOC content by PEF. In addition, chromatographic analyses allowed the identification of aliphatic and cyclic by-products formed during the cleavage of the AR1 structure.

Experimental

Chemicals

AR1 (disodium 8-acetamido-1-hydroxy-2-phenylazonaphthalene-3,6-disulfonate, $C_{18}H_{13}N_3Na_2O_8S_2$, CI 18050, 60% purity) was purchased from Sigma-Aldrich. Anhydrous sodium sulfate used as background electrolyte and iron(II) sulfate heptahydrate used as catalyst in EF and PEF were of analytical grade from Merck. Solutions were prepared with bidistilled water and their pH was adjusted to 3 before the electrolyses with analytical grade sodium hydroxide or sulfuric acid from Merck. Other chemicals were obtained from Merck and Sigma-Aldrich.

Electrolytic system

The electrolyses were performed in an open, undivided cell containing a 100 mL solution and featuring a double jacket for circulation of external thermostated water at 35 °C (WiseCircu® WCB-11 water bath). The solution was stirred with a magnetic bar at 800 rpm to ensure good mixing and transport of reactants. The cell contained a 2.5 cm^2 Si/BDD thin-film electrode from Adamant® (500 ppm B) as the anode and a 2.5 cm^2 carbon-PTFE air-diffusion cathode from Electrocell®. The cathode was fed with compressed air flowing at 1 L min^{-1} for H_2O_2 generation. The trials were performed at constant current provided by an MCP M10-QD305 power supply. In PEF, the solution was irradiated with a Black Ray B100AP lamp.

Equipment and analytical procedures

The solution pH was measured on an Extech 321990 pH-meter. Samples withdrawn at regular time intervals from electrolyzed solutions were neutralized at pH 7-8 to stop the degradation process and filtered with 0.45 μm PTFE filters from Whatman before analysis. The decolourization of AR1 solutions was monitored from the absorbance (A) decay at the maximum visible wave-length (λ_{max}) of 520 nm, measured from the spectra recorded on a Cary 1E UV/Vis spectrophotometer (Varian). The percentage of colour removal or decolourization efficiency was then determined as follows [21]:

$$\text{Colour removal, \%} = 100\frac{A_0 - A_t}{A_0} \tag{6}$$

where A_0 and A_t denote the absorbance at initial time and after an electrolysis time t, respectively.

The mineralization of solutions was monitored from their total organic carbon (TOC) abatement, determined on a Vario TOC Select analyzer. From these data, the mineralization current efficiency (MCE) at a given current (I / A) and electrolysis time (t / h) were estimated as follows [1]:

$$\text{MCE, \%} = 100\frac{nFV_s\Delta(TOC)_{exp}}{4.32\times10^7 mIt} \tag{7}$$

where F is the Faraday constant (96487 C mol^{-1}), V_s is the solution volume (L), $\Delta(TOC)_{exp}$ is the experimental TOC decay (mg L^{-1}), 4.32×10^7 is a conversion factor (3600 s h^{-1} × 12000 mg mol^{-1}) and m is the number of carbon atoms of AR1 (18 atoms). The number of electrons (n) consumed per each dye molecule was taken as 98 considering that its mineralization leads to carbon dioxide and nitrate and sulfate ions as follows:

$$C_{18}H_{13}N_3Na_2O_8S_2 + 45\ H_2O \rightarrow 18\ CO_2 + 3\ NO_3^- + 2\ SO_4^{2-} + 2\ Na^+ + 103\ H^+ + 98\ e^- \tag{8}$$

AR1 decay was followed by reversed-phase high performance liquid chromatography (HPLC) with a Waters 625 LC fitted with a Hibar® RP-18e 5 µm, 150 × 4 mm, column at 25 °C and coupled with a photodiode array detector set at λ = 520 nm. The mobile phase was a 70:30 (v/v) acetonitrile/1.0 mM ammonium acetate (pH 4) mixture at 0.5 mL min^{-1}. Generated carboxylic acids were detected by ion-exclusion HPLC using the same LC fitted with a Bio-Rad Aminex HPX 87H, 300 × 7.8 mm, column at 25 °C and setting the array detector at λ = 210 nm. The isocratic elution at 0.6 mL min^{-1} with 4 mM H_2SO_4 as the mobile phase yielded good peaks for maleic (t_r = 8.8 min), oxamic (t_r = 10.6 min), malic (t_r = 11.3 min), formic (t_r = 14.9 min) and acetic (t_r = 16.1 min).

The cyclic and/or aromatic intermediates were analyzed by gas chromatography coupled to mass spectrometry (GC-MS). Several electrolyses were carried out under different experimental conditions for short and long times. The final solutions were collected together until reaching 500 mL, which were then extracted three times with 30 mL CH_2Cl_2. The resulting organic solution (90 mL) was dried with anhydrous Na_2SO_4, then filtered and completely evaporated in a rotary to obtain a pale yellow solid that was further analyzed.

Results and Discussion

Influence of the experimental parameters on the degradation of Acid Red 1 by EF process

Solutions containing 300 mg L^{-1} AR1 (i.e., 0.59 mM AR1 or 100 mg L^{-1} TOC) were electrolyzed at 60 mA cm^{-2} in the presence of different amounts of Fe^{2+} as catalyst. As can be seen in Fig. 1, the absence of Fe^{2+} (so-called EO process) caused the slowest decolourization and TOC abatement, only reaching 70 % colour removal after 70 min and 25 % TOC decay after 180 min. Under the present EO conditions, given the weak oxidation power of H_2O_2, the organic matter can be mainly degraded by BDD($^\bullet$OH) formed via reaction (5). This radical tends to be very active towards the initial pollutants and their by-products because it is weakly physisorbed on the anode surface and it is generated at a very positive potential. Moreover, it is known that hydroxyl radicals can react at high rate with all double bonds in the aromatic rings and, especially, with the –N=N– bond. However, since BDD($^\bullet$OH) is confined to the anode vicinity, the degradation process becomes severely limited by mass transport, thus being needed a much longer electrolysis time to effectively destroy the molecules in a batch system without recirculation like the one tested here.

In contrast to the previous finding, within the same time period the presence of Fe^{2+} allowed the complete decolourization as well as a greater mineralization in all cases, which can be accounted for by the crucial contribution of $^{\bullet}OH$ formed in the bulk from Fenton's reaction (2). As can be seen in Fig. 1a, the increase in Fe^{2+} concentration from 0.1 to 0.5 and then to 1.0 mM clearly accelerated the colour removal, being necessary 70, 60 and 40 min, respectively, to get colourless solutions.

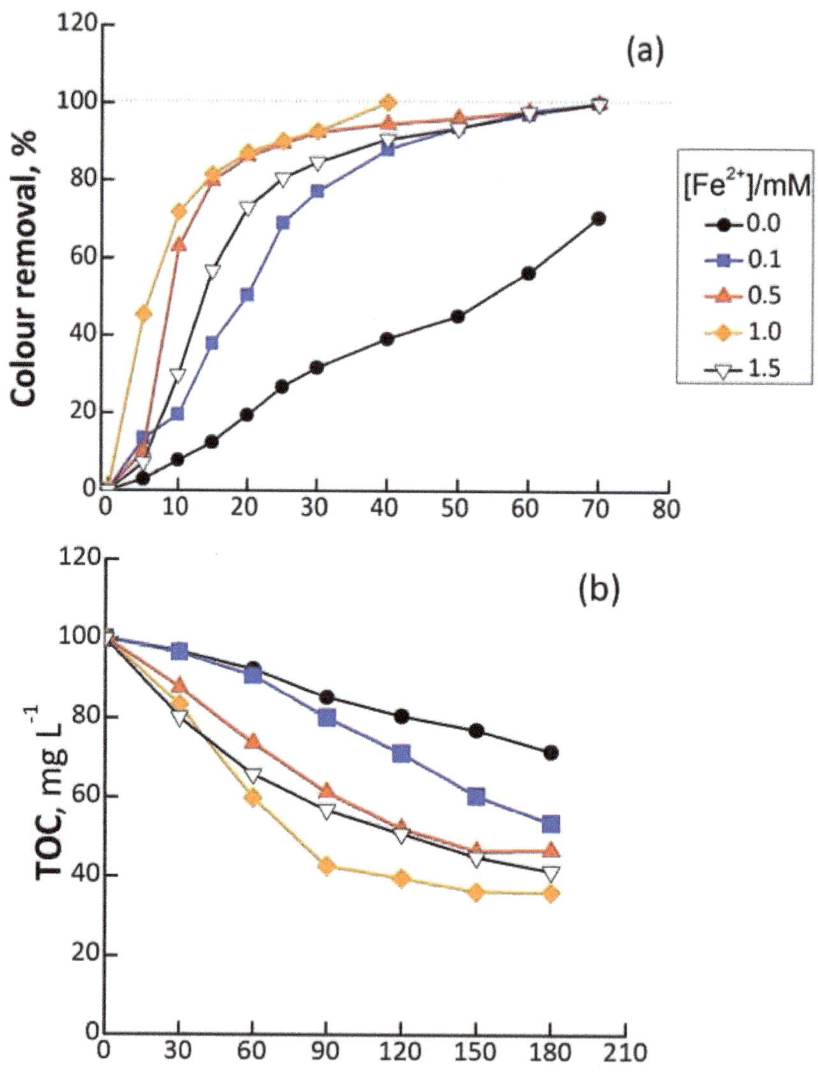

Figure 1. *Effect of Fe^{2+} concentration on **(a)** decolourization efficiency at 520 nm and **(b)** TOC removal with electrolysis time for the electro-Fenton treatment of solutions of 300 mg L^{-1} AR1 in 0.05 M Na_2SO_4 at pH 3.0, 35 °C and 60 mA cm^{-2}.*

The rising catalyst content had a positive effect on the mineralization profiles as well, since 47, 54 and 63 % TOC removal was attained after 180 min using 0.1, 0.5 and 1.0 mM Fe^{2+}. In contrast, further increase to 1.5 mM was detrimental, eventually leading to slower colour removal and only 59 % TOC abatement at 180 min. This phenomenon can be mainly explained by the larger extent of parasitic reactions causing the consumption of $^{\bullet}OH$, particularly by Fe^{2+}. It must be noted that the mineralization was always partial, with a tendency to reach a plateau owing to the plausible accumulation of refractory intermediates (as confirmed later on) that could not be oxidized by $^{\bullet}OH$ in the bulk and were very slowly destroyed by BDD($^{\bullet}OH$). In conclusion, 1.0 mM was chosen as the optimum Fe^{2+} concentration for subsequent tests.

The effect of current density within the range 8-80 mA cm^{-2}, carried out under conditions shown in Fig. 1 with 1 mM Fe^{2+} as the optimized amount, is depicted in Fig. 2. These trials aimed at exploring the possibility of enhancing the decolourization and mineralization kinetics, which is based on the fact that the applied current determines the yield of BDD($^{\bullet}OH$) formed via reaction (5) as well as that of $^{\bullet}OH$ via reaction (2) because it depends on the H_2O_2 formation rate and Fe^{2+} regeneration rate.

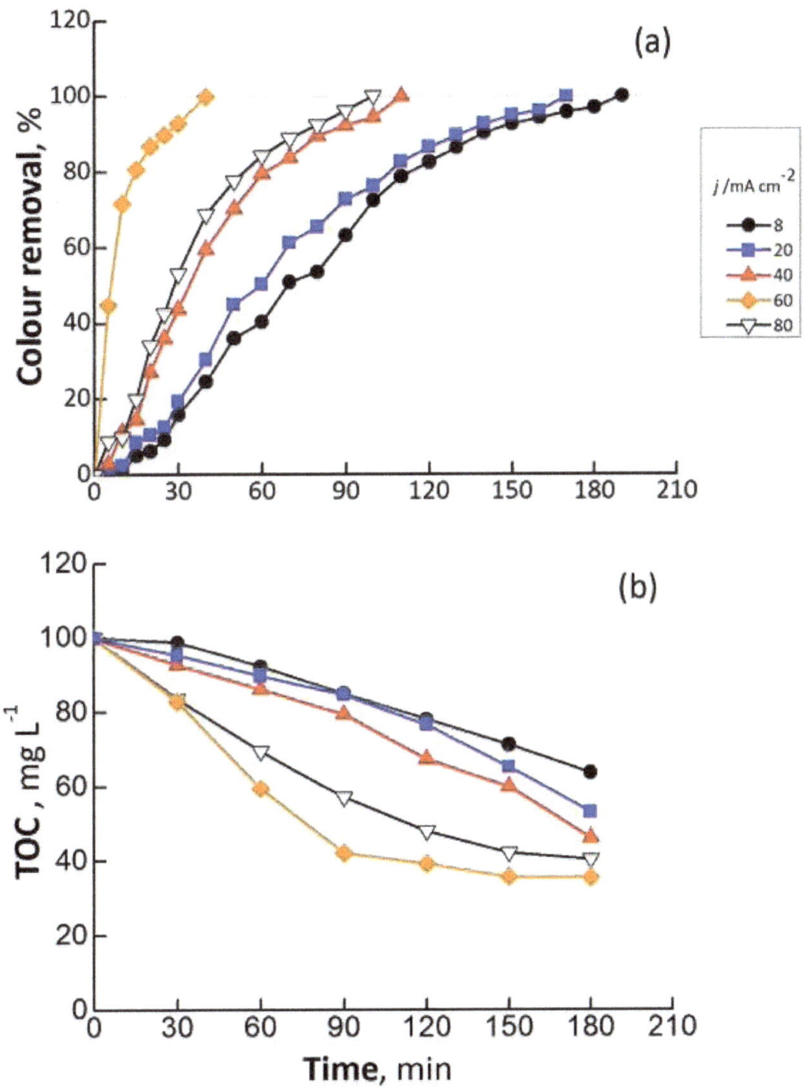

Figure 2. *Effect of current density on* ***(a)*** *decolourization efficiency at 520 nm and* ***(b)*** *TOC removal with electrolysis time for the electro-Fenton treatment of solutions of 300 mg L^{-1} AR1 in 0.05 M Na_2SO_4 with 1.0 mM Fe^{2+} at pH 3.0 and 35 °C.*

A progressive increase in current density from 8 to 60 mA cm^{-2} caused the acceleration of both, decolourization and mineralization. This can be easily explained by the faster generation of BDD($^{\bullet}OH$) on the anode and $^{\bullet}OH$ in the bulk. Note that even the lowest current densities were able to yield the complete decolourization at long electrolysis time. However, further increase to 80 mA cm^{-2} was detrimental since it caused a slower colour removal and led to a lower TOC removal. This negative effect arises from the shift in cathode potential to unfavourable values that promoted the reduction of O_2 to H_2O over reaction (1) and hindered the conversion of Fe^{3+} to Fe^{2+}. As a matter of fact, H_2O_2 concentration analyzed during the electrolyses at 60 and 80 mA cm^{-2} reached 25 mM and 15 mM, respectively. Thus, 60 mA cm^{-2} was chosen as the optimum value.

Effect of UVA light

As discussed in the Introduction, for most of the contaminated solutions studied in the past it was possible to enhance the degradation process by irradiating them with UV light, which favoured the oxidation of pollutants and their by-products due to the action of reaction (3) and (4). In the present study, no direct photolysis of AR1 by UV light was observed, since its peak during the HPLC analyses remained unchanged. This ensured the photostability of AR1 during the electrolyses run under PEF conditions. For this purpose, the AR1 solutions were treated as done in the previous EF experiments but incorporating the UV lamp near the cell.

As shown in Fig. 3a, solutions of 300 mg L^{-1} AR1 treated by PEF under the optimized conditions described before (*i.e.*, 1.0 mM Fe^{2+} and 60 mA cm^{-2}) were more quickly decolourized compared to EF trials, being required 50 min instead of 60 min to become colourless (see Fig. 1a for comparison). The key contribution of photoreduction reaction (3) favoured the faster regeneration of Fe^{2+}, which then was able to accelerate the production of $^{\bullet}OH$ from reaction (2). On the other hand, PEF also yielded a much larger mineralization after 180 min, reaching 90 % owing to photodecarboxylation reaction (4). As reported elsewhere [1], some of the reaction intermediates can form stable complexes with Fe(III) that can be effectively degraded only upon action of UV photons since $^{\bullet}OH$ and BDD($^{\bullet}OH$) are much less effective. Fig. 3a also depicts the decay of AR1 monitored by HPLC during the same experiment. Its profile is quite similar to colour removal profile, which means that no other coloured by-products were formed during the treatment. Assuming a pseudo-first-order kinetics, an apparent rate constant (k_{app}) of 2.74×10^{-3} s^{-1} for AR1 decay was determined. The decay of the dye was also similar to the colour removal trend for EF treatment (not shown), revealing a much smaller k_{app} = 0.59×10^{-3} s^{-1}. Therefore, the beneficial synergy between $^{\bullet}OH$, BDD($^{\bullet}OH$) and UV light for the decontamination of AR1 solutions is demonstrated.

Due to production needs, actual wastewaters may present a significant variation in the dye content over time and thus, it is mandatory that the water treatment technology is flexible enough to be adapted to such changes. The effect of AR1 concentration on the mineralization profile *vs.* time is shown in Fig. 3b. A similar TOC removal was attained after 180 min for solutions containing 30 and 100 mg L^{-1} TOC. In contrast, only 75 % mineralization was reached for solutions with 200 mg L^{-1} TOC, which is simply due to the much larger number of organic molecules to be degraded in the latter case. But, an important feature to be highlighted is the progressively increasing slope of the curves (*i.e.*, larger mineralization rate) upon increase of AR1 concentration, which can be related to the more efficient reaction between $^{\bullet}OH$/BDD($^{\bullet}OH$) and the organic molecules. Indeed, low AR1 concentrations cause the waste of radicals in self-destruction and other side reactions, whereas high organic contents lead to effective oxidation reactions. This is clearly demonstrated in Fig. 3c, which compares the evolution of MCE vs time for several EAOPs. For solutions with 100 mg L^{-1} TOC, the efficiency increases in the sequence EO < EF < PEF, with maximum values of 10, 25 and 55 %, respectively. As discussed before, this can be related to the more favorable synergy between different oxidants in the case of PEF. In addition, a greater MCE resulted from the treatment of larger AR1 concentrations in PEF, reaching 85 % during the treatment of 200 mg L^{-1} TOC, thus confirming the lower extent of parasitic reactions that cause radical waste. Note that MCE tends to decrease at long electrolyses time, which can be explained by (i) the formation of more resistant intermediates and (ii) the mass transport limitations related to low organic loads.

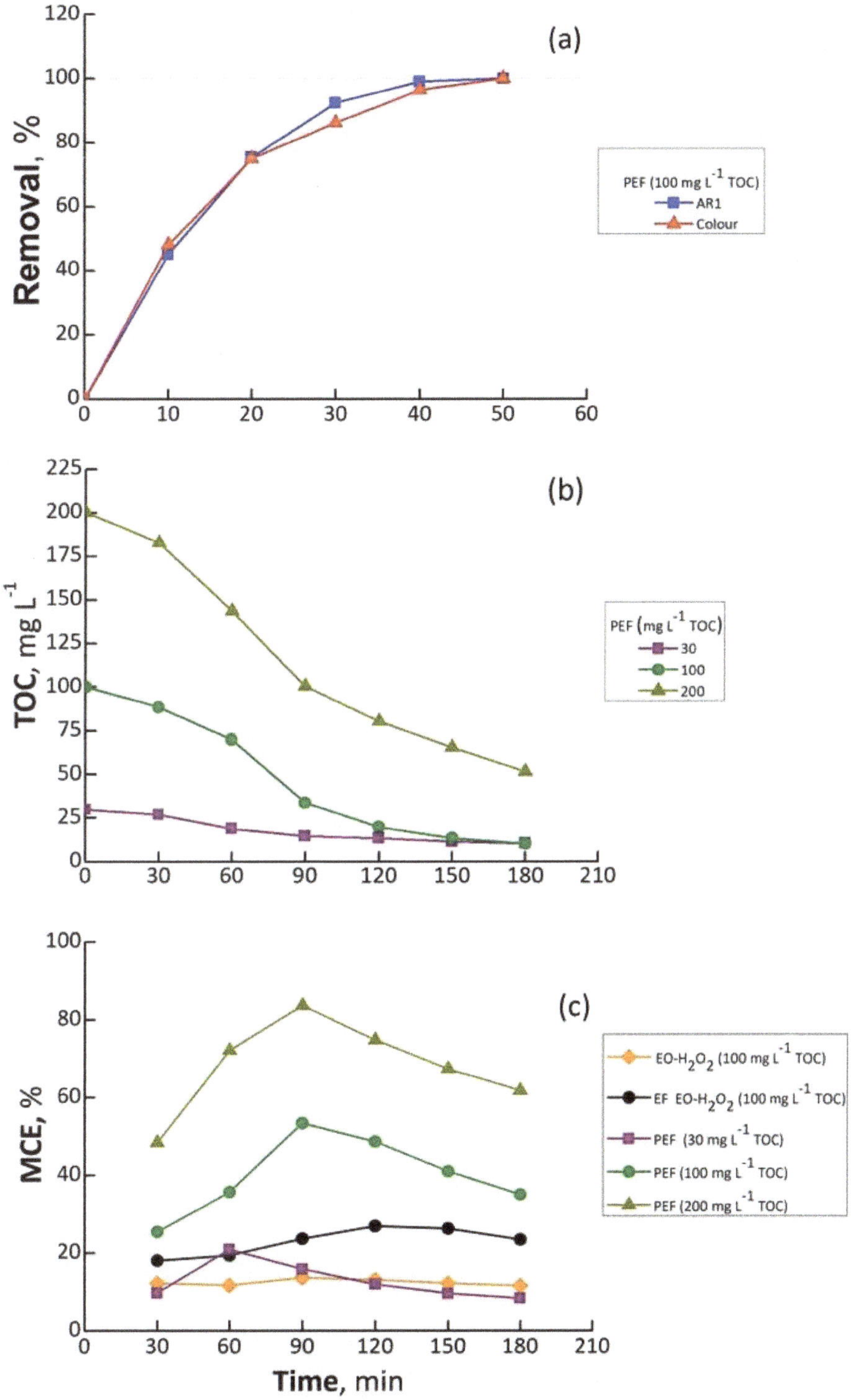

Figure 3. (a) *Decolourization efficiency at 520 nm and percentage of AR1 removal with electrolysis time for the photoelectro-Fenton treatment of solutions of 300 mg L^{-1} AR1 in 0.05 M Na_2SO_4 with 1.0 mM Fe^{2+} at pH 3.0, 35 °C and 60 mA cm^{-2}.* ***(b)*** *TOC abatement* vs. *time for the same experiment compared to trials at different AR1 concentrations.* ***(c)*** *MCE for trials shown in (b) compared to EO (0 mM Fe^{2+}) and EF (1.0 mM Fe^{2+}) treatments shown in Fig. 1b.*

Identification of reaction intermediates

Table 1 summarizes the seven aromatic intermediates identified during PEF treatments.

Table 1. *Structures of Acid Red 1 and its degradation intermediates identified by GC-MS analysis.*

Chemical name	Structure	*m/z*
Acid Red 1	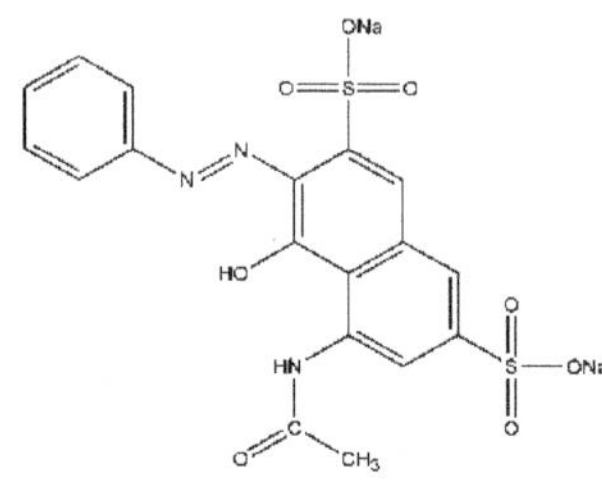	509.42
Naphthalene derivatives		
Disodium 8-acetamido-1-hydroxy-2-(4-hydroxyphenylazo)naphthalene-3,6-disulfonate	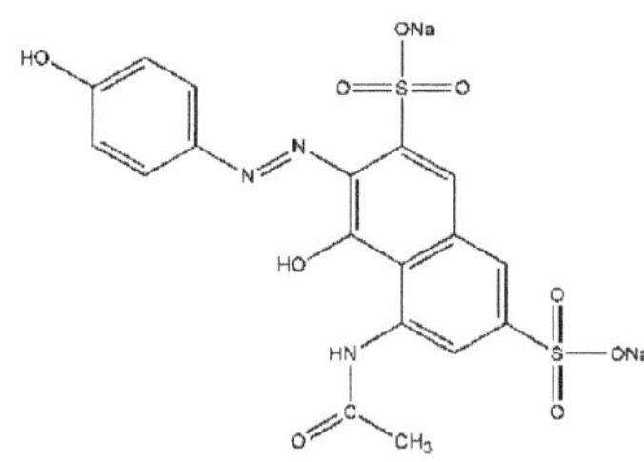	524.12
Disodium 1,2-dihydroxy-3-phenylazonaphthalene-4,7-disulfonate	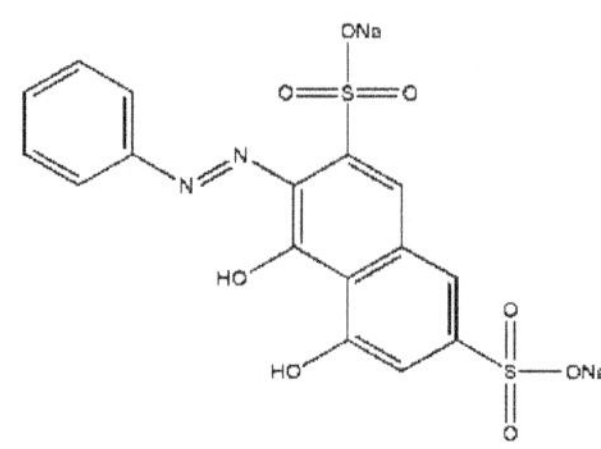	468.37
(5-Phenylazo-3,4,6-trihydroxynaphthalene)sulfonic acid	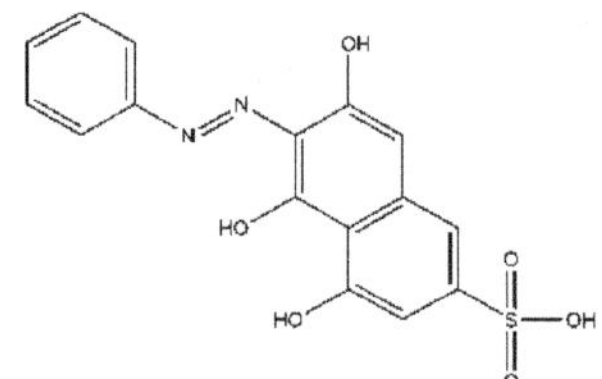	362.33
Benzene derivatives		
Sodium (3,5-dihydroxy-2-phenylazo)benzenesulfonate	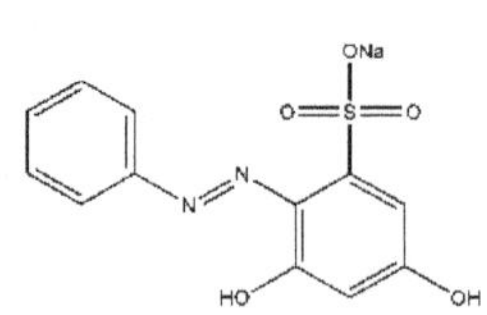	316.26
Sodium (1-hydroxy-2-[4-hydroxyphenylazo]-2-oxo)ethanesulfonate	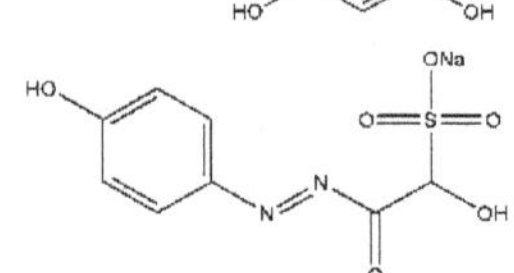	282.21
2-(3,5-dihydroxyphenylazo)-1-hydroxy-2-oxo-ethanesulfonic acid	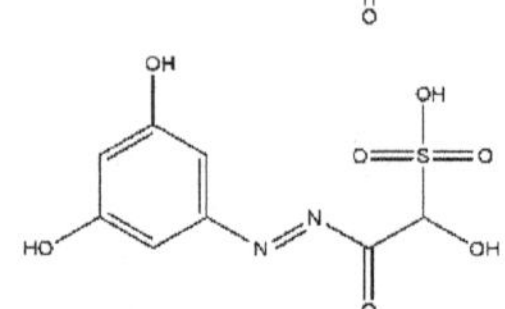	276.22
Hydroquinone	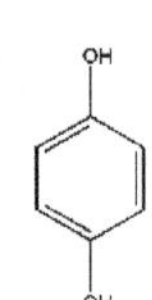	110.11

As can be seen, •OH and BDD(•OH) led to the hydroxylation of the benzenic and naphthalenic rings of AR1 to yield three naphthalene derivatives. The subsequent action of radicals onto AR1 and/or onto those intermediates led to the formation of four benzene derivatives. The accumulation of these aromatic intermediates would be dangerous due to their inherent high toxicity and thus, PEF treatments had to be prolonged until their complete disappearance.

The progressive cleavage and oxidation of the aromatic intermediates gave rise to the formation of short-chain aliphatic carboxylic acids, as also described elsewhere for other pollutants [22]. Five C_1-C_4 acids were identified by ion-exclusion HPLC, namely maleic, oxamic, malic, formic and acetic. As explained from Fig. 3, PEF ensured the almost complete removal of all these acids since the TOC in the final solutions was <10 % after 180 min.

Conclusions

PEF technology is confirmed as a very powerful alternative for giving response to environmental concerns related to water contamination by organic pollutants. This process allows a much faster destruction of AR1 as well as a more significant and efficient TOC removal compared to EO and EF, thus becoming a promising technology for the treatment of industrial wastewaters containing this azo dye. The toxic intermediates formed during the first degradation stages are completely transformed into aliphatic molecules, which are slowly converted to CO_2 and H_2O. The use of renewable energy such us sunlight in sunny countries like Chile and Spain, giving rise to the so-called solar photoelectro-Fenton (SPEF) process, would be an interesting feature for real-scale application.

Acknowledgements*: The authors thank CONICYT (Chile) for support under FONDECYT grant 1130391 and DICYT-USACh, as well as for the PhD fellowship N° 21130071 awarded to C. González-Vargas.*

References

[1] E. Brillas, I. Sirés, M. A. Oturan, *Chemical Reviews* **109** (2009) 6570–6631
[2] A. Dirany, S. Efremova Aaron, N. Oturan, I. Sirés, M. A Oturan, J. J. Aaron, *Analytical and Bioanalytical Chemistry* **400** (2011) 353–360
[3] A. El-Ghenymy, N. Oturan, M. A. Oturan, J.A. Garrido, P. L. Cabot, F. Centellas, R. M. Rodríguez, E. Brillas, *Chemical Engineering Journal* **234** (2013) 115–123
[4] S. Loaiza-Ambuludi, M. Panizza, N. Oturan, A. Özcan, M. A. Oturan, *Journal of Electroanalytical Chemistry* **702** (2013) 31–36
[5] N. Oturan, M. Zhou, M. A. Oturan, *Journal of Physical* Chemistry A **114** (2010) 10605–10611
[6] R. Salazar, M. S. Ureta-Zañartu, *Water Air Soil Pollution* **223** (2012) 4199–4207
[7] J. Urzúa, C. González-Vargas, F. Sepúlveda, M. S. Ureta-Zañartu, R. Salazar, *Chemosphere* **93** (2013) 2774–2781
[8] M. Zhou, Q. Yu, L. Lei, *Dyes and Pigments* **77** (2008) 129–136
[9] M. Panizza, M. A. Oturan, *Electrochimica Acta* **56** (2011) 7084–7087
[10] E. J. Ruiz , A. Hernández-Ramírez, J. M. Peralta-Hernández, C. Arias, E. Brillas, *Chemical Engineering Journal* **171** (2011) 385–392
[11] R. Salazar, E. Brillas, I. Sirés, *Applied Catalysis B: Environmental* **115-116** (2012) 107–116
[12] M. Panizza, G. Cerisola, *Chemical Reviews* **109** (2009) 6541–6569
[13] C. A. Martínez-Huitle, E. Brillas, *Applied Catalysis B: Environmental* **87**(2009) 105–145
[14] European Food Safety Authority, *The EFSA Journal* **515** (2007) 1-28

[15] Official Journal of the European Union L 195/8, 27.7.2007. Commission regulation (EC) No 884/2007 of July 2007 on emergency measures suspending the use of E 128 Red 2G as food colour

[16] A. F. Villa , F. Conso, *EMC - Toxicologie-Pathologie* **1** (2004) 161-177

[17] Cs. M. Földváry, L. Wojnárovits, *Radiation Physics and Chemistry* **78** (2009) 13–18

[18] N. K. Daud, M.A. Ahmad, B.H. Hameed, *Chemical Engineering Journal* **165** (2010) 111–116

[19] S. Thomas, R. Sreekanth, V. A. Sijumon, U. K. Aravind, C. T. Aravindakumar, *Chemical Engineering Journal* **244** (2014) 473–482

[20] Z. Shen, W. Wang, J. Jia, J. Ye, X. Feng, A. Peng, *Journal of Hazardous Materials* **B84** (2001) 107–116.

[21] A. Thiam, M. Zhou, E. Brillas, I. Sirés, *Applied Catalysis B: Environmental* **150-151**(2014) 116–125

[22] M. A. Oturan, M. Pimentel, N. Oturan, I. Sirés, *Electrochimica Acta* **54** (2008) 173–182

10

Wastewater treatment by multi-stage batch adsorption and electrochemical regeneration

FADHIL M. MOHAMMED[‡], EDWARD P. L. ROBERTS[✉], ANDREW K. CAMPEN* and NIGEL W. BROWN*

School of Chemical Engineering and Analytical Science, University of Manchester. The Mill, Oxford Road, Manchester M13 9PL, UK; [‡]Current address: Ministry of Science and Technology, Baghdad, Iraq

**Arvia Technology Ltd, Daresbury Innovation Centre, Keckwick Lane, Daresbury, Cheshire WA4 4FS, UK.*

[✉]Corresponding Author: E-mail: edward.roberts@manchester.ac.uk

---

Abstract

The removal and destruction of a tri-phenyl methane dye, Acid Violet 17 (AV17), from aqueous solution by adsorption and electrochemical regeneration was studied using a graphite intercalation compound (GIC) adsorbent. It was demonstrated that the adsorbent could be regenerated by anodic oxidation of the adsorbed dye in a simple electrochemical cell. The GIC adsorbent recovered its initial adsorption capacity after 40 to 60 min of treatment at a current density of 10 mA cm^{-2}, corresponding to a charge of 12 to 18 C g^{-1} of adsorbent. The charge passed is consistent with that expected for mineralisation of the dye, suggesting that the dye was removed and destroyed with high charge efficiency. The energy cost of the regeneration was found to be around 120 J per g of adsorbent regenerated or 115 J per mg of the AV17 dye removed and destroyed. A model describing the process of wastewater treatment by multiple cycles of adsorption and electrochemical regeneration, based on adsorption isotherm data, has been developed and validated. It was found that relatively modest improvements in the adsorption capacity of the adsorbent material could significantly improve the process performance.

Keywords

Adsorption; electrochemical regeneration; graphite intercalation compound; tri-phenyl methane dye; acid violet.

Introduction

Dyestuff removal has been one of the most persistent problems in wastewater treatment in the last thirty years. Recent methods which have been investigated for dye removal include adsorption, ion exchange, chemical oxidation, precipitation and biological treatment [1,2]. There are ever more dyes available commercially, and due to their complex structure and synthetic origin, most are difficult to decolorize [3]. Therefore, it is necessary to remove them from liquid wastes at least to a limit accepted by national and international regulatory agencies before discharge. Adsorption is a widely used technique for the removal of dyes from wastewater and can be effective for overall treatment, particularly if the sorbent is cheap, does not require a pre-treatment step before application and is easy to regenerate [4]. Adsorption on activated carbon is a technology which has been widely studied to remove dyes from wastewater but the high capital and regeneration cost of activated carbon has led to the search for low cost materials. In the last twenty years, many investigators have studied the feasibility of inexpensive, commercially available materials, that are easy to regenerate and re-utilized as many times as possible [5]. In recent years many inexpensive, widely available materials have been investigated as adsorbents to remove dyes from contaminated water. Wood, fly ash, coal, zeolite, silica, and agricultural wastes have all been tried with varying degrees of success [6-14].

The adsorption of dyes from aqueous solutions onto these inexpensive adsorbent materials focused on the determination of the capacity, kinetics, equilibrium isotherms and the effect of different parameters such as thermodynamics, pH, bonding mechanisms, and desorption [14-16]. However, only limited application of such data has been directed to the modelling of the batch adsorption / regeneration process for wastewater treatment.

Electrochemical regeneration of activated carbon adsorbents was demonstrated by Narbaitz and Cen in 1994 [17] and there have been several recent studies of this process [18-20]. However, the process is not economic as regeneration is slow and the energy cost is high [21], due to the poor conductivity of the activated carbon bed. Recent work has shown the potential of graphite intercalation compound (GIC) adsorbent materials that can be electrochemically regenerated very rapidly and cheaply [21-24].

Conventional adsorption processes are designed based on the single use of a batch of adsorbent. The rapid electrochemical regeneration process discussed above opens up the possibility of using a batch of adsorbent for multiple cycles of adsorption with regeneration carried out on-site. In this way the adsorbent can be reused many times. In addition, a treatment process can be envisaged where the water is treated by the same batch of adsorbent through several cycles of adsorption and regeneration. This approach has been used to treat radioactive organic waste in the nuclear industry [25]. In this study we aim to carry out the first detailed study of this multi-stage adsorption and electrochemical regeneration process.

Experimental

Materials

The adsorbate used in this study was Acid Violet 17, a powdered, anionic tri-phenyl methane dye, which has three substituent groups and is a mono sodium salt. It was supplied by KEMTEX Educational Supplies Ltd under the trade name Kenanthrol Violet 2B at a dye content of about 22 %. Analysis is by UV/Vis spectroscopy at a λ_{max} of 542 nm, as described in our earlier paper [26]. The adsorbent used was a graphite intercalated compound (GIC), Nyex^{TM} 1000, supplied by Arvia Technology Ltd., which contains around 94 wt. % carbon and has a particle size range of 100 – 700 μm [26].

Adsorption/electrochemical regeneration methodology

A laboratory scale sequential batch rig (see Fig. 1) was used for both adsorption and electrochemical regeneration, which were carried out at the ambient laboratory temperature of 20°C. In each experiment 100 g of GIC adsorbent was mixed with 1 L of water in the adsorption zone by sparging air into the bottom of the rig.

After adsorption, the adsorbent was allowed to settle into the anodic compartment of the electrochemical regeneration zone. The adsorbent bed was in contact with a graphite anode and was separated from the perforated stainless steel cathode (316L with open area 33%, 3 mm holes) by a microporous polyethylene separator (Daramic 350, Grace GmbH). The separator acts as a barrier to minimize the transport of electrolyte through the membrane and to ensure that the GIC bed does not contact the cathode, as this would short circuit the electrochemical cell. The cathode compartment was filled with 0.3 wt% NaCl solution to provide good conductivity, and the pH of this solution was adjusted to below 2 using hydrochloric acid to ensure that the membrane was stable. The anode, cathode and membrane of the electrochemical cell had dimensions of 10 cm by 7 cm, and the gap between the anode current feeder and the membrane was 2.2 cm. The 100 g of adsorbent used formed a bed of depth 5 cm in the anode compartment. A DC power supply was used to apply a current of 0.5 A to the cell, corresponding to a current density (from studies by Brown, 2005) of 10 mA cm^{-2} (based on the membrane area).

Ideally electrochemical regeneration should occur throughout the bed depth in the anode compartment. Although it is difficult to measure the current distribution in the packed bed electrode, previous studies with GIC adsorbents have examined the effect of bed thickness [22] or separated and tested layers of adsorbent in the regeneration bed [24]. These studies have indicated that complete regeneration can be achieved with bed depths of up to around 20 mm.

Prior to each experiment 100 g of fresh adsorbent was mixed with 1 litre of clean water for 30 min before being allowed to settle into the electrochemical regeneration cell. The water was drained off and a current of 0.5 A was applied for 30 min in order to oxidise any organic impurities present on the surface of the adsorbent. A volume of 1 L of solution containing 120 mg L^{-1} of AV17 was then added to the cell. This concentration was selected in order to saturate the adsorbent and to give a significant equilibrium concentration. This was necessary to ensure that the equilibrium concentration achieved with the regenerated adsorbent can be accurately measured.

The adsorbent and AV17 solution were mixed by sparging air into the cell for 120 min. This adsorption time was found to be sufficient to ensure that equilibrium was reached at these conditions. Once the adsorption stage was complete, the air was switched off and the adsorbent particles settled into the anodic compartment of the electrochemical cell. The treated liquid was drained off and a sample was taken and analysed by UV/visible spectroscopy in order to determine the loading of AV17 dye on the adsorbent.

The bed was regenerated for a period of between 10 and 120 min. After regeneration the supernatant was drained from above the bed of GIC adsorbent and the adsorption stage was repeated using a fresh AV17 solution (120 mg L^{-1}) which was added to the cell. The adsorption stage was repeated by sparging with air for a period of 120 min. A sample of the solution obtained after adsorption was analysed by UV/vis spectroscopy, as described above. The performance of the regeneration was characterised using the 'regeneration efficiency', obtained by comparing the equilibrium adsorbent loading achieved before and after regeneration:

$$\eta_r \,/\, \% = \frac{q_r}{q_i} \times 100 \qquad (1)$$

where q_i and q_r are the equilibrium loading (q_e) of AV17 on the adsorbent (mg g^{-1}) obtained before and after regeneration, respectively, calculated using Equation (2).

$$q_e = \frac{V(C_0 - C_e)}{W} \quad (2)$$

where C_0 and C_e are the initial and equilibrium concentration of AV17 in solution respectively (mg L^{-1}), V is the volume of solution (L), and W is the mass of adsorbent used (g).

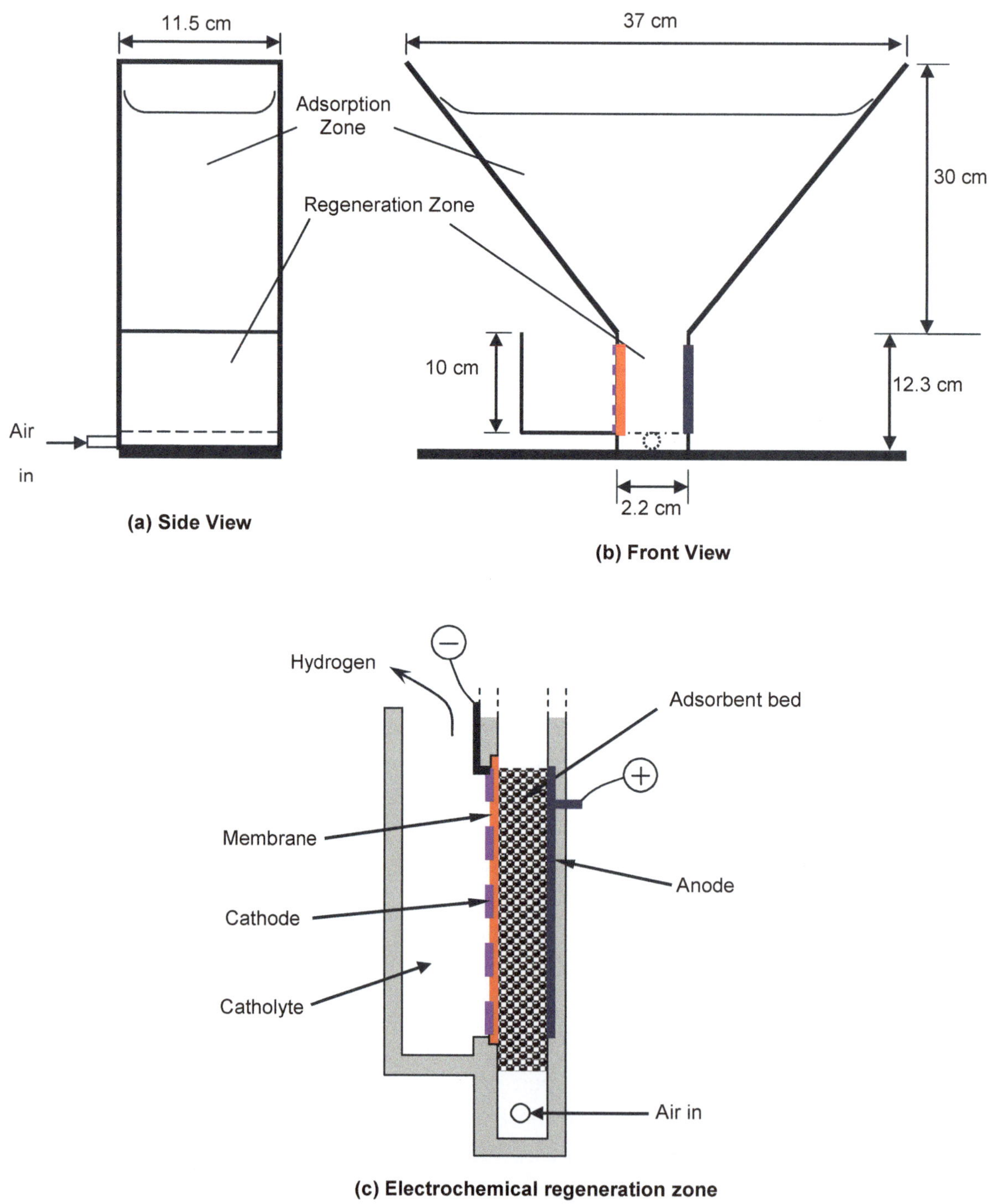

Figure 1. *Laboratory scale sequential batch rig for electrochemical regeneration of the GIC adsorbent: schematic diagram of the side* ***(a)*** *and front* ***(b)*** *elevation of the rig, and* ***(c)*** *schematic diagram showing a cross section of the electrochemical regeneration zone.*

A similar technique was used for multi-stage batch adsorption and electrochemical regeneration, but in this case the solution was not drained off after the adsorption cycle. In this case, the initial concentration of AV 17 was much higher, 668 mg L^{-1} as this solution was treated by a series of adsorption/regeneration cycles. For these experiments, the amount of GIC adsorbent was increased to 125 g, which formed a bed of 7 cm depth in the anode compartment. In addition, the adsorption time was reduced to 60 min and the regeneration time was 30 min.

Results and Discussion

Adsorption isotherm

The design of a batch adsorber system requires knowledge of the equilibrium isotherm to understand the adsorption process behaviour and provide fundamental physiochemical data for evaluating the maximum capacity of the adsorbent [27]. Batch sorption studies have been performed previously to investigate the kinetic and equilibrium isotherm of the adsorption of the Acid violet 17 [26]. The experimental data were found to fit the Langmuir isotherm model, Equation (3).

$$q_e = \frac{bk_L C_e}{1 + bC_e} \qquad (3)$$

where k_L and b are the Langmuir constants related to the capacity of the adsorbent (mg g^{-1}) and the intensity of adsorption, (L mg^{-1}), respectively.

There is little data available in the literature on the adsorption of AV17 by activated carbon, but there is some data on other dyes such as methylene blue [28]. The comparison of Langmuir isotherm constants for adsorption of organic dye calculated in this work with those determined by [28] for activated carbon are shown in Table 1. The low surface area of the adsorbent leads to a relatively low adsorption capacity, as indicated by the relatively low value of k_L. However, if the adsorptive capacity (k_L) is normalised with the specific surface area, it is found that the GIC adsorbent is able adsorb a higher mass of dye per unit area than the activated carbons studied (albeit for a different adsorbate). This suggests that either the GIC has more adsorption sites per unit area or (more likely) that some of the activated carbon surface area in the micropores and is not accessible to the dye molecules.

Table 1. *Comparison of Langmuir constants and surface area for adsorption of methylene blue (MB), and AV17 onto activated carbons [28] and the GIC adsorbent Nyex™ 1000 [29], respectively, at room temperature (23°C) and normal pH.*

Adsorbent	Adsorbate	Langmuir constant		Surface area (BET), m^2 g^{-1}	Adsorptive capacity, mg m^{-2}
		k_L / mg g^{-1}	b / L mg^{-1}		
GIC (Nyex™ 1000)	AV17	0.987	0.31	1.0	0.987
Activated carbon (PAC1)	MB	307	0.12	863.50	0.36
Activated carbon (PAC2)	MB	345	0.15	857.14	0.4
Activated carbon (F400)	MB	455	0.2	1216.4	0.37

The essential characteristic of the Langmuir isotherm shape on whether adsorption is ‘favourable’ or ‘unfavourable’ can be classified in terms of dimensionless separation factor or equilibrium parameter, R_L, Equation (4) [30].

$$R_L = \frac{1}{1+bC_{mx}} \quad (4)$$

where C_{mx} is the maximum solution concentration studied. The separation factor R_L indicates the isotherm shape according to Table 2.

Table 2. *Effect of separation factor on isotherm shape [30].*

Value of R_L	**Type of isotherm**
$R_L>1$	Unfavourable
$R_L=1$	Linear
$0<R_L<1$	Favourable
$R_L=0$	Irreversible

The separation factor for the GIC adsorbent (b = 0.31 L mg^{-1}) has been calculated at range of maximum concentrations of AV17 as shown in Figure 2. It can be observed from this figure that the value of R_L lies between 0 and 1 at all initial dye concentrations, confirming the favorable uptake of the AV17 by the GIC adsorbent.

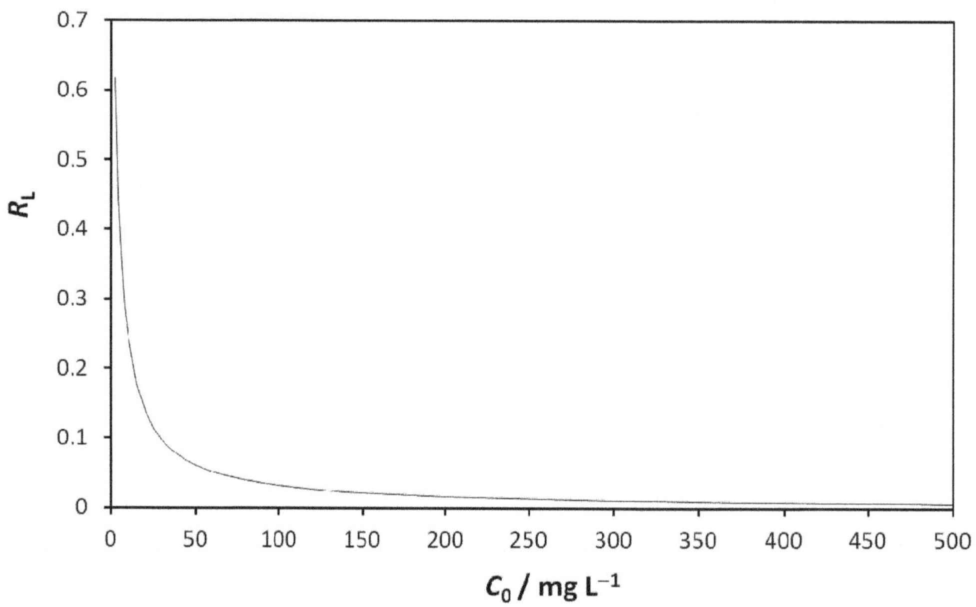

Figure 2. *Separation factor for Acid Violet 17 onto the GIC adsorbent Nyex™ 1000 at 20°C.*

Using the Langmuir model to determine the solid phase loading q_{mx} in equilibrium with C_{mx}:

$$q_{mx} = \frac{bk_L C_{mx}}{1+bC_{mx}} \quad (5)$$

Thus:

$$\frac{q_e}{q_{mx}} = \frac{C_e}{C_{mx}} \frac{1+bC_{mx}}{1+bC_e} \quad (6)$$

Defining a dimensionless solid phase concentration as $q = (q_e/q_{mx})$, and a dimensionless liquid phase concentration as $C = (C_e/C_{mx})$, we obtain:

$$q = \frac{C}{R_L + (1 - R_L)C} \tag{7}$$

Measured values of q for a range of values of C are plotted in Figure 3, where the maximum equilibrium concentration was C_{mx} = 46 mg L^{-1}. The experimental data was fitted to Equation (7) by finding the value of R_L that gave a minimum in the sum of the absolute error (SAE) between the (q, C) data and Equation (7), indicating a value of R_L = 0.065. Similar results have been reported for separation factors for the adsorption of the dye AV17 onto activated carbon prepared from sunflower seed hull [16] and orange peel [14].

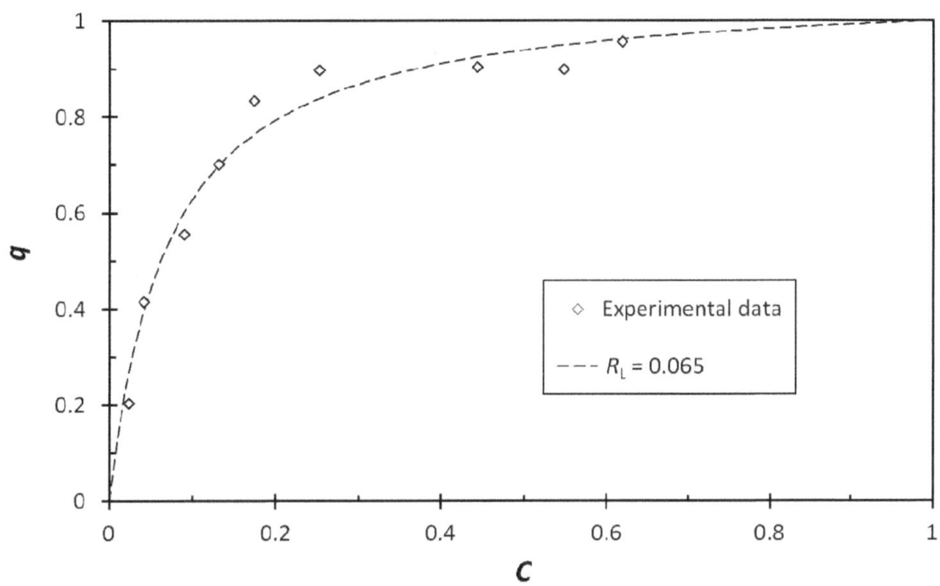

Figure 3. *Isotherm shape for adsorption of AV17 onto the GIC adsorbent Nyex^{TM} 1000 at room temperature (23°C) and the curve obtained using Equation (7) with* R_L *= 0.065.*

Adsorption and electrochemical regeneration

The liquid phase concentration of AV17 dye after adsorption was found to be in the range 13.2 to 15.4 mg L^{-1}, corresponding to a loading of AV17 on the GIC adsorbent of 1.07 to 1.05 mg g^{-1}. This was slightly higher than the value of k_L (0.987 mg g^{-1}) from the Langmuir adsorption isotherm [29], possibly due to natural variability of the adsorbent, attrition or surface modification of the adsorbent during the washing procedure.

It was found that the full adsorption capacity could be recovered after 40 min of electrochemical regeneration, corresponding to a charge passed of 12 C g^{-1} (Figure 4). Further increases in the regeneration time led to an increase in the adsorption capacity to a value around 10 % higher than the original capacity. There are a number of possible reasons for this increase in adsorption capacity after regeneration. Previous studies [24] have suggested that the electrochemical regeneration can lead to a slight roughening of the adsorbent surface, and thus an increase in the specific surface area available for adsorption. Another possibility is the presence of oxidising species formed in the adsorbent bed during regeneration which may react with the organic dye leading to an apparent increase in adsorption capacity.

For regeneration times less than 40 min, the full capacity of the adsorbent was not recovered. This suggests that some AV17 or its breakdown products remained adsorbed on the surface of the GIC adsorbent. When the full adsorption capacity is recovered, the AV17 dye is either fully mineralised, or breakdown products are formed which do not adsorb on the adsorbent and are thus released into the water. Chemical oxygen demand (COD) analysis of treated water (unpublished data from Arvia Technology Ltd) have not found organic breakdown products in solution, suggesting that breakdown products remain adsorbed until full mineralisation is achieved.

It is possible to estimate the charge required to fully mineralise the dye; however this will depend on the products formed. The maximum charge required can be estimated by assuming complete anodic oxidation of the AV17 to carbon dioxide, sulphate and nitrate:

$$C_{41}H_{44}N_3NaO_6S_2 + 93H_2O \longrightarrow 41CO_2 + 230H^+ + Na^+ + 2SO_4^{2-} + 3NO_3^- + 224e^- \quad (8)$$

Alternatively a lower estimate of the charge required for mineralisation can be obtained by assuming that the products were carbon monoxide, sulphide and nitrogen:

$$C_{41}H_{44}N_3NaO_6S_2 + 35H_2O \longrightarrow 41CO + 114H^+ + Na^+ + 2S^{2-} + 1.5N_2 + 111e^- \quad (9)$$

The theoretical charge, Q, for mineralisation of the AV17 dye can be estimated from:

$$Q = \frac{q_e n F}{M_w} \quad (10)$$

where n is the number of electrons required per molecule of dye oxidised, F is Faraday's constant (96487 C mol^{-1}), and Mw is the molecular weight of the dye (761.9 g mol^{-1}). The value of q_e was 1.05 mg g^{-1}, and using n = 111 and 224 [based on Equations (4) and (5) respectively] the theoretical charge required for mineralisation of the AV17 would be between 15 and 30 C g^{-1}. Comparison of these values with the data in Fig. 4 suggests that the charge efficiency for the anodic oxidation is high, but further work is needed to determine whether any breakdown products are present in the treated water.

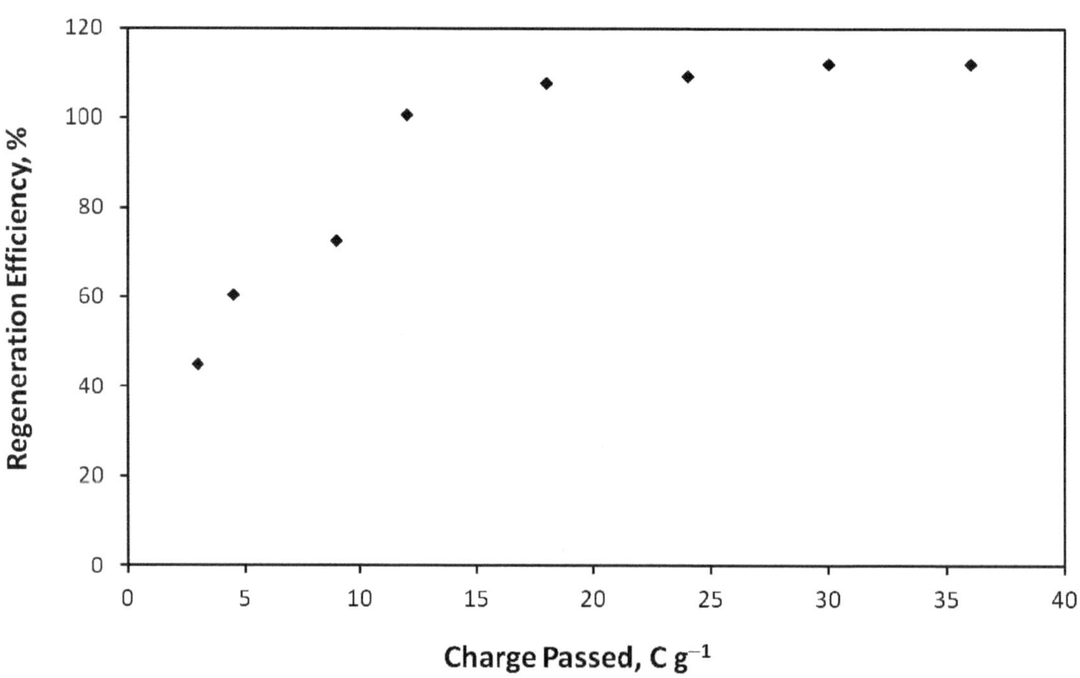

Figure 4. *Regeneration efficiency as a function of charge passed during electrochemical regeneration of the GIC adsorbent Nyex™ 1000 loaded with 1.06 mg g^{-1} of the organic dye AV17, using a current density of 10 mA cm^{-2} and a bed depth 2.2 cm.*

The cell voltage was not continuously monitored, but was observed to be relatively stable at around 6 V throughout regeneration. The cell voltage is dominated by ohmic losses in the membrane and in the water in the anode compartment. Slight variations in the cell voltage were observed, presumably associated with changes in the conductivity of the water in the anode compartment (for example due the decrease in pH caused by the electrochemical processes) or the formation of gas bubbles leading to a decrease in effective conductivity. It was thus not possible to use the cell voltage to determine the completion of the regeneration. Based on a cell voltage of 6 V, the energy cost of the regeneration was 120 J per g of adsorbent regenerated or 115 J per mg of AV17 removed and destroyed.

Multi-stage batch design model

Batch adsorption and regeneration process are usually carried on small volumes of wastewater. However, the efficiency of dye removal can be improved by carrying out the treatment using multi-stage adsorption/regeneration system. An adsorption isotherm can be used to predict the design of a single stage batch adsorption system [31-36]. A wastewater treatment process consisting of a series of adsorption and regeneration stages can be considered to be a multi-stage equilibrium operation. The adsorption isotherm can thus be used to design a multi-stage adsorption / regeneration process as shown in Figure 5.

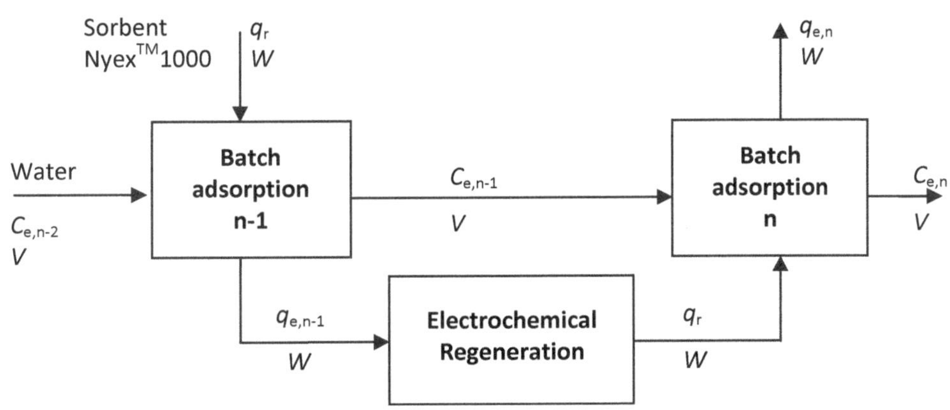

Figure 5. *Schematic diagram of two stages in a multi-stage batch adsorption and regeneration system, considered as a multi-stage equilibrium process.*

The design objective is to reduce the dye concentration in the feed (of volume V) from C_0 to C_n (mg L^{-1}), reusing the adsorbent (of mass W) in each adsorption stage after regeneration. The dye mass balance for batch adsorber system (n) in Figure 5 can be written as:

$$V\,C_{e,n-1} + W\,q_r = V\,C_{e,n} + W\,q_{e,n} \qquad (11)$$

For the electrochemical regeneration process used in this study, we can assume that the regeneration efficiency was 100% (based on the data shown in Figure 4), so $q_r = q_0 = 0$, and Equation (11) thus becomes:

$$V\,(C_{e,n-1} - C_{e,n}) = W\,q_{e,n} \qquad (12)$$

The experimental adsorption data were found to fit the Langmuir isotherm model (Equation 3) for adsorption of AV17 on the GIC adsorbent. Thus the Langmuir equation can be used to substitute for $q_{e,n}$, giving:

$$\frac{W}{V}=\frac{(C_{e,n-1}-C_{e,n})}{\left(\dfrac{bk_L C_{e,n}}{1+bC_{e,n}}\right)} \tag{13}$$

Equation (13) can be rearranged to a quadratic form and solved to give:

$$C_{e,n}=\frac{-\left(\dfrac{W}{V}bk_L-bC_{e,n-1}+1\right)\mp\sqrt{\left(\dfrac{W}{V}bk_L-bC_{e,n-1}+1\right)^2+4bC_{e,n-1}}}{2b} \tag{14}$$

For the first adsorption cycle (n = 1), $C_{e,n-1}$ in Equation (14) is replaced by the initial concentration C_0.

Equation (14) was used to determine the concentration after each stage of treatment for a given feed concentration C_0 and mass of adsorbent (W) using the Langmuir isotherm constants for the GIC / AV17 system: b = 0.31 L mg^{-1} and k_L = 0.987 mg g^{-1}. For a given feed concentration, number of adsorption stages and required percentage dye removal, the solution was iterated numerically using a forward derivative Newton's method with a convergence limit of 10^{-4} in order to calculate the amount of adsorbent, W (g), required.

The amount of adsorbent W was estimated for a five stage adsorption regeneration system to achieve AV17 dye removals ranging from 10 to 99 %, for a range of initial concentrations (55 - 680 mg L^{-1}), and the results obtained are plotted in Figure 6.

The results indicate that the amount of adsorbent required increases linearly with the percentage dye removal, up to removals of around 90 to 95 %. This can be explained by observation that the adsorbent approaches its maximum loading in each of the adsorption cycles, so the amount of dye removed per gram of adsorbent is approximately constant. However, to achieve very high removals the dye concentration in the final adsorption stage will be low, so that the adsorbent used in this stage will not be fully saturated. The amount of adsorbent required thus begins to increase more rapidly as the final concentration decreases towards zero.

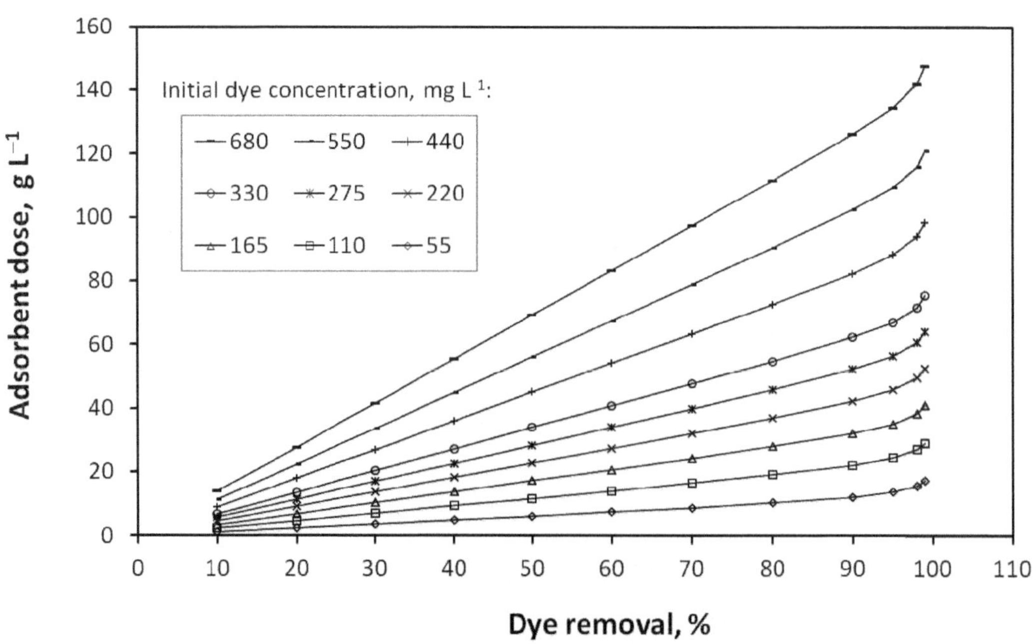

Figure 6. *The adsorbent dose required as a function of the dye removal for a range of initial dye concentrations using a five stage batch adsorption and regeneration system.*

For validation of the model, an experimental study of a five stage adsorption – regeneration process was carried out for the adsorption of AV17 (with an initial concentration of C = 668 mg L^{-1}) onto the GIC adsorbent (with an adsorbent dose of 125 g L^{-1}). The values of the percentage removal achieved (R_n) after each stage obtained from the model and the experiment (Equations 14 and 15) were compared, as shown in Figure 7. The percentage removal of AV17 was observed to increase from 19 to 93 % in stages 1 to 5, and there was good agreement between the experimental data and the model. From the experimental data, the adsorbent loading after each adsorption cycle was found to be approximately 1±0.05 mg g^{-1}, consistent with the capacity of adsorbent obtained from the isotherm study [24]. This confirms that for the conditions used in this experiment the adsorbent was approaching saturation with AV17 after each adsorption cycle.

$$R_n / \% = \frac{C_0 - C_n}{C_0} 100 \tag{15}$$

The adsorbent material investigated in this study has a very low adsorbent capacity, and there is significant potential to develop new adsorbent materials suitable for electrochemical regeneration but with increased adsorptive capacity compared to the GIC adsorbent Nyex™ 1000. Recent studies have shown that higher specific surface area adsorbent materials similar to the GIC used in this study can offer improved adsorption capacity [37,38]. In order to evaluate the benefits of increased adsorptive capacity, the effect of the value of the Langmuir k_L on the multi-stage adsorption-regeneration process was investigated. In this case, the number of adsorption/regeneration stages required to achieve 99.9 % AV17 removal (with an initial concentration of 1 g L^{-1}) was calculated for an adsorbent dose of 125 g L^{-1} and for a range of values of k_L (see Figure 8). As k_L was increased, it was found that stage number of stages of adsorption-regeneration required decreased such that a doubling of the adsorbent capacity leads to a reduction in the number of stages required from 10 to 5. A doubling of the adsorbent capacity has been demonstrated [37,38] and an order of magnitude increase in k_L can readily be envisaged, which would reduce the number of stages required to only 2. This would have a significant impact on the economics of the process.

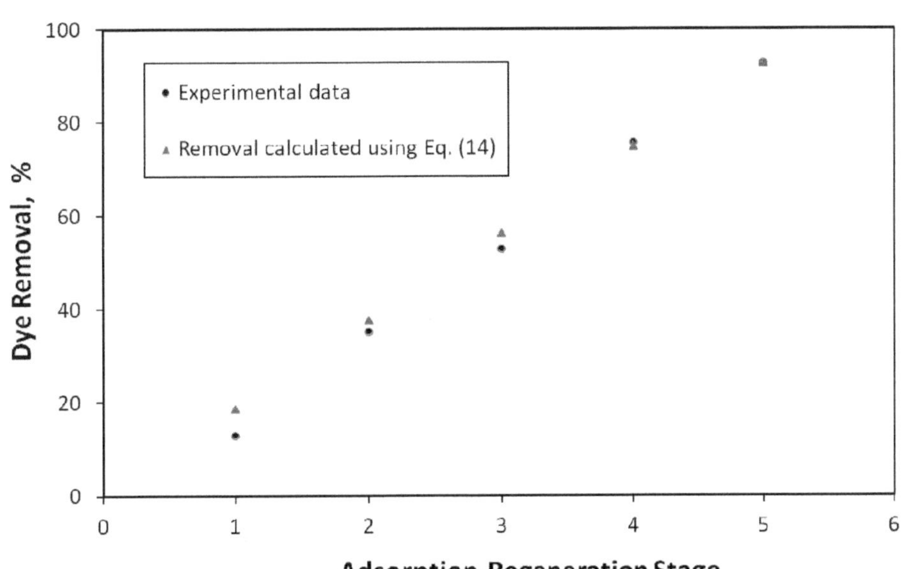

Figure 7. *Performance of batch adsorption and electrochemical regeneration for removal of AV17 using the GIC adsorbent Nyex™ 1000 with five stages adsorption and regeneration. The initial AV17 concentration was 680 mg L^{-1} and the adsorbent dose was 125 g L^{-1}.*

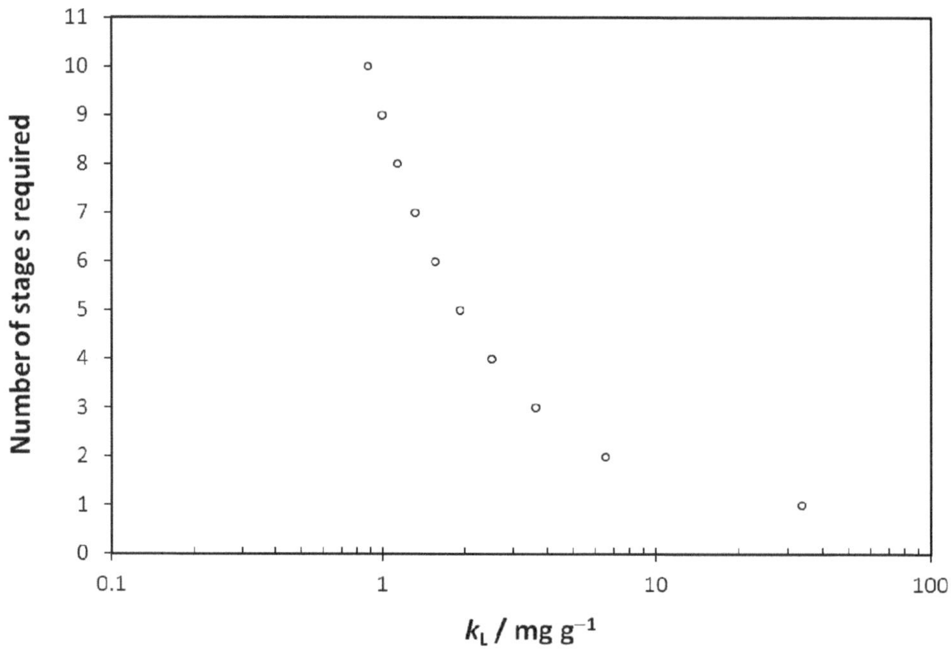

Figure 8. *Effect of adsorptive capacity (k_L) on the number of stages required for 99.9% removal of AV17 (with an initial concentration of 1 g L^{-1}) using multi-stage adsorption-regeneration with an adsorbent dose of 125 g L^{-1}. 125 g L^{-1}.*

Conclusions

It has recently been shown that low capacity unexpanded GIC adsorbents can be regenerated rapidly and cheaply, so that they can be regenerated and reused on-site [23,25,26]. This study describes the first detailed study of the adsorption/regeneration characteristics of a graphite intercalation compound. The laboratory scale batch regeneration tests demonstrated that the GIC adsorbent loaded with the organic dye Acid Violet 17 could be fully regenerated with a charge of around 15 C g^{-1}. This charge is of the correct order of magnitude to suggest that the dye was mineralised during regeneration. The energy cost of the regeneration was found to be around 120 J per g of adsorbent regenerated or 115 J per mg of AV17 removed and destroyed.

The effective treatment of wastewater containing a dissolved organic contaminant by multi-stage batch adsorption with electrochemical regeneration has been demonstrated for the first time. Multistage adsorption / regeneration was shown to be effective for the removal of high concentrations of AV17 dye using the GIC adsorbent, exploiting its rapid electrochemical regeneration. Removal of 93 % of the dye from a solution containing 668 mg L^{-1} of AV17 was achieved after five cycles of treatment with 125 g L^{-1} of the GIC adsorbent. A design model for the multi-stage adsorption and electrochemical regeneration process based on the Langmuir equilibrium isotherm has been developed and validated for removal of a dye. This model was used to predict the dose of adsorbent required to achieve a range of dye removals for a given number of adsorption / regeneration cycles. The model was found to be in good agreement with the experimental results obtained for five stages of adsorption / regeneration. It was found that the predicted number of stages of batch adsorption / regeneration required to achieve 99.9 % AV17 removal was halved when the adsorptive capacity of the adsorbent was doubled. This finding confirms that development of adsorbents which can be regenerated electrochemically with increased capacity compared to the GIC used in this study could significantly improve the economics of the process.

Acknowledgements: *The authors would like to thank the UK Engineering and Physical Sciences Research Council, ARVIA Technology Ltd and the Iraqi Ministry of Higher Education and Scientific Research for their financial support for this research.*

References

[1] P.A. Pekakis, N.P. Xekoukoulotakis, D. Mantzavinos, *Water Res.* **40** (2006) 1276-1286.
[2] J. Mo, J.-E Hwang, J. Jegal, J. Kim, *Dyes Pigments* **72** (2007) 240-245.
[3] P. Nigam, G. Armour, I.M. Banat, D. Singh, R. Marchant, *Bioresource Technol.* **72** (2000) 219-226.
[4] S.B. Wang, Y. Boyjoo, A. Choueib, Z.H. Zhu, *Water Res.* **39** (2005) 129-138.
[5] G. San Miguel, S.D. Lambert, N.J.D. Graham, *Water Res.* **35** (2001) 2740-2748.
[6] N. Kannan, M.M. Sundaram, *Dyes Pigments* **51** (2001) 25-40.
[7] Y. Guo, S. Yang, W. Fu, J. Qi, R. Li, Z. Wang, H. Xu, *Dyes Pigments* **56** (2003) 219-229.
[8] C. Namasivayam, M. Dinesh Kumar, K. Selvi, R. Ashruffunissa Begum, T. Vanathi, R.T. Yamuna, *Biomass Bioenerg.* **21** (2001) 477-483.
[9] C. Namasivayam, D. Kavitha, *Dyes Pigments* **54** (2002) 47-58.
[10] G. McKay, M. El Geundi, M.M. Nassar, *Water Res.* **21** (1987) 1513-1520.
[11] M. Özacar, I.A. Sengil, *J. Hazard. Mater.* **98** (2003) 211-224.
[12] G. Annadurai, R.-S. Juang, D.-J. Lee, *J. Hazard. Mater.* **92** (2002) 263-274.
[13] G.S. Gupta, G. Prasad, V.N. Singh, *Water Res.* **24** (1990) 45-50.
[14] R. Sivaraj, C. Namasivayam, K. Kadirvelu, *Waste Manage.* **21** (2001) 105-110.
[15] R.A. Shawabkeh, M.F. Tutunji, *Appl. Clay Sci.* **24** (2003) 111-120.
[16] N. Thinakaran, P. Baskaralingam, M. Pulikesi, P. Panneerselvam, S. Sivanesan, *J. Hazard. Mater.* **151** (2008) 316-322.
[17] R.M. Narbaitz, J.Q. Cen, *Water Res.* **28** (1994) 1771-1778.
[18] P. Canizares, J. Lobato, J. Garcia-Gomez, M.A. Rodrigo, *J. Appl. Electrochem.* **34** (2004) 111-117.
[19] R.M. Narbaitz, A. Karimi-Jashni, *Environ. Technol.* **30** (2009) 27-36.
[20] R. Berenguer, J.P. Marco-Lozar, C. Quijada, D. Cazorla-Amoros, E. Morallon, *Carbon* **48** (2010) 2734-2745.
[21] N.W. Brown, E.P.L. Roberts, A. Chasiotis, T. Cherdron, N. Sanghrajka, *Water Res.* **38** (2004) 3067-3074.
[22] N.W. Brown, E.P.L. Roberts, A.A. Garforth, R.A.W. Dryfe, *Electrochim. Acta* **49** (2004) 3269-3281.
[23] N.W. Brown, E.P.L. Roberts, A.A. Garforth, R.A.W. Dryfe, *Water Sci. Technol.* **49** (2004) 219-225.
[24] N.W. Brown, Adsorption and electrochemical regeneration for wastewater treatment: development of a process, PhD Thesis, UMIST, Manchester, UK, 2005.
[25] N.W. Brown, D.A. Wickenden, E.P.L. Roberts, *Waste Management Symposium*, Proceedings of WM2012, Phoenix, Arizona, USA, 2012.
[26] F.M. Mohammed, E.P.L. Roberts, A. Hill, A.K. Campen, N.W. Brown, *Water Res.* **45** (2011) 3065-3074.
[27] S. Ismadji, S.K. Bhatia, *Can. J. Chem. Eng.* **78** (2000) 892-901.
[28] E.N. El Qada, S.J. Allen, G.M. Walker, *Chem. Eng. J.* **135** (2008) 174-184.
[29] F.M. Mohammed, Modelling and design of water treatment processes using adsorption and electrochemical regeneration, PhD Thesis, University of Manchester, Manchester UK, 2011.
[30] M. Gordon, *J. Chem. Technol. Biot.* **32** (1982) 759-772.
[31] G. McKay, M.S. Otterburn, J.A. Aga, *Water Air Soil Poll.* **24** (1985) 307-322.

[32] M. Dogan, M. Alkan, Y. Onganer, *Water Air Soil Poll.* **120** (2000) 229-248.
[33] M. Özacar, I.A. Sengil, *Environ. Geol.* **45** (2004) 762-768.
[34] M. Özacar, I.A. Sengil, *Biochem. Eng. J.* **21** (2004) 39-45.
[35] V. Vadivelan, K.V. Kumar, K. V. *J. Colloid Interf. Sci.* **286** (2005) 90-100.
[36] E.I. Unuabonah, K.O. Adebowale, A.E. Ofomaja, *Water Air Soil Poll.* **200** (2009) 133-145.
[37] H.M.A. Asghar, Development of graphitic adsorbents for water treatment using adsorption and electrochemical regeneration, PhD Thesis, University of Manchester, Manchester, UK, 2011.
[38] H.M.A. Asghar, E.P.L. Roberts, S.N. Hussain, A.K. Campen, N.W. Brown, Submitted to *J. Appl. Electrochem.* (2012).

11

Didanosine determination in diluted alkaline electrolyte by adsorptive stripping voltammetry at the mercury film electrode

PERCIO AUGUSTO MARDINI FARIAS✉, ARNALDO AGUIAR CASTRO*, ANA ISA PEREZ CORDOVES*

Department of Chemistry, Pontifícia Universidade Católica do Rio de Janeiro, 22451-900, Rio de Janeiro, RJ, Brazil

**Fac. Química, Química Analítica, Universidad de La Habana, Cuba*

✉Corresponding Author

Abstract

A stripping method for the determination of didanosine at the submicromolar concentration levels is described. The method is based on controlled adsorptive accumulation of didanosine at thin-film mercury electrode followed by a linear cyclic scan voltammetry measurement of the surface species. Optimum experimental conditions include a NaOH solution of 2.0×10^{-3} mol L^{-1} (supporting electrolyte), an accumulation potential of -0.20 V, and a scan rate of 100 mV s^{-1}. The response of didanosine is linear over the concentration range 0.01 – 0.10 ppm. For an accumulation time of 7 minutes, the detection limit was found to be 0.43 ppb (1.0×10^{-9} mol L^{-1}). The more convenient relation to measuring the didanosine in presence of the metals ions, efavirenz, acyclovir, nevirapine, indinavir, nelfinavir, saquinavir, lamivudine and zidovudine were also investigated. The utility of the method is demonstrated by the presence of didanosine together with hypoxanthine, ATP or DNA

Keywords
Didanosine determination; metals ions; antiretroviral drugs; hypoxanthine; ATP; DNA; thin-film mercury electrode; linear cyclic scan stripping voltammetry.

Introduction

Didanosine (Figure 1) is a nucleoside analogue of adenosine. It differs from other nucleoside analogues, because it does not have any of the regular bases, instead it has hypoxanthine attached to the sugar ring. Within the cell, didanosine is phosphorylated to the active metabolite of dideoxyadenosine triphosphate (ddATP) by cellular enzymes. Like other anti-HIV nucleoside analogs, it acts as a chain terminator by incorporation and inhibits viral reverse transcriptase by competing with natural dATP. [1-3].

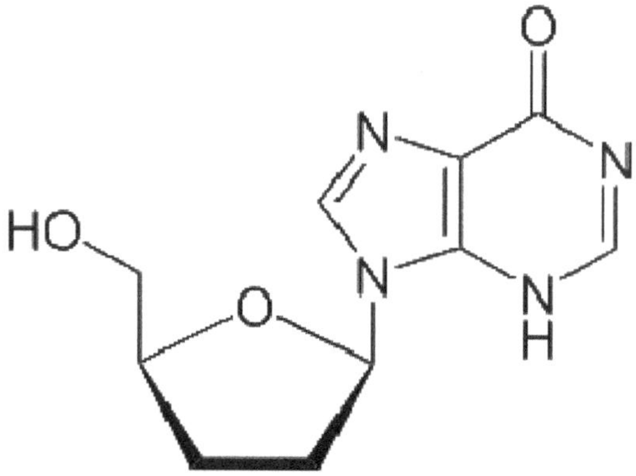

Figure 1. *Didanosine; [9-[(2R,5S)-5-(hydroxymethyl)oxolan-2-yl]-6,9-dihydro-3H-purin-6-one]*

Several chromatographic [4-11] analytical methods have been developed for the determination of didanosine but only one electroanalytical work was reported [12]. With the recent advances in capabilities of the adsorptive stripping voltammetry, new methodologies have been developed for adenine, thymine, guanine, ATP, DNA and antiretroviral drugs using alkaline solution with low ionic strength as the supporting electrolyte [13-20]. The present work reports a new stripping voltammetric procedure for trace detection of didanosine based on its adsorption at the thin film mercury electrode. The advantages, instrumental parameters, and possible limitations of this procedure are also explained in this paper. Furthermore, the effects of a wide range of potentially interfering compounds such as efavirenz, acyclovir, nevirapine, indinavir, nelfinavir, saquinavir, lamivudine, zidovudine, some metal ions, hypoxanthine, ATP and DNA were examined.

Experimental

Linear cyclic voltammograms were obtained with an EG&G PAR model 384-B Polarographic Analyser (Princeton Applied Research, Princeton, NJ, USA), equipped with an external cell and a Houston Ametek-DMP-40 series digital plotter. The working electrode was a glassy carbon electrode (GCE, 3.0 mm diameter, BAS- Bioanalytical Systems, Inc., West Lafayette, Indiana 47906, USA) with thin-film mercury, an Ag/AgCl reference electrode with vycor tip and a platinum auxiliary electrode. A magnetic stirrer and stirring bar (Nalgene Cat. No. 6600-0010, Rochester, NY, USA) provided convective transport during the accumulation step.

Stock solutions of 10^{-2} mol L^{-1} $Hg(NO_3)_2$ were initially prepared by dissolution of 0.4 g of mercury (II) nitrate in 100 mL of an acidified Milli-Q water (5 % of HNO_3). The thin mercury film was formed using a glassy carbon electrode (GCE, BAS) initially polished with alumina (PK-4, BAS) and then mounted with the help of a Teflon holder in a voltammetric cell provided with an Ag/AgCl reference electrode, a platinum auxiliary electrode, and containing 1 mL of stock mercury(II) nitrate solution, 1 mL of 10^{-1} mol L^{-1} potassium nitrate solution and 8 mL of purified water. The solution was purged with nitrogen for 240 s to eliminate the oxygen initially present. Mercury plating was carried over for 5 min at a cell of -0.9 V. After checking that the electrode was plated properly the set of electrodes was rinsed with water and a new clean cell containing the analyte solution was fitted. Each thin mercury film formed on glassy carbon electrode is stable during two hours [18-20].

Water purified in a Milli-Q purification system (Millipore, Billerica, MA, USA) was used for all dilutions and sample preparations. All chemicals were analytical reagent grade. Didanosine standard was used as received by the Instituto Nacional de Controle de Qualidade em Saúde - Brazil (INCQS – lot L1, 100.2%). Stock solutions of 1000 ppm were prepared dissolving 50 mg of the target reagent didanosine into 5 mL of 2 mol L^{-1} NaOH and water until a volume of 50 mL was reached. Diluted didanosine solutions of 100 or 10 ppm were prepared daily using 5 mL of 1000 or 100 ppm didanosine into water until 50 mL was reached. Stock solutions of acyclovir were prepared in accordance with didanosine procedure. Stock solutions of 1000 ppm of other HIV drugs were prepared dissolving 50 mg of the target reagent into 5 mL of 2 mol L^{-1} NaOH, 5 mL of ethylic alcohol and water until a volume of 50 mL was reached. A 1000 ppm copper and other metals ions stock solutions (atomic absorption standard solution, Sigma-Aldrich Brasil Ltda.) were used, and diluted as required for standard additions. Sigma Chemicals (Sigma-Aldrich Brasil Ltda., São Paulo-SP, Brasil) stock solutions of 1000 ppm were prepared by dissolving 25 mg of the target reagent hypoxanthine plus solid NaOH with 25 mL of water (to achieve a final concentration of 0.1 mol L^{-1} NaOH). Stock solutions of 1000 ppm of adenosine 5'-triphosphate, disodium salt hydrate (ATP) was prepared by dissolving 10 mg of the target reagent in 2 mL of diluted perchloric acid (10^{-1} mol L^{-1}) with an subsequent solution heating at 70 °C during 30 seconds. After heating, the sample was cooled down, and diluted to 10 mL with water. Single-stranded calf thymus DNA (Cat. No. D-8899; Lot 43H67951) was used as received from Sigma. A 500 μg DNA mL^{-1} stock solution (around 5 mg 10 mL^{-1}; Lyophilized powder containing 63% DNA) was prepared according to ATP procedure. The solutions were stored in the dark at 4 °C.

A known volume (10 mL) of the supporting electrolyte solution (2.0×10^{-3} mol L^{-1} sodium hydroxide) was added to the voltammetric cell and degassed with nitrogen for 8 min (and for 60 seconds before each adsorptive stripping cycle). Initially, the condition potential (usually -0.9 V) was applied to the electrode for a selected time (usually 60 s). The conditioning potential is used for cleaning the surface of the thin mercury electrode. Afterwards the initial potential (usually -0.20 V) was applied to the electrode with a selected time (usually 240 s), while the solution was slowly stirred. The stirring was then stopped, and after 30 s, the voltammogram was recorded by applying a negative-going potential scan. The scan (usually at 100 mV s^{-1}) was terminated at -0.90 V, and the adsorptive stripping cycle was repeated with the same thin-film mercury. After the background stripping voltammograms had been obtained, aliquots of the didanosine standards were introduced. The entire procedure was automated, as controlled by 384-B Polarographic Analyser. Throughout this operation, nitrogen was passed over the solution surface. All data were obtained at room temperature (25 °C).

Results and discussion

Figure 2 compares differential-pulse, linear-scan and linear cyclic adsorptive stripping voltammograms for 0.05 ppm didanosine in a 2.0×10^{-3} mol L^{-1} NaOH solution, after 120 seconds preconcentration with stirring at -0.20 V. A mercury film was use as work electrode. After equilibrium time of 30 s the differential pulse (A), linear (B) or cyclic cathodic voltammogram (C) were recorded at 25 (A) or 100 (B,C) mV s^{-1}. Both scan modes offer excellent signal-to-background characteristics but linear scan offers higher current peak and greater speed, and is recommended for the determination of didanosine. The linear cyclic stripping mode yield well-defined peak, with half-width (b½) of 130 mV, and was used throughout the further study. The didanosine cyclic cathodic peak of 2407 nA (I_p) appears at -0.56V (E_p). No anodic peaks were observed in the first scan.

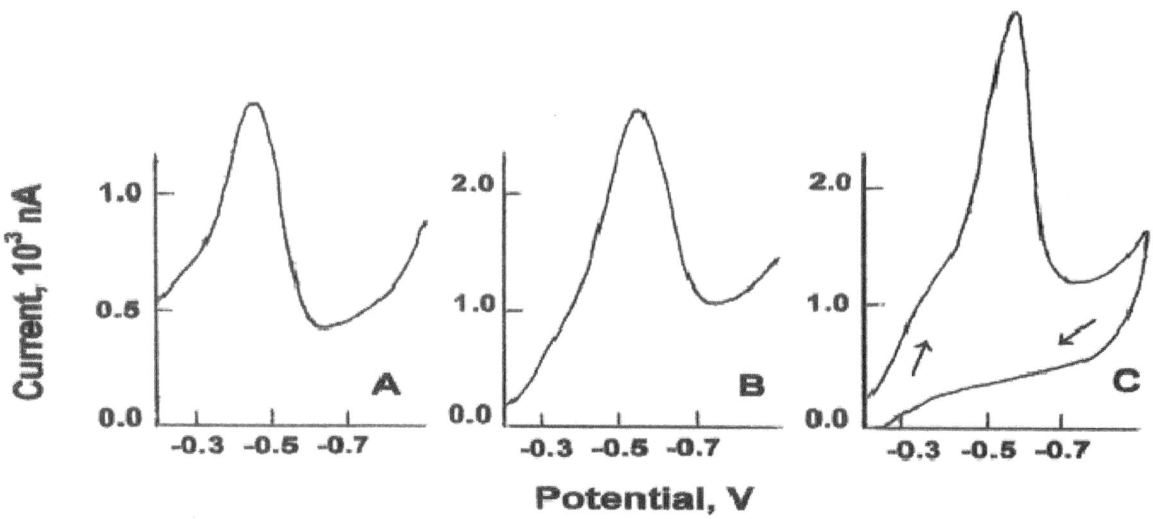

Figure 2. *Differential-pulse (A), linear-scan (B) and adsorptive cyclic (C) voltammograms of 0.05 ppm didanosine in 2.0 x 10^{-3} mol L^{-1} NaOH. Conditioning time, 60 s at -0.90 V; accumulation time, 120 s at -0.20 V with stirring; amplitude pulse, 50 mV (A); scan rate, 25 (A) and 100 (B,C) mV s^{-1}; thin-film mercury electrode*

Several other chemical and instrumental parameters such as supporting electrolyte, pH, accumulation potential and time, and scan rate, which directly affect the didanosine adsorptive stripping peak response, were optimized. The adsorption properties of the didanosine can vary with the composition of the supporting electrolyte. Various electrolytes, *e.g.* Briton-Robinson buffer, phosphate buffer and NaOH solution, were evaluated as suitable media for the adsorptive stripping measurement of didanosine. Best results (with respect to signal enhancement and reproducibility) were obtained in NaOH electrolyte. This alkaline medium was employed throughout this study.

The adsorptive stripping signal of didanosine depends on the sample pH. Figure 3A, shows the dependence of the didanosine peak current on the solution pH (from 2 to 12). No response to didanosine was observed in solutions more acidic than pH 7. Increasing pH from 8 to 11 resulted in a rapid increase in the didanosine peak current. However, the stability of didanosine peak in aqueous solutions decreases with pH above of 11. Accordingly, the solution of approximately pH 11 was used throughout to satisfy the sensitivity and stability requirements.

The effect of accumulation potential was observed from +0.05 to -0.40 V and the highest didanosine peak currents were observed at the accumulation potential of -0.20 V (Figure 3B).

The peak current (I_p) for the surface-adsorbed didanosine is directly proportional to the scan rate (υ) (figure 3C). A plot of I_p *vs.* υ was linear (correlation coefficient, 0.998), with a slope of 54.4, over the 10-200 mV s^{-1}.

The dependence of the linear cyclic current peak with the pre-concentration time was verified. For longer accumulation time, more didanosine is adsorbed and higher peaks current appears. Such time-dependent profiles represent the corresponding adsorption isotherms because the peak current depends on the amount adsorbed. Initially the current increases linearly (0-180 s), and then drops at longer accumulation times. The resulting plot of peak current vs. accumulation time (0-180 s) show a slope of 36.03 nA s^{-1} and a correlation coefficient of 0.996. Figure 4 shows two didanosine voltamogramms with accumulation time of 120 (A) and 240 (B) seconds. With 240 s of pre-concentration time the peak current for 1.0 ppm of didanosine was about 2.1 times larger than the corresponding peak obtained with a 120 s response.

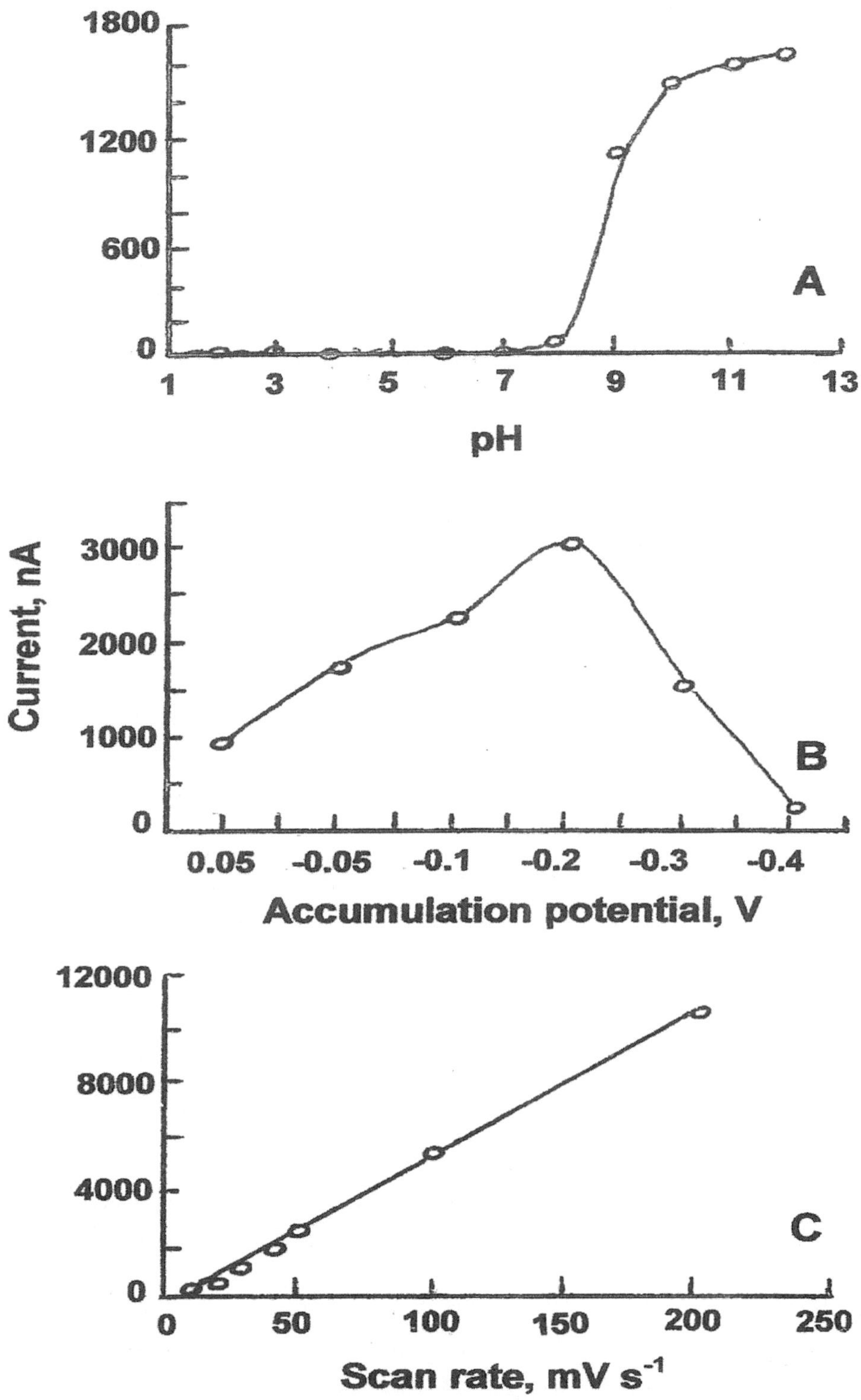

Figure 3. *A - Effect of pH on the current peak height of the linear cyclic adsorptive stripping voltammograms of 1.0 ppm of didanosine in solution. Condition time, 60 s at -0.90 V; accumulation time 240 s at -0.30 V; final potential, -0.9 V; scan rate, 50 mV s^{-1}; equilibrium time, 30 s; thin-film mercury electrode (5 min at -0.9 V). B - Effect of accumulation potential on the current peak height of the linear cyclic adsorptive stripping voltammograms of 0.05 ppm didanosine in 2.0×10^{-3} mol L^{-1} NaOH. Accumulation time, 120 s; scan rate, 100 mV s^{-1}. Other conditions same as 3A. C - Effect of scan rate on the current peak height of the linear cyclic voltammograms of 1.0 ppm didanosine in 2.0×10^{-3} mol L^{-1} NaOH. Acculutation time, 240 s at -0.20 V. Other conditions same as 3A.*

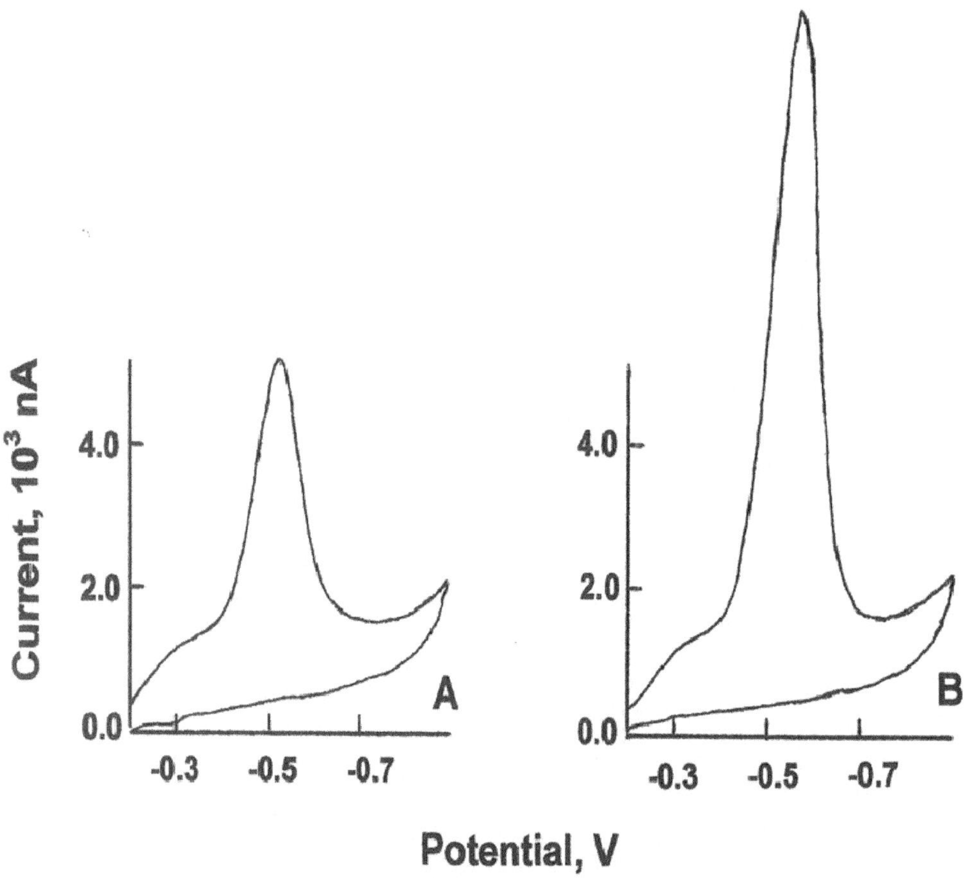

Figure 4. *Increasing current effects of accumulation time at -0.20 V on the linear cyclic adsorptive stripping voltammograms of 1.0 ppm of didanosine in 2.0 x 10^{-3} mol L^{-1} NaOH. The (A) and (B) curve is relative at accumulation time of 120 and 240 s. Conditioning time, 60 s at -0.90 V; scan rate, 100 mV s^{-1} ; equilibrium time, 30 s; thin-film mercury electrode (5 min at -.9 V).*

The effective pre-concentration (accumulation time) associated with the adsorption process results in significant lowering of the detection limit compared to the corresponding solution measurements. A detection limit of 0.43 ppb (1.0×10^{-9} mol L^{-1}) was estimated from quantification of 0.01 ppm after 7 min accumulation (S/N = 2). Thus, 4 ng could be detected in the 10 mL of solution used.

The reproducibility was estimated by ten successive measurements in stirred 0.05 ppm didanosine solution (other conditions: supporting electrolyte, 0.002 mol L^{-1} NaOH; conditioning time, 60 s at -0.9 V; accumulation time, 180 s at 0.2 V; final potential, -0.9 V; scan rate, 100 mV s^{-1}; equilibrium time, 30 s and thin-film mercury electrode). The mean peak current was 3255 nA with a range of 3111 - 3481 nA and a relative standard deviation of 2.6 %. Such precision compares similarly with those reported for other compounds measured by adsorptive stripping analysis [13-20]. The E_p and the $b_{1/2}$ remain the same at -0.56 V and 130 mV, respectively.

Figure 5 shows voltammograms for increasing didanosine concentration (A: 0.005, B: 0.030 and C: 0.050 ppm) after 300 s accumulation. Well-defined stripping peaks (at -0.56 V) were observed from 0.005 to 0.050 didanosine concentration range. The resulting plot of peak current versus concentration is linear (slope 22550 nA ppm^{-1}; correlation coefficient, 0.992). Such linearity prevails as long as linear isotherm conditions (low surface coverage) exist. A separate experiment was performed to test linearity at high concentration range resulting also in well-defined stripping peaks between 0.2 and 0.4 ppm didanosine (accumulation time, 180 s at -0.3 V; scan rate, 50 mV s^{-1}; final potential, -0.9 V; other conditions same as in Figure 5). The resulting plot of peak current versus concentration also showed linearity (slope 751 nA ppm^{-1}; correlation coefficient, 0.997).

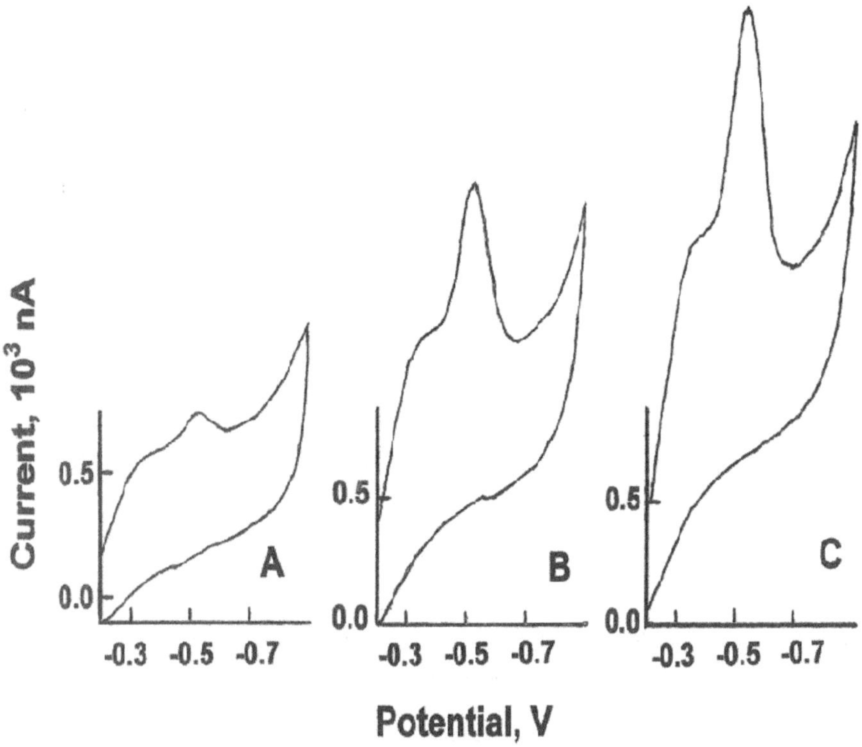

Figure 5. *Linear cyclic scan adsorptive stripping voltammograms obtained after increasing the didanosine concentration in a solution of 2.0 x 10^{-3} mol L^{-1} NaOH (with 1% v/v of ethylic alcohol). The (A), (B) and (C) curve is relative at 0.005, 0.030 and 0.050 ppm of didanosine concentration. Accumulation time, 300 s at -0.20 V. Condition time, 60 s at -0.9 V. Scan rate, 100 mV s^{-1}. Final potential at -0.9 V. Equilibrium time, 30 s. Thin-film mercury electrode (5 min at -0.9 V).*

Practical applications of differential pulse adsorptive stripping analysis may be affected by interferences due to the presence of ions and/or surface active compounds. With respect to the surface reaction, double layer charge or direct interactions among these substances may inhibit or aid in the accumulation of the analyte. Measurements of 0.05 ppm didanosine were not affected by addition of up to 0.01 cobalt(II); up to 0.06 ppm nickel(II) and up to 0.10 ppm cadmium(II), lead(II), zinc(II), copper(II) and iron(III). Using higher concentration a perceptible negative shift of didanosine peak was observed by presence of nickel (II) or cobalt (II) indicating a possible formation of Ni- or Co-didanosine complex. More dramatic interference was observed with cobalt(II). The addition of 0.03 ppm cobalt(II) resulted in a 5-fold increase of didanosine peak height. The metal ions chosen to be used in this study were based on their importance to human life. Cu, Fe, Co and Zn are essential while Pb and Cd are considered toxic to human life, since Ni is a genetic modifier. Preliminary studies were carried out for the determination of didanosine in presence of others antiretroviral drugs routinely used for the treatment of a human immunodeficiency virus (HIV): efavirenz, acyclovir, nevirapine, indinavir, nelfinavir, saquinavir, lamivudine and zidovudine. These antiretroviral drugs were chosen because they are produced in Brazil and have low costs. Measurements of 0.05 ppm didanosine were not affected by addition of up to 0.01 ppm of lamivudine or zidovudine; up to 0.03 ppm of neviparine; up to 0.04 ppm of indinavir; up to 0.08 ppm of nelfinavir; up to 0.10 ppm of acyclovir and saquinavir. The more dramatic interference was observed with efavirenz. The addition of 0.03 ppm efavirenz resulted in a twofold increase in didanosine peak height. A linear scan adsorptive stripping voltammogram for 0.1 ppm of free efavirenz in a 2×10^{-3} mol L^{-1} NaOH solution showed a cathodic peak at -0.36V. A perceptible negative shift of didanosine peak was observed by presence of nelfinavir.

Figure 6 illustrates the method's suitability for the determination of didanosine through linear cyclic adsorptive stripping voltammetry in a synthetic sample contained contain several antiretroviral drugs (efavirenz, nevirapine, didanosine, lamivudine, zidovudine and acyclovir; all with 2.0 ppm of concentration). Four successive standard additions to sample resulted in well-shaped adsorptive stripping peaks. The didanosine peak in the original sample (curve A) can, thus, be quantified based on the resulting standard addition plot (also show in Figure 6). Because of the inherent sensitivity of the method, short (60 s) accumulation times could be used. Five consecutive analysis of sample yielded an average value of 2.5 ppm with standard deviation of the 0.4 ppm. This is in "relative" agreement with the didanosine value (2.0 ppm).

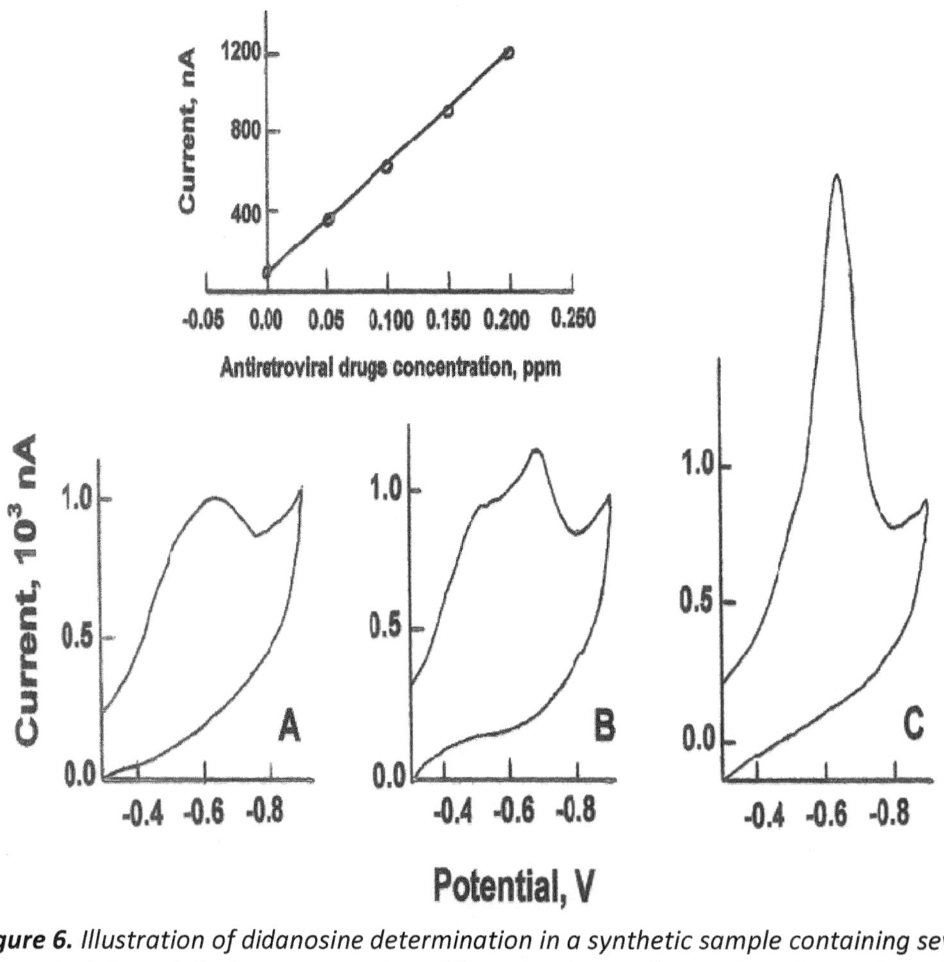

Figure 6. *Illustration of didanosine determination in a synthetic sample containing several antiretroviral drugs (efavirenz, nevirapine, didanosine, lamivudine, zidovudine and acyclovir; all with 2 ppm of concentration) by linear cyclic adsorptive stripping voltammetry. Supporting electrolyte, 10 mL of 2.0×10^{-3} mol L^{-1} NaOH. A - addition of 40 µL of synthetic sample; B -0.025 ppm and C - 0.20 ppm addition of standard didanosine. Condition time, 60 s at -0.9 V. Accumulation time, 60 s at -0.30 V. Final potential, -0.9 V. Scan rate, 100 mV s^{-1}. Equilibrium time, 30 s. Thin-film mercury electrode (5 min at -0.9 V).*

Preliminary studies were also carried out for the determination of didanosine in the presence of hypoxanthine, ATP and DNA. The didanosine is a nucleoside analogue of adenosine [1-3]. The current measurements of 0.05 ppm didanosine (other conditions: supporting electrolyte,

2×10^{-3} mol L^{-1} NaOH; condition time, 60 s at -0.9 V; accumulation time, 120 s at -0.2 V; final potential, -1.10 V; scan rate, 100 mV s^{-1}; the equilibrium time, 30 s and thin-film mercury electrode) were not affected by the addition of up to 0.10 ppm of DNA. In another experiment, after the addition of 0.10 ppm of DNA and copper(II) in 0.05 ppm of didanosine (with a potential peak at -0.56 V), a Cu(II) anodic and DNA cathodic peaks were observed at -0.20 and -0.75 V, respectively (Figure 7).

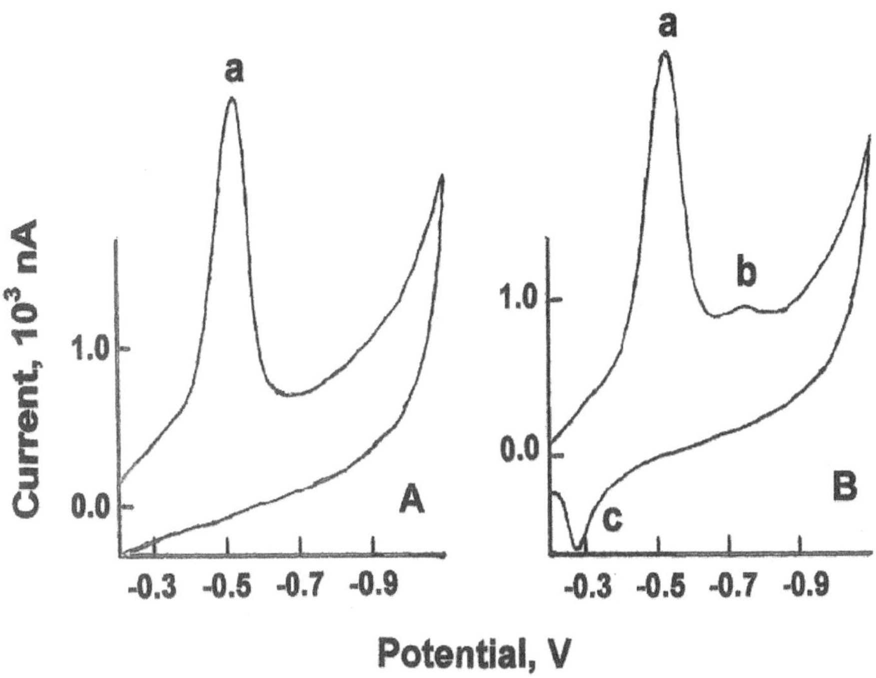

Figure 7. *Linear cyclic pulse adsorptive stripping voltammogram of 0.05 ppm didanosine (A) in 0.10 ppm copper(II) and DNA presence (B) in a solution of 2.0 x 10^{-3} mol L^{-1} NaOH. (a) Represent the didanosine, (b) the DNA and (c) the copper (II) peak. Accumulation time, 120 s at - 0.20 V. Condition time, 60 s at -0.9 V. Scan rate, 100 mV s^{-1}. Final potential at -1.1 V. Equilibrium time, 30 s. Thin-film mercury electrode (5 min at -0.9 V).*

The current measurements of 0.05 ppm didanosine (other conditions: same as DNA experiment) were not affected by addition of up to 0.10 ppm of ATP. In another experiment, after the addition of 0.10 ppm of ATP and copper(II) in 0.05 ppm of didanosine (with a potential peak at -0.52 V), a Cu(II) anodic and ATP cathodic peak were observed in -0.15 and -0.75 V, respectively. The current measurements of 0.05 ppm didanosine (final potential, -1.20 V; other conditions the same as DNA experiment were not affected by addition of up to 0.10 ppm of hypoxanthine. In another experiment, after the addition of 0.10 ppm of hypoxanthine and copper(II) in 0.05 ppm of didanosine (with a potential peak at -0.52 V), a Cu(II) anodic and hypoxanthine cathodic peaks were observed in -0.30 and -1.10 V, respectively. Without the didanosine presence, the hypoxanthine, ATP and DNA also showed similar behavior in presence of copper(II) [13-20].

Conclusions

An effective means for the determination of trace levels of didanosine has been described. The use of a diluted alkaline electrolyte provided a sensitive and selective adsorptive stripping voltammetric method for didanosine determination. Interference studies indicate the possibility of the determination of didanosine in presence of others antiretroviral drugs for the treatment of a human immunodeficiency virus (HIV), like acyclovir, nevirapine, indinavir, nelfinavir and saquinavir. In particular, this approach offers similar efficiency to chromatographic methods. The didanosine peak increases in the presence of nickel(II) or cobalt(II) indicating a possible formation of Ni or Co-didanosine complex. The didanosine with the DNA peak are separated by 0.19 V and with the ATP by 0.23 V. Also, further studies using diluted alkaline solution as supporting electrolyte and film mercury electrode modified *in situ* by metallic ions can be used for detection of other drugs, and DNA-intercalating dyes, as well as amino-acids, peptides and proteins determinations.

Acknowledgements: *The authors gratefully acknowledge the CAPES-Brazil and MES-Cuba for their support of this work. In addition, we thank Dr. Katia Cristina Leandro of Fundação Oswaldo Cruz for generously supplying the sample of didanosine.*

References

[1] E. De Clercq, *J. Clin. Virol.* **30** (2004) 115-133.
[2] E. De Clercq, *Int. J. Biochem. Cell B.* **36** (2004) 1800-1822.
[3] Sweetman, In: S.C. Sweetman, Editor, *Martindale the Complete Drug Reference* (35 th ed.), Pharmaceutical Press (2007).
[4] R.D.E. Estrela, M.C. Salvadori, R.S.L. Raices, G. Suarez-Kurtz, *J.Mass. Spectrom.* **38** (2003) 378-385.
[5] Y.Huang, E. Zurlinden, E. Lin, X.H. Li, J. Tokumoto, J. Golden, A. Murr, J. Engstrom, J. Conte, *J. Chromatogr.B* **799** (2004) 51-61.
[6] A.M.C. de Oliveira, T.C.R. Lowen, L.M. Cabral, E.M. dos Santos, C.R. Rodrigues, H.C. Castro, T.C. dos Santos, *J.Pharmaceut. Biomed.* **38** (2005) 751-756.
[7] C.P.W.G.M. Verweij-van Wissen, R.E. Aarnoutse, D.M. Burger, *J. Chromatogr.B* **816** (2005) 121-129.
[8] T.N. Clark, C.A. White, M.G. Bartlett, *Biomed. Chromatogr.* **20** (2006) 605-611.
[9] A. S. Kumari, K. Prakash, K. E. V. Nagoji, M. E. B. Rao, *Asian J. Chem.* **19** (2007) 2633-2636.
[10] S. Notari, C. Mancone, T. Alonzi, M. Tripodi, P. Narciso, P. Ascenzi, *J. Chromatogr.B* **863** (2008) 249-257.
[11] C.V. Kumar, D. Anantantahakumar, K. Madhuri, J.V.L.N.S. Rao, V.L.N. Seshagiri, *Asian J. Chem.* **23** (2011) 583-586.
[12] K.I. Ozoemena, R.I. Stefan-van Staden, T. Nyokong, *Electroanal.* **21** (2009) 1651-1654.
[13] P.A.M. Farias, A. de L.R. Wagener, A.A. Castro, *Anal. Let.* **34** (2001) 2125-2140.
[14] P.A.M. Farias, A. de L.R. Wagener, A.A. Castro, *Anal. Let.* **34** (2001) 1295-1310.
[15] P.A.M. Farias, A. de L.R. Wagener, A.A. Junqueira, A.A. Castro, *Anal. Let..* **40** (2007) 1779-1790.
[16] P.A.M. Farias, A.A. Castro, Wagener, A.de L.R., A.A. Junqueira, *Electroanal.* **19** (2007) 1207-1212.
[17] P.A.M. Farias, K.C. Leandro, J.C. Moreira, *Anal. Let..* **43** (2010) 1951-1957.
[18] A.A.Castro, M.V.N. de Souza, N.A. Rey, P.A.M. Farias, *J. Braz.Chem. Soc.* **22** (2011) 1662-1668.
[19] A.I.P. Cordoves, P.A.M. Farias, Curr. Pharm. Anal. **7** (2011) 71-78.
[20] A.A. Castro, R.Q. Aucelio, N.A. Rey, E.M. Miguel, P.A.M. Farias, *Comb. Chem. High T.Scr.* **14** (2011) 22-27.

12

Mathematical modeling and reliability analysis of a 3D Li-ion battery

RICHARD HONG PENG LIANG, TANGSHENG ZOU, KARTHIK SOMASUNDARAM*✉, WEI TONG*, ERIK BIRGERSSON**

Raffles Science Institute, Raffles Institution, One Raffles Institution Lane, Singapore 575954

**Department of Mechanical Engineering, National University of Singapore, Singapore 117576*
***Department of Chemical and Biomolecular Engineering, National University of Singapore, Singapore 117576*

✉Corresponding Authors

Abstract
The three-dimensional (3D) Li-ion battery presents an effective solution to issues affecting its two-dimensional counterparts, as it is able to attain high energy capacities for the same areal footprint without sacrificing power density. A 3D battery has key structural features extending in and fully utilizing 3D space, allowing it to achieve greater reliability and longevity. This study applies an electrochemical-thermal coupled model to a checkerboard array of alternating positive and negative electrodes in a 3D architecture with either square or circular electrodes. The mathematical model comprises the transient conservation of charge, species, and energy together with electroneutrality, constitutive relations and relevant initial and boundary conditions. A reliability analysis carried out to simulate malfunctioning of either a positive or negative electrode reveals that although there are deviations in electrochemical and thermal behavior for electrodes adjacent to the malfunctioning electrode as compared to that in a fully-functioning array, there is little effect on electrodes further away, demonstrating the redundancy that a 3D electrode array provides. The results demonstrate that implementation of 3D batteries allow it to reliably and safely deliver power even if a component malfunctions, a strong advantage over conventional 2D batteries.

Keywords
3D batteries, Li-ion battery, mathematical model, reliability analysis, thermal model

Introduction

The demand for small scale, high power density sources has increased with the advent of miniaturized electronic devices such as micro-electromechanical systems (MEMS), micro-robots, micro-sensors and implantable medical devices. The lithium-ion battery is considered as a viable energy storage system that can cater to many of these applications, as it is able to attain the high energy densities required.

Conventional batteries with planar cathode and anode layers arranged in a parallel-plate configuration with a separator in between are known as two-dimensional (2D) cells, in which transport of Li-ions between the electrodes is one-dimensional (1D) in nature. This 2D cell currently used in commercial applications still face several constraints, especially regarding power limitations [1–3] and reliability which affect millions of industrial and small-scale consumers [4,5].

As a result, three-dimensional (3D) architectures have been developed for the lithium-ion battery [1,3,6] partly to ameliorate some of these concerns and further harness its potential as a key energy solution for the future. This nascent concept describes cells with key structural features extending in and fully utilizing 3D space. As shown in Fig. 1a, a 3D cell typically consists of anodes and cathodes which have active surface areas exposed in three dimensions in closely-spaced arrays in a 2D plane. With this, we have to reconsider the phenomena of mass and charge transport, electronic and ionic conductivity and electron-transfer kinetics in the form of 3D batteries.

The 3D cell promises many benefits: it can attain enhanced energy capacity without compromising on power density, while maintaining the same areal footprint [6–8]; it enables us to take advantage of more extensive interactions between the active materials [1,9,10], allowing us to adopt a design that improves reliability in the event that an active component ceases to function. The flow of energy and current in the conventional parallel-plate design essentially stops when an intermediate component fails. On the other hand, even if one of the electrodes in a 3D design fails, the battery can potentially continue to operate (albeit with reduced capacity and performance) as repeating units of electrodes arranged in a tessellation can provide redundancy.

Experiments have been able to produce working precursors to fully functional 3D batteries through the use of a variety of electrochemical deposition techniques. For instance, C-MEMS (Carbon-Microelectromechanical Systems), [11–13] a solution for miniaturization, uses photolithography to implement photoresist arrays on a SiO_2 surface, followed by pyrolysis at high temperatures in an oxygen-free environment. Changing conditions under which these steps are carried out allows one to vary design shapes, and also mechanical and transport properties [2,9,14,15]. Lithographic techniques vapour deposition techniques are being used to prepare independent arrays of the electrodes [1,2,16–21]. Recently, interdigitated Li-ion microbatteries are prepared using 3D printing techniques [22].

However, computational studies and mathematical modelling have yet to be entirely developed for the 3D Li-ion cell, as they either do not completely solve for coupled electrochemistry, transport phenomena and heat generation, or consider only electrochemical phenomena for optimization studies. Hart *et al.* [6] modelled and estimated current densities and potentials for arrays of electrodes with different geometries, while Zadin *et al.* [23–25] focused on simulating the ionic transport mechanisms in liquid and polymer electrolytes inside a 3D microbattery assuming non-porous solid electrodes to show how cell geometry can give rise to qualitatively non-uniform current densities and thus suboptimal surface utilization. However, neither model considers the electrochemical activity or thermal behaviour inside the electrodes. On the other hand, detailed mathematical models have been formulated to predict transient local electrochemical and thermal changes in a Li-ion cell in a rectilinear geometry [26–29] and a spiral-wound geometry [30,31], both essentially 2D-collapsible geometries.

By virtue of the lack of a detailed resolution for modelling and simulation of coupled electrochemistry, transport phenomena and heat generation for the 3D Li-ion cell, the aim of this paper is twofold: First, to employ a thermal-electrochemical model for studying the behaviour of a 3D liquid electrolyte Li-ion cell applicable to various geometries; second, to apply this model to a planar tessellated electrode geometry and conduct a reliability analysis to demonstrate the redundancy and longevity that a 3D battery can attain. Our mathematical model will investigate the transient conservation of charges, species and energy; it couples the electrochemical and thermal behaviour through the heat generation arising from reversible, irreversible and ohmic heating as well as through the temperature-dependent transport and electrochemical parameters.

Mathematical Formulation

The 3D battery has the advantage of having a larger areal energy capacity than the conventional 2D design, but also has a disadvantage of having non-uniform current density. This would lead to the poor utilization of the electrode materials, resulting in lower cell efficiency, non-uniform heat dissipation *etc.* Studies have shown that a checkerboard cathode/anode array configuration, where each electrode is surrounded by four nearest neighbour opposite electrodes, provides a more uniform current output around every electrode compared to that of other arrays. Current uniformity in this 3D design would render it more useful in a wider variety of applications [6].

As a result, we have selected a 3D battery that consists of positive and negative electrodes arranged in square planar tessellation as shown in Fig. 1 for our study. Each electrode is adjacently bounded by four electrodes of opposite sign. Current collectors plates are present at either end of the electrodes; one for the anode and another for the cathode.

For the individual electrodes, we consider two shapes of different extremes - circles and squares. Square electrodes provide a cleaner tessellation in the array, shorter average distances between electrodes and a higher packing efficiency. Circular electrodes are however much more feasible to implement and representative of real-life manufacturing processes. Hence, we consider both the square and circular electrode arrays in this paper, and compare the differences in performance between these two extremes.

In order to ensure that the comparison between both square and circular arrays is fair, a few restrictions have been imposed. The minimum distance between each electrode (w_{se}), as well as the thickness of both electrodes (w_{pe}, w_{ne}) i.e. the diameter of the circular cross-section or the edge length of the square cross-section, is kept constant in both arrays. The dimensions are provided in Tables 1 and 2.

The electrochemical and thermal behaviour of a three-dimensional (3D) Li-ion cell consisting of a graphite negative electrode (ne) and a manganese oxide spinel positive electrode (pe) as shown in Fig. 1 is studied. The electrodes and the spaces between the electrodes are filled with an electrolyte solution (el) of $LiPF_6$ salt in 1:2 ethylene carbonate: dimethyl carbonate solvent. The materials considered here are the same as that used for a conventional cell that is commercially available with the assumption that these can be utilized to fabricate 3D batteries using the conventional techniques.

As depicted in Fig. 1, there are two main scales involved in the modeling of a Li-ion cell: the macro- and the micro-scale. In short, the transport of ions and electrons in the cell between the electrodes is referred as the transport at the macroscale, which includes species transport in the liquid electrolyte, electronic charge conduction in the solid phase and ionic charge conduction in the liquid electrolyte; and the diffusion of ions in the active material present in the electrodes is referred to as transport at the microscale, which includes diffusion of lithium in the active material of the porous electrodes.

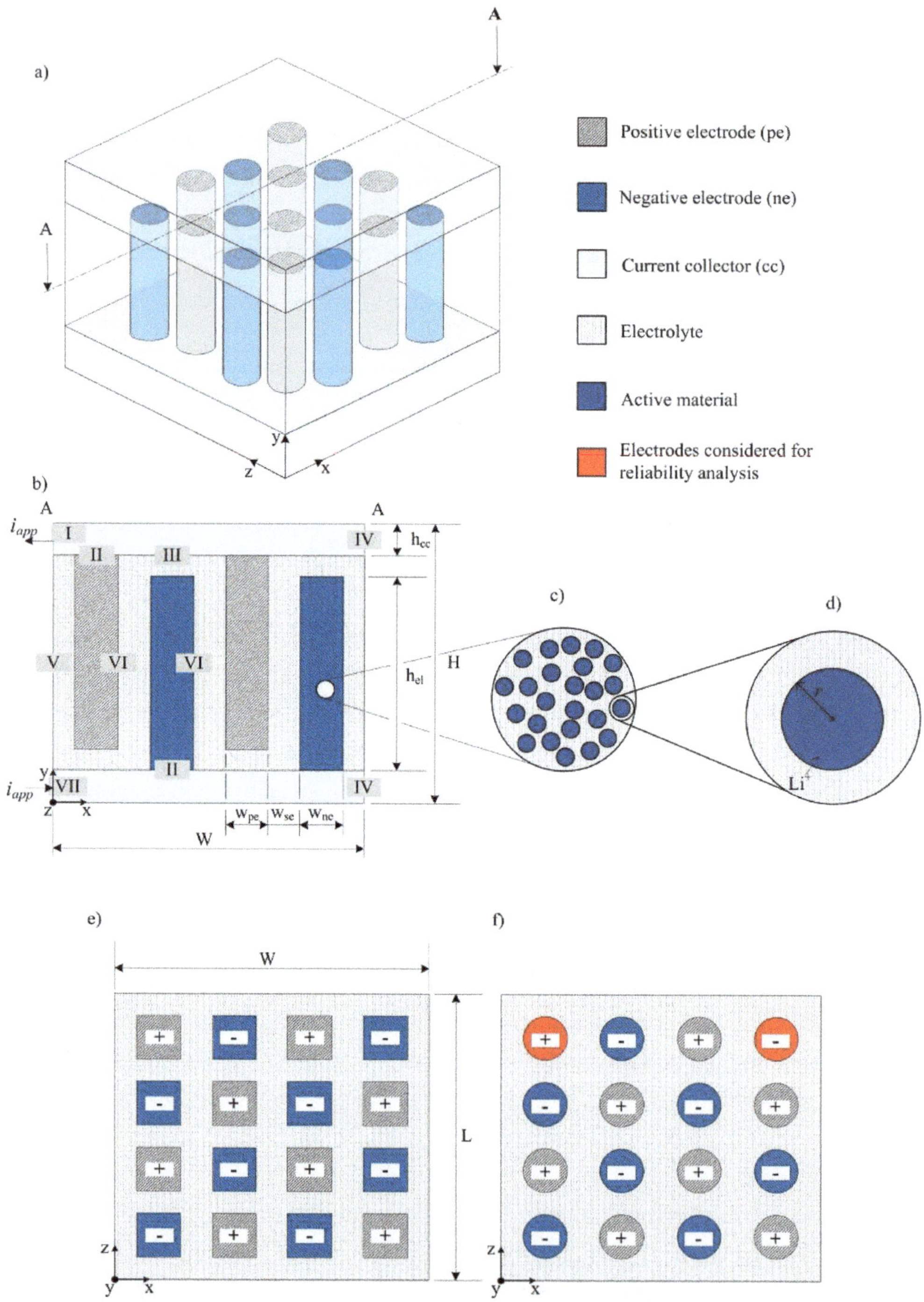

Figure 1. *(a) Schematics of* ***(a)*** *3D Li-ion battery,* ***(b)*** *section AA showing the various functional layers in the battery with the roman numerals indicating the interfaces of these layers and the boundaries,* ***(c)*** *agglomerate structure in the negative electrode (positive electrode also exhibits similar structure),* ***(d)*** *diffusion of lithium in active material in the electrodes on the microscale,* ***(e)*** *top view of the battery (xz-plane) with square cross-section, and (f) circular cross-section*

Table 1. *Physical properties and design parameters of battery components*

Parameter	Unit	cc (-)	ne	el	pe	cc (+)	Ref.
c_l^0	mol m^{-3}	-	2.0×10^3			-	27
C_p	J kg^{-1} K^{-1}	3.8×10^2	7.0×10^2	7.0×10^2	7.0×10^2	8.7×10^2	34,37
c_s^0	mol m^{-3}	-	1.5×10^4	-	3.9×10^3	-	27
c_s^{max}	mol m^{-3}	-	2.6×10^4	-	2.3×10^4	-	27
D_l	m^2 s^{-1}	-	7.5×10^{-11}			-	27
D_s	m^2 s^{-1}	-	3.9×10^{-14}	-	1.0×10^{-13}	-	27
E_{a,D_l}	kJ mol^{-1}	-	10			-	34
E_{a,D_s}	kJ mol^{-1}	-	4	-	20	-	34
E_{a,σ_l}	kJ mol^{-1}	-	20			-	34
h_i	m	10×10^{-6}	5×10^{-4}	-	5×10^{-4}	10×10^{-6}	-
k	W m^{-1} K^{-1}	3.8×10^2	0.05×10^2	0.01×10^2	0.05×10^2	2.0×10^2	34,37
k_0	mol$^{2.5}$ m$^{-0.5}$ s^{-1}	-	2×10^{-11}	-	2×10^{-11}	-	-
R_s	m	-	12.5×10^{-6}	-	8.5×10^{-6}	-	27
W_i	m	-	10×10^{-5}	-	10×10^{-5}	-	-
α_a, α_c	-	-	0.5	-	0.5	-	27
ε_p	-	-	0.14	-	0.19	-	27
ε_l	-	-	0.36	-	0.44	-	27
ε_f	-	-	0.03	-	0.07	-	27
ϑ_i^0	-	-	0.56	-	0.17	-	27
ρ	kg m^{-3}	9.0×10^3	1.9×10^3	1.2×10^3	4.1×10^3	2.7×10^3	27,34,37
σ_s	S m^{-1}	6.0×10^7	1.0×10^2	-	3.8	3.8×10^7	34

Table 2. *Other model parameters*

Parameter	Unit	Value
H	m	5.7×10^{-4}
h_t	W m^{-2} K^{-1}	5
i_{app}	A m^{-2}	7×10^2 (circular); 9×10^2 (square)
L	m	6.6×10^{-4}
T_a, T_{ref}	K	298.15
W	m	6.6×10^{-4}
w_{se}	m	5.2×10^{-5}

The model is based on the porous-electrode theory developed by Newman and Tiedemann [32,33] and embodies the following main assumptions:

1. Isotropic material properties;
2. Uniform distribution of active materials of the same size in the electrodes;
3. The active material is assumed to be spherical; *i.e.*, we only need to consider the radial direction at the microscale;
4. Side reactions are assumed negligible.

The mathematical formulation consists of the conservation equations of species and charge, together with conservation of energy at macroscale [27,34]; the diffusion length or the polynomial approximation approach is employed for the conservation of lithium inside the active material at microscale. For the sake of brevity, the governing equations, initial conditions, and constitutive relations are provided in tables in the appendix and the details can be found in our earlier work [31].

The physical properties and design adjustable parameters pertaining to the geometry studied in this work are given in Tables 1 and 2.

The current density is prescribed at the positive current collector at the boundary I (see Fig. 1b for placement of roman numerals) and also Newton's law of cooling is specified here. At the interface between the current collector/electrode or the current collecting tab/current collector, continuity of energy flux and solid-phase current is specified; insulation is specified for the ionic flux and current. At the current collector/electrolyte, insulation is defined for the solid phase current and continuity for the energy flux. At the boundaries IV and V, there is no flow of ions/electrons as well as energy and hence insulation is specified here as well for solid phase current and energy. At the electrode/electrolyte interfaces, continuity of energy flux and ionic flux as well as ionic current is defined and since there is no flow of electrons across the interface, insulation for solid phase current is defined. The current is collected from the negative current collector at the boundary VII or otherwise this end is grounded and also Newton's law of cooling is specified here.

Numerics

The commercial finite-element solver, COMSOL Multiphysics 3.5a [35], was employed to solve the 3D model. Linear elements were implemented for all dependent variables φ_s, φ_l, c_l, c_s^{surf}, T and c_s^{avg} the direct solver UMFPACK was chosen as linear solver with a relative convergence tolerance of 10^{-4}, and solutions for all models were tested for mesh independence. All computations were carried out on a workstation with dual-core processors (2.33 GHz) and a total of 64 GB random access memory (RAM).

Charge and discharge currents, i_{app}, were applied at the respective boundaries with a smoothed Heaviside function. For both 3D arrays, there were 3×10^4 elements and 1.6×10^5 degrees of freedom (DoF) which required around 45 GB of memory for solution under 5 C discharge with a solution time of around 1 hr. The reliability analysis was carried out by targeting one of the electrodes – either positive or negative – as marked in our computational cell in Fig. 1f. We assume that there will be no exchange current in the malfunctioning electrode as no reaction occurs; to reflect this, the transfer current per unit volume, J, is set to zero in the simulation for that electrode alone.

Results and Discussion

In this section, we begin by studying the behaviour during discharge of a 3D lithium-ion battery with tessellated electrode geometry, for both the square and circular electrode arrays. A reliability analysis is then carried out for the circular electrode array to study the behaviour of the battery when one of the electrodes in the array malfunctions. Only the discharge process will be illustrated as charging exhibits a similar behaviour.

Standard Discharge Behaviour

Discharge curves

First, we shall explore the global behaviour of the 3D battery in terms of potential during discharge for different C-rates, as shown in Fig. 2. The drop in cell voltage during the discharge in both the square and circular electrode arrays is similar. Both arrays exhibit a gentle and constant decrease in potential for most of the discharge period except for the sharp drop towards the end of discharge, similar to that of the 2D cell. Under all discharge rates, there is a sudden drop in the potential from the initial value of 4.2 V during the initial period of discharge that corresponds to ohmic losses. At 1 C-rate, there is a gradual drop in the potential from 3.9 V to 3 V until the end of discharge at 3600 s when a sudden drop in the potential is seen as in Fig. 2. While a similar trend is observed for discharge at 2 C-rate, the array experiences a much faster drop in potential at 5 C-rate due to a higher discharge current density.

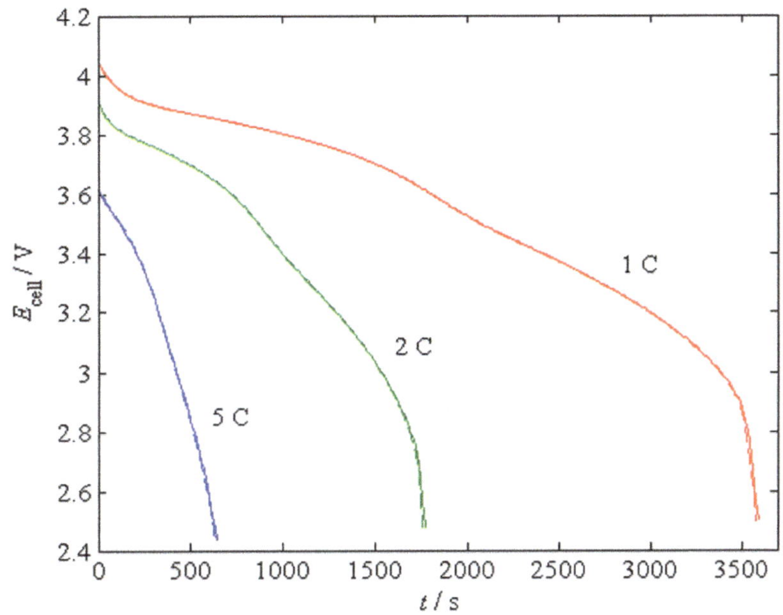

Figure 2. *Battery voltage during discharge at various discharge rates for square (continuous) and circular (dashed) electrodes*

Because electrodes in a 3D array are arranged in a parallel configuration, an array containing a large number of electrodes will have a greater capacity compared to a single pair of electrodes as compared to the 2D cell. Having more electrodes in an array would increase the total current delivered but keeping the potential unchanged. The array capacity can also be increased by elongating the electrodes into the plane, instead of increasing the capacity by making the electrodes thicker as in 2D batteries. As such, power density is not sacrificed for an increase in the array capacity in 3D batteries since the distance between electrodes remains the same. On the other hand, array capacity is gained at the expense of power density in 2D cells as an increase in capacity is attained by increasing the thickness of electrodes.

Electrochemical behaviour

During discharge, lithium ions deintercalate from the active material in the negative electrode and enter the electrolyte; the reverse process happens in the positive electrode. Hence, the concentration of lithium ions in the electrolyte increases in the negative electrode and decreases in the positive electrode, as depicted in Fig. 3. The time constant for diffusion is around 100 s (w_i^2 / D_l^{eff}) after which the concentration profile reaches a pseudo-steady state. The lithium-ion concentration reaches a maximum of 2100 mol m^{-3} at 1 C-rate and 2500 mol m^{-3} at 5 C-rate in the cell, as well as a minimum of 1900 mol m^{-3} at 1 C-rate and 1700 mol m^{-3} at 5 C-rate. These maximum and minimum concentration values are found in the negative and positive corner electrodes respectively, since the corner electrodes are only surrounded by two nearest neighbours: a negative corner electrode has fewer adjacent positive electrodes to consume fewer lithium ions, while a positive corner electrode has fewer adjacent negative electrodes to produce lithium ions.

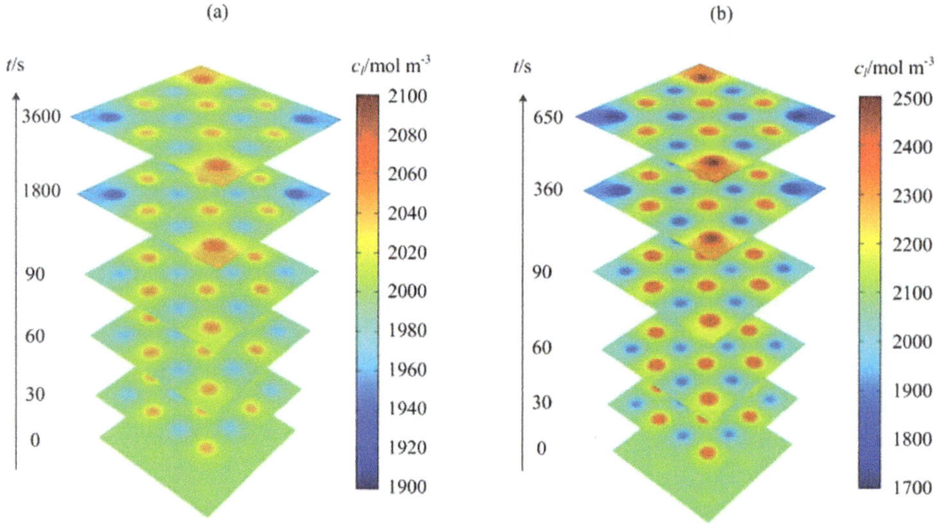

Figure 3. *Variation of concentration of lithium ions in the electrolyte along the xz-plane at $y=2.8\times10^{-4}$ m at various times during discharge at 1 C and 5 C-rates for the circular electrodes*

Moreover, there is variation in the lithium ion concentration in the electrolyte along the height of the electrodes, as illustrated in Fig. 4a. Both arrays display similar behaviour, with the square electrode array attaining a slightly higher concentration of 2105 mol m^{-3} in a negative corner electrode compared to 2090 mol m^{-3} in the equivalent circular electrode. Extreme values are observed in the areas nearer the current collector: negative electrodes have the highest lithium ion concentrations near the negative current collector as consumption of lithium ions by the positive electrode is lowest near the negative current collector; positive electrodes have the lowest lithium ion concentrations near the positive current collector as production of lithium ions in the negative electrode is lowest near the positive current collector.

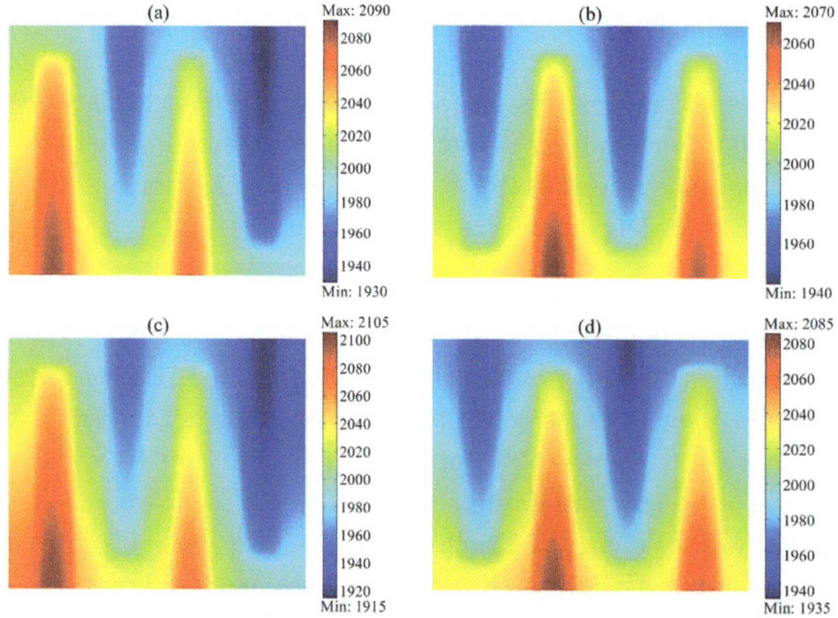

Figure 4a. *Concentration profile of lithium ions in the electrolyte at the end of discharge at 1 C-rate along the xy-plane at $z=10^{-4}$ m (a, c) and at $z=2.5\times10^{-4}$ m (b, d) for circular and square electrodes respectively*

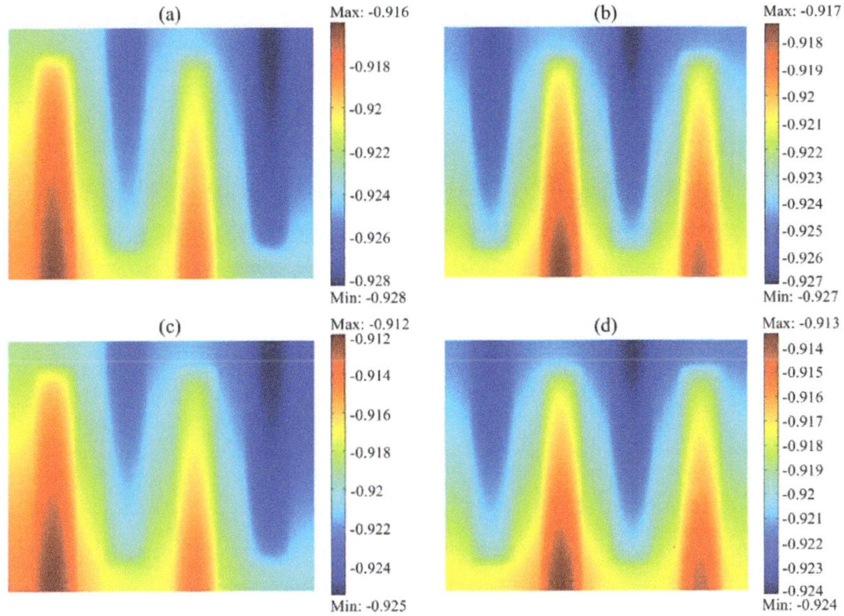

Figure 4b. *Liquid phase potential profile in the electrolyte at the end of discharge at 1 C-rate along the xy-plane at z=10^{-4} m (a, c) and at z=2.5×10^{-4} m (b, d) for circular and square electrodes respectively*

The unequal distribution of lithium ions also results in a variation in the liquid phase potential in the cell, as can be seen in Fig. 4b. Again, behaviour of both arrays are similar, with a maximum potential drop of 12×10^{-3} V in the circular electrode array compared to 13×10^{-3} V in the square electrode array. This potential drop, which can be attributed to ohmic losses and the concentration overpotential, gradually accentuates as the concentration gradient steepens during discharge. Extreme values of potential drop are observed nearest to the current collectors, because of similar reasons that cause extreme values in lithium ion concentration. First, we shall explore the global behaviour of the 3D battery in terms of potential during discharge for discharge curves.

Heat generation and thermal behaviour

The temperature of both circular and square arrays increases over time during discharge due to heat generation, as shown in Fig. 5. For the square electrode array, the temperature increases by 12K, 20 K and 30 K above the ambient temperature for discharge rates of 1 C, 2 C and 5 C, respectively.

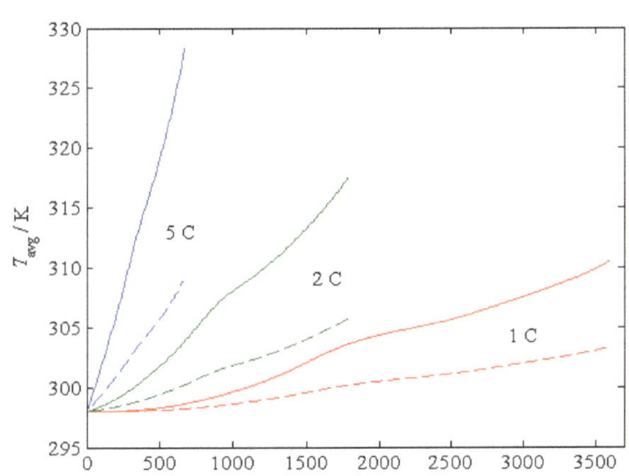

Figure 5. *Average battery temperature of the battery during discharge at various discharge rates for square (continuous) and circular (dashed) electrodes*

On the other hand, for the circular electrode array, the temperature increase is smaller, being 5 K, 8 K and 11 K for same discharge rates. Such a difference in array temperature over time is due to the higher volume of the square electrodes as compared to the circular ones leading to more heat generation in the square electrodes than the circular ones.

A comparison of the contribution from various layers towards heat generation is provided in Figs. 6a-b. The negative electrode is the highest contributor, amounting to 60 % in both discharge rates, due to its lower ionic conductivity and less porous nature. The positive electrode is the second highest contributor. Heat generation in the current collector and electrolyte is purely by ohmic heating and remains almost constant throughout the discharge.

a b

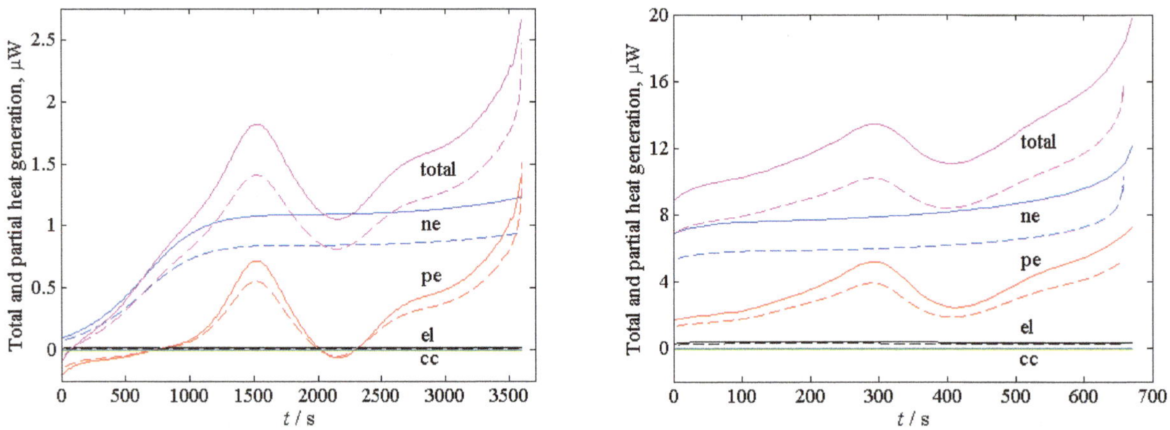

Figure 6. *Time history of total heat generation and heat generation in various layers during discharge for square (continuous) and circular (dashed) electrodes at* ***a*** *– 1 C-rate, and* ***b*** *– 5 C-rate*

Reliability Analysis

Due to the similarity in the behaviour of both the circular and square electrode arrays in the reliability analysis, we will only consider the circular electrode array. The behaviour of neighbouring electrodes is studied when either a target anode or a cathode malfunctions/fails. Given the extreme discharge conditions of 5 C discharge for lithium-ion batteries, we shall discuss the results for 5 C discharge rates as a worst-case scenario.

Discharge curves

The variation of cell voltage with time during discharge at 5 C-rate under both perturbed cases is presented in Fig. 7. As expected, there is a decrease in the discharge time compared to the normal case due to the loss in the energy capacity of the battery because of the malfunctioning of the electrodes. When the positive electrode malfunctions, there is a decrease of 60 s in the discharge time corresponding to a 10 % decrease in the energy capacity. Similarly, for the malfunctioning negative electrode, there is a decrease of 110 s in the discharge time corresponding to a 17 % decrease in the energy capacity. The negative electrode has higher theoretical capacity than the positive electrode (defined as C_i in the constitutive relations) and hence the capacity of the battery is reduced more when the negative electrode malfunctions compared to the positive electrode malfunctioning. Thus, unlike the conventional parallel-plate design, the 3D array is still able to safely generate power when an electrode malfunctions, though with a lower energy capacity.

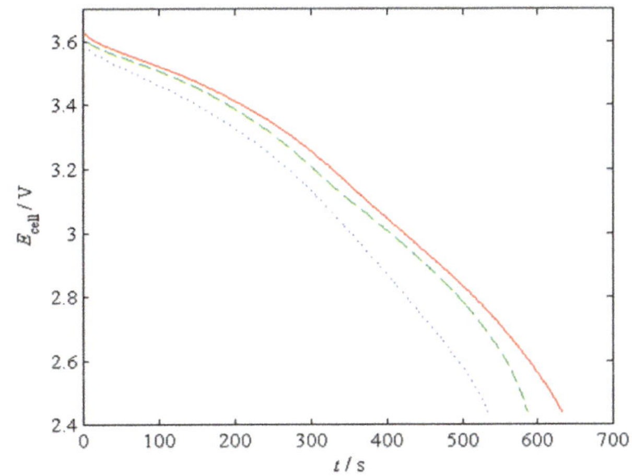

Figure 7. *Battery voltage during discharge at 5 C-rate under normal (continuous) and perturbed cases of malfunctioning either positive (dashed) or negative (dotted) electrodes as marked in Fig. 1*

Electrochemical behaviour

When a positive electrode malfunctions, the concentration distribution of lithium ions in the rest of the array away from the perturbed cell remains largely similar throughout high discharge rates relative to that in a non-perturbed array, as shown by comparing Figs. 3b and 8a. However, the concentration of lithium ions in the electrolyte in the malfunctioning positive electrode increases continuously from the initial value of 2×10^3 mol m^{-3} to 2.3×10^3 mol m^{-3} at the end of discharge, compared to a decrease to 1.8×10^3 mol m^{-3} on average in the electrolyte in the other positive electrodes. Because there is an inflow of ions from the adjacent negative electrodes that act as source of lithium ions, but no reaction in the active material taking place to consume them, lithium ions accumulate within the electrolyte in the malfunctioning positive electrode. This effect is concentrated locally in and around the malfunctioning electrode as shown in Fig. 8a and is insignificant towards the other electrodes in the array.

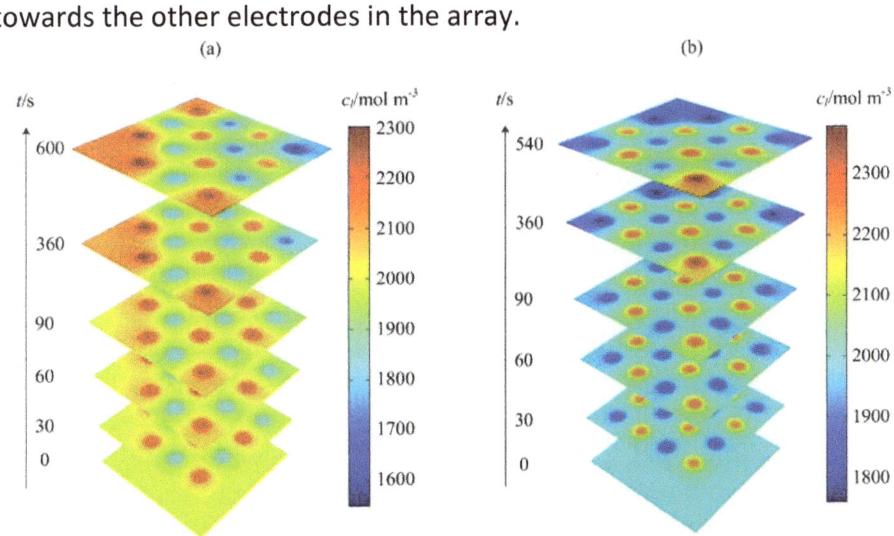

Figure 8. *Variation of concentration of lithium ions in the electrolyte along the xz-plane at y=2.8 × 10^{-4} m at various times during discharge at 5 C-rate under malfunctioning (a) positive and (b) negative electrodes of circular cross-section*

When a negative electrode malfunctions, the concentration distribution of lithium ions in the rest of the array also remains largely similar throughout high discharge rates relative to that in a non-perturbed array, as shown by comparing Figs. 3b and 8b. However, the concentration of lithium ions in the electrolyte in the malfunctioning negative electrode decreases continuously from the initial value of 2 x 10^3 mol m^{-3} to 1.8 x 10^3 mol m^{-3} at the end of discharge, compared to an increase to 2.3 x 10^3 mol m^{-3} on average in the other negative electrodes. As the adjacent positive electrodes act as sink for lithium ions due to the reduction reaction, a concentration gradient develops between these electrodes and the malfunctioning electrode. Lithium ions diffuse into the positive electrodes from the electrolyte and the malfunctioning negative electrode does not give out the lithium ions due to the absence of reaction in the active material, resulting in the depletion of the ions within the electrolyte there. This effect is also concentrated locally in and around the malfunctioning electrode as shown in Fig. 8b and is insignificant towards the rest of the array.

The cessation of function of any electrode would directly impact the electrochemical behaviour on neighbouring electrodes to a significant degree, due to diffusion of lithium ions in and out of the malfunctioning electrode which otherwise would not occur in a non-perturbed array. However, this effect is negligible for electrodes further away during both the reaction phase and the diffusion phase, due to the presence of many other functioning electrodes in an otherwise intact array.

Heat generation and thermal behaviour

The rise in average temperature of the array in the perturbed cases in similar to that in the standard discharge, in that the average temperature increases steadily at first before becoming more gradual towards the end of discharge. However, there are some minor differences. When a positive electrode malfunctions, the increase in temperature is smaller by 0 - 2 K compared to the normal case during a 5 C discharge, and only 0 - 0.2 K during 1 C discharge rates, due to the decrease in the number of heat sources. On the other hand, when an anode malfunctions, the increase in temperature is smaller by 0 - 1 K compared to the normal case during a 5 C discharge, and 0 - 0.2 K during 1 C discharge, due to the decreased heat generation in the negative electrodes which is highest contributor to heat generation as seen before. Differences in thermal behavior are negligible for electrodes further away from the malfunctioning electrode.

Conclusions

This paper presents a thermal-electrochemical coupled model for next-generation 3D Li-ion batteries applied to two different electrode geometries - square and circular. A reliability analysis was also conducted to analyse the effect of a single malfunctioning electrode on the rest of the array.

In summary, the performance of the 3D cell during discharge under normal conditions was similar for both the square and circular electrode arrays; potential, thermal behaviour and electrochemical behaviour also did not show marked differences at any discharge rates under the selected design parameters. Furthermore, the cell capacity can be increased by simply adding more electrodes in the plane of the array or increasing the height of the electrodes, without compromising on power density unlike in the conventional 2D design.

The performance of the 3D electrode array during perturbed conditions, in terms of the changes in the potential and the concentration distribution, was only significant in and around the malfunctioning electrode. A maximum change of around 17 % in the energy capacity and 10 % in lithium ion concentration in the electrolyte in and around the malfunctioning electrode under a 5 C discharge was seen. Deviation in electrochemical behaviour is negligible more than one cell away from the malfunctioning electrode. The model can also easily be extended to account for various types of 3D designs and conditions.

This design demonstrates a clear advantage in terms of reliability over the 2D battery, as the 3D array does not stop functioning even if one of the electrodes fail, unlike the conventional 2D parallel-plate design. This allows power to be continuously delivered in a safe manner until the battery is eventually replaced, as there is only minor deviation in thermal behaviour of the cell. As the array behaviour is likely to be significantly affected only when many electrodes malfunction, actual implementation of 3D batteries with full-size arrays is viable. With good performance due to the redundancy provided by the array, the 3D cell will be able to deliver reliability benefits which are crucial in many modern applications. Further, the model can be extended to study the transport in solid polymer electrolytes as well. Also, the model can be employed to study the behaviour of the battery when there is short-circuiting of the electrodes which seems to be a common problem in microbatteries.

Nomenclature

A_s	specific surface area for the faradaic reaction per unit volume, m^2/m^3	C_s	concentration of lithium in active material in the electrodes, mol m^{-3}
C_l	electrolyte concentration, mol m^{-3}	C_p	specific heat capacity, J kg^{-1} K^{-1}
c_s^{avg}	average concentration of Li in the active material, mol m^{-3}	c_s^{surf}	surface concentration of Li in the active material, mol m^{-3}
D_l	diffusion coefficient of electrolyte, m^2 s^{-1}	D_s	diffusion coefficient of Li in the active material in the electrodes, m^2 s^{-1}
E_a	activation energy for a variable, kJ mol^{-1}	F	Faraday's constant, 96487 C mol^{-1}
H	height of the battery, m	h_i	height of the functional layers in the battery, m
h_t	heat transfer coefficient, W m^{-2} K^{-1}	i_{app}	applied current density, A m^{-2}
i_0	exchange current density, A m^{-2}	i_l	liquid phase current density, A m^{-2}
i_s	solid phase current density, A m^{-2}	i_f	faradaic transfer current density, A m^{-2}
J	local charge transfer current per unit volume, A m^{-3}	k	thermal conductivity, W m^{-1} K^{-1}
k_0	reaction rate constant, $mol^{2.5}$ $m^{-0.5}$ s^{-1}	L	length of the battery, m
l_s	diffusion length, m	N_l	species (lithium ion) flux, mol m^{-2} s^{-1}
n	normal vector	Q	volumetric heat generation, W m^{-3}
q	conductive heat flux, W m^{-2}	R	gas constant, J mol^{-1} K^{-1}
R_s	radius of active material, m	r	radial coordinate
t	time, s	t_+^0	transference number of cation
T	temperature, K	T_a, T_0	ambient and initial temperature, K
T_{ref}	reference temperature, 298.15 K	U_{ref}	open circuit potential of the electrode, V
V_i	volume of the electrode i, m^3	W	width of the battery, m
W_i	thickness of the layer i, m		

Greek

α_a, α_c	anodic/cathodic transfer coefficient	ε_l	volume fraction of the electrolyte in the electrodes
ε_f	volume fraction of the conductive filler additive in the electrodes	ε_p	volume fraction of the polymer in the electrodes
η	overpotential, V	γ	Bruggeman constant (= 1.5)
ρ	density, kg m^{-3}	σ_l	ionic conductivity of electrolyte, S m^{-1}
σ_s	electronic conductivity of solid matrix, S m^{-1}	φ_l	liquid phase potential, V
φ_s	solid phase potential, V	υ	local state of charge of the electrodes

Subscripts

cc	current collector	ne	negative electrode
pe	positive electrode	el	electrolyte
l	liquid/ electrolyte		

Superscripts

0	initial values	eff	effective values
max	maximum values		

References

[1] J. W. Long, B. Dunn, D. R. Rolison, H. S. White, *Chem. Rev.* **104** (2004) 4463–4492

[2] M. Beidaghi, C. Wang, *Micro- and Nanotechnology Sensors, Systems, and Applications II*. Edited by George, Thomas; Saif Islam, M.; Dutta, Achyut K. Proceedings of the SPIE, Volume 7679, 2010

[3] M. Roberts, P. Johns, J. Owen, D. Brandell, K. Edstrom, G. E. Enany, C. Guery, D. Golodnitsky, M. Lacey, C. Lecoeur, H. Mazor, E. Peled, E. Perre, M. M. Shaijumon, P. Simon, P.-L. Taberna, *J. Mater. Chem.* 21 (2011) 9876–9890

[4] J. Weinert, A. Burke, X. Wei, *J. Power Sources* 172 (2) (2007) 938–945

[5] J. Tarascon, M. Armand, *Nature* **414** (2001) 359–367

[6] R. W. Hart, H. S. White, B. Dunn, D. R. Rolison, *Electrochem. Commun.* **5** (2003) 120–123

[7] B. Dunn, J. W. Long, D. R. Rolison, *The Electrochem. Soc. Interface* **17(3)** (2008) 49–53

[8] R. Salot, S. Martin, S. Oukassi, M. Bedjaoui, J. Ubrig, *Appl. Surf. Sci.* **256(3)** (2009) S54–S57

[9] H.-S. Min, B. Y. Park, L. Taherabadi, C. Wang, Y. Yeh, R. Zaouk, M. J. Madou, B. Dunn, *J. Power Sources* **178** (2008) 795–800

[10] J. F. M. Oudenhoven, L. Baggetto, P. H. L. Notten, *Adv. Energy Mater.* **1(1)** (2011) 10–33

[11] J. Kim, X. Song, K. Kinoshita, M. Madou, R. White, *J. Electrochem. Soc.* **145(7)** (1998) 2314–2319

[12] R. Kostecki, X. Song, K. Kinoshita, *J. Electrochem. Soc.* **147(5)** (2000) 1878–1881.

[13] S. Ranganathan, R. McCreery, S. M. Majji, M. Madou, *J. Electrochem. Soc.* **147(1)** (2000) 277–282

[14] C. Wang, L. Taherabadi, G. Jia, Y.Yeh, B. Dunn, M. Madou, *Electrochem. Solid-State Lett.* **7(11)** (2004) A435–A438

[15] D. R. Rolison, J. W. Long, J. C. Lytle, A. E. Fischer, M. E. B. Christopher P. Rhodes and, Todd M. McEvoy, A. M. Lubers, *Chem. Soc. Rev.* **38(1)** (2009) 226–252

[16] Y. Shao-Horn, C. Hidrovo, S. Jurga, H. Smith, G. Barbastathis, *204th meeting of the Electrochemical Society,* Pennington, NJ, abstr. 1279, 2003

[17] K. Dokko, J. Sugaya, H. Nakano, T. Yasukawa, T. Matsue, K. Kanamura, *Electrochem. Commun.* **9** (2007) 857–862

[18] G. T. Teixidor, R. B. Zaouk, B. Y. Park, M. J. Madou, *J. Power Sources* **183** (2008) 730–740

[19] H.-S. Min, B. Y. Park, L. Taherabadi, C. Wang, Y. Yeh, R. Zaouk, M. J. Madou, B. Dunn, *J. Power Sources* **178(2)** (2008) 795–800

[20] M. Fleischauer, J. Li, M. Brett, *J. Electrochem. Soc.* **156(1)** (2009) A33–A36

[21] M. Kotobuki, Y. Suzuki, H. Munakata, K. Kanamura, Y. Sato, K. Yamamoto, T. Yoshida, *J. Electrochem. Soc.* **157(4)** (2010) A493–A498

[22] K. Sun, T.-S. Wei, B. Y. Ahn, J. Y. Seo, S. J. Dillon, J. A. Lewis, *Adv. Mater.* **5** (2013) 1–5

[23] V. Zadin, H. Kasemagi, A. Aabloo, D. Brandell, *J. Power Sources* **195** (2010) 6218–6224

[24] V. Zadin, D. Brandell, H. Kasemgi, A. Aabloo, J. O. Thomas, *Solid State Ionics* **192** (2011), 279–283

[25] V. Zadin, D. Brandell, *Electrochim. Acta* **57** (2011) 237–243

[26] T. Fuller, M. Doyle, J. Newman, *J. Electrochem. Soc.* **141(1)** (1994) 1–10

[27] M. Doyle, J. Newman, *J. Electrochem. Soc.* **143(6)** (1996) 1890–1903

[28] P. M. Gomadam, J. W. Weidner, R. A. Dougal, R. E. White, *J. Power Sources* **110** (2002) 267–284

[29] C.-W. Wang, A. M. Sastry, *J. Electrochem. Soc.* **154(11)** (2007) A1035–A1047

[30] X. Zhang, *Electrochim. Acta* **56** (2011) 1246–1255

[31] K. Somasundaram, E. Birgersson, A. S. Mujumdar, *J. Power Sources* **203** (2012) 84–96

[32] J. Newman, W. Tiedemann, *AIChE J.* **21(1)** (1975) 25–41

[33] J. Newman, K. E. T. Alyea (Eds.), *Electrochemical Systems*, 3rd Edition, John Wiley and Sons, Inc., 2004

[34] V. Srinivasan, C. Y. Wang, *J. Electrochem. Soc.* **150(1)** (2003) A98–A106

[35] Comsol Multiphysics 3.5a, http://www.comsol.com, visited May 11, 2009

[36] B. A. Johnson, R. E. White, *J. Power* Sources *70* (1998) 48–54

[37] D. R. Baker, M. W. Verbrugge, *J. Electrochem. Soc.* **146(7)** (1999) 2413-2424

Appendix A

***Table A.1** Governing equations*

Governing equation	Flux
$\nabla\cdot\mathbf{i}_s = -J$ (pe,ne,cc)	$\mathbf{i}_s = -\sigma_s^{eff}\nabla\varphi_s$
$\nabla\cdot\mathbf{i}_l = J$ (pe,ne,el)	$\mathbf{i}_l = -\sigma_l^{eff}\nabla\varphi_l + \frac{2RT\sigma_l^{eff}}{F}(1-t_+^0)\nabla(\ln c_l)$
$\varepsilon_l\frac{\partial c_l}{\partial t} + \nabla\cdot\mathbf{N}_l = \frac{J}{F}$ (pe,ne,el)	$\mathbf{N}_l = -D_l^{eff}\nabla c_l + \frac{\mathbf{i}_l t_+^0}{F}$
$\frac{D_s}{l_s}(c_s^{surf} - c_s^{avg}) = -\frac{i_f}{F}, \frac{dc_s^{avg}}{dt} = -\frac{3i_f}{FR_s}$, (ne,pe)	-
$(\rho C_p)^{eff}\frac{\partial T}{\partial t} + \nabla\cdot\mathbf{q} = Q$ (pe,ne,el,cc)	$\mathbf{q} = -k^{eff}\nabla T$

Appendix B

***Table B.1** Constitutive relations*

$J =$	$\begin{cases} A_s i_f & \text{(ne,pe)} \\ 0 & \text{(el,cc)} \end{cases}$
$i_f =$	$i_0\left\{\exp\left(\frac{\alpha_a\eta F}{RT}\right) - \exp\left(-\frac{\alpha_c\eta F}{RT}\right)\right\}$
$i_0 =$	$Fk_0\sqrt{c_l(c_s^{\max} - c_s^{surf})c_s^{surf}}$
$\eta =$	$\varphi_s - \varphi_l - U_{ref,\,i}^{eff}$, i=ne,pe
$A_s =$	$\frac{3(1-\varepsilon_l-\varepsilon_f-\varepsilon_p)}{R_s}$
$Q =$	$J\eta + JT\frac{\partial U_{ref,\,i}}{\partial T} + \sigma_s^{eff}(\nabla\varphi_s)^2 + \sigma_l^{eff}(\nabla\varphi_l)^2 + \frac{2RT\sigma_l^{eff}}{F}(1-t_+^0)\nabla(\ln c_l)\cdot\nabla\varphi_l$, i=ne,pe
$\vartheta_{ne}, \vartheta_{pe} =$	$\frac{c_s^{surf}}{c_s^{\max}}$
$C_i =$	$V_i(1-\varepsilon_l-\varepsilon_f-\varepsilon_p)\rho_i C_{th}$, i=ne,pe
$U_{ref,\,i}^{eff} =$	$U_{ref,\,i} + (T-T_{ref})\frac{\partial U_{ref,\,i}}{\partial T}$, i=ne,pe
$\sigma_s^{eff} =$	$\sigma_s(1-\varepsilon_l-\varepsilon_f-\varepsilon_p)$
$\sigma_l^{eff} =$	$\sigma_l\varepsilon_l^{\gamma}$
$k_i^{eff} =$	$k_i(1-\varepsilon_l) + k_l\varepsilon_l$, i=ne,pe
$(\rho C_p)_i^{eff} =$	$(\rho C_p)_i(1-\varepsilon_l) + (\rho C_p)_l\varepsilon_l$, i=ne,pe
$D_l^{eff} =$	$D_l\varepsilon_l^{\gamma}$
$\sigma_l =$	$-1.172\times10^{-14}c_l^4 + 1.3605\times10^{-10}c_l^3 - 5.2245\times10^{-7}c_l^2 + 6.7461\times10^{-4}c_l + 1.0793\times10^{-2}$
$\Theta(T) =$	$\Theta(T_{ref})\exp\left[\frac{E_{a,\Theta}}{R}\left(\frac{1}{T_{ref}} - \frac{1}{T}\right)\right], \Theta = D_s, D_l, \sigma_l$

$l_s =$	$\frac{R_s}{5}$
$\varphi_s^0 =$	$U_{ref,\,pe}(\vartheta_{pe}^0) - U_{ref,\,ne}(\vartheta_{ne}^0)$
$\varphi_l^0 =$	$-U_{ref,\,ne}(\vartheta_{ne}^0)$
$E_{cell} =$	$\varphi_s\|_{I} - \varphi_s\|_{VII}$
$\frac{\partial U_{ref,\,ne}}{\partial T} =$	$\frac{344.1347\exp(-32.9633\vartheta_{ne}+8.3167)}{1+749.0756\exp(-34.7909\vartheta_{ne}+8.8871)} - 0.852\vartheta_{ne} + 0.3622\vartheta_{ne}^2 + 0.2698$
$\frac{\partial U_{ref,\,pe}}{\partial T} =$	$-4.1453 + 8.1471\vartheta_{pe} - 26.0645\vartheta_{pe}^2 + 12.766\vartheta_{pe}^3 + 4.3127\exp(0.5715\vartheta_{pe}) - 0.1842\exp\left(-\left(\frac{\vartheta_{pe}-0.5169}{0.0462}\right)^2\right)$ $+1.2816\sin(-4.9916\vartheta_{pe}) - 0.0904\sin(-20.9669\vartheta_{pe} - 12.5788) + 0.0313\sin(31.7663\vartheta_{pe} - 22.4295)$
$U_{ref,\,ne} =$	$-0.16 + 1.32\exp(-3\vartheta_{ne}) + 10\exp(-2000\vartheta_{ne})$
$U_{ref,\,pe} =$	$4.1983 + 0.0565\tanh(-14.5546\vartheta_{pe} + 8.6094) - 0.0275\left(\frac{1}{(0.9984-\vartheta_{pe})^{0.4924}} - 1.9011\right)$ $-0.1571\exp(-0.0474\vartheta_{pe}^8) + 0.8102\exp(-40(\vartheta_{pe} - 0.1339))$

Appendix C

***Table C.1** Boundary conditions*

I	$\mathbf{n}\cdot\mathbf{i}_s = -i_{app}, \mathbf{n}\cdot\mathbf{q} = h_t(T - T_a)$
II	$\mathbf{n}\cdot\mathbf{i}_s\|_{II_+} = \mathbf{n}\cdot\mathbf{i}_s\|_{II_-}, \mathbf{n}\cdot\mathbf{q}\|_{II_+} = \mathbf{n}\cdot\mathbf{q}\|_{II_-}, \mathbf{n}\cdot\mathbf{i}_l = \mathbf{n}\cdot\mathbf{N}_l = 0$
III	$\mathbf{n}\cdot\mathbf{i}_s = 0, \mathbf{n}\cdot\mathbf{q}\|_{III_+} = \mathbf{n}\cdot\mathbf{q}\|_{III_-}$
IV, V	$\mathbf{n}\cdot\mathbf{i}_s = 0\,(IV), \mathbf{n}\cdot\mathbf{i}_l = 0\,(V), \mathbf{n}\cdot\mathbf{q} = 0$
VI	$\mathbf{n}\cdot\mathbf{i}_s = 0, \mathbf{n}\cdot\mathbf{i}_l\|_{VI_+} = \mathbf{n}\cdot\mathbf{i}_l\|_{VI_-}, \mathbf{n}\cdot\mathbf{q}\|_{VI_+} = \mathbf{n}\cdot\mathbf{q}\|_{VI_-}, \mathbf{n}\cdot\mathbf{N}_l\|_{VI_+} = \mathbf{n}\cdot\mathbf{N}_l\|_{VI_-}$
VII	$\varphi_s = 0, \mathbf{n}\cdot\mathbf{q} = h_t(T - T_a)$

***Table C.2** Initial conditions*

$c_s^{surf} = c_s^{avg} = c_s^0$
$c_l = c_l^0$
$\varphi_s = \begin{cases} 0 & (\text{ne, cc(-)}) \\ \varphi_s^0 & (\text{pe, cc(+)}) \end{cases}$
$\varphi_l = \varphi_l^0$ (ne, pe, el)
$T = T_0$

13

A review of liquid metal anode solid oxide fuel cells

ALIYA TOLEUOVA*,**, VLADIMIR YUFIT***, STEFAAN SIMONS*,****, WILLIAM C. MASKELL*,***, DANIEL J. L. BRETT*,✉

*Electrochemical Innovation Lab, Department of Chemical Engineering, University College London, Torrington Place, London WC1E 7JE, UK
**School of Engineering, Nazarbayev University, 53 Kabanbay Batyr Ave, Astana 010000, Republic of Kazakhstan
***Department of Earth Science and Engineering, Imperial College London, London SW7 2AZ, UK
****International Energy Policy Institute, UCL Australia, 220 Victoria Square, Adelaide, Australia
✉Corresponding Author

Abstract
This review discusses recent advances in a solid oxide fuel cell (SOFC) variant that uses liquid metal electrodes (anodes) with the advantage of greater fuel tolerance and the ability to operate on solid fuel. Key features of the approach are discussed along with the technological and research challenges that need to be overcome for scale-up and commercialisation.

Keywords
Solid oxide fuel cell; liquid metal anode; direct carbon fuel cell.

Introduction

The world's growing energy demands have led to increasing environmental and resource availability concerns. In this regard, stabilization of increasing anthropogenic CO_2 emissions is one of the most urgent issues associated with global climate change. According to the International Energy Agency (IEA), fossil fuels will remain the primary energy sources, representing more than 75% of the overall growth in energy use from 2007 to 2030 [1]. Moreover, carbon-free energy technologies are unable to meet global demands in the foreseeable future [2]. Therefore, continued use of fossil fuels in a 'low-carbon economy' necessitates the reduction of CO_2 emissions by improving process efficiency. Notwithstanding the reduction in greenhouse gas emissions brought about by high efficiency, fuel cells generate carbon dioxide free from diluting nitrogen which can be sequestered at lower cost compared with the exhaust products from most electricity generation based upon combustion of fossil fuels.

While coal remains the most available and abundant fuel source around the world (60% of all global sources), highly efficient fuel technologies are required to minimize the severe environmental impact of electric energy production from coal [3]. In this regard, several studies [4–7] have presented novel concepts for the direct electrochemical oxidation of carbon (coal or biomass) for highly efficient electricity generation using Liquid Metal Electrode Solid Oxide Fuel Cells (LME SOFCs). This new class of fuel cell offers the advantage of high efficiency with the benefits of greater fuel tolerance, the concept being particularly well suited to the direct use of solid fuel.

High temperature solid oxide fuel cells (SOFC) are highly efficient electrochemical energy conversion devices capable of converting the chemical energy stored in gaseous fuels (including hydrogen, alkanes, alcohols, alkenes, alkynes, ketones, *etc.*) into electrical power [8]. The operational efficiency of such systems is higher than conventional heat engines since the process is electrochemical and not constrained by the Carnot efficiency limitation. Typical efficiencies obtained from SOFCs are in the range of 50-60% (LHV). The technology is highly scalable with systems in the hundreds of kilowatts demonstrated. The SOFC is increasing attention as a clean and efficient power-generating technology. However, durability and performance degradation due to impurities contained in hydrocarbon fuels, such as sulfur and other trace elements, continue to be a concern [4]. Contaminants may react with the anode materials via various mechanisms to decrease electrochemical reaction rates by increasing charge transfer resistance and result in mechanical failure of materials [9]. In order to mitigate the influence of fuel impurities on cell degradation, the supplied fuel should undergo pretreatment using a variety of adsorbents and filters [10]. Alternatively, substantial improvements in anode materials are necessary to make the anode more tolerant of fuel contaminants. One approach to achieving this is to use a liquid metal anode solid oxide fuel cell (LMA SOFC). This allows operation on a variety of carbonaceous fuels – gaseous or solid, while remaining highly tolerant towards fuel contaminants.

Some of the earliest work on liquid metal electrodes was reported by Cell Tech Power, who used liquid tin as their anode material of choice [5,6,11,12] and achieved a power density of 170 mW cm^{-2} on hydrogen and JP-8 fuel [4]. The projected efficiency of such a system operating on coal is 61 % [13].

However, LMA SOFC technology is not yet well-understood. The nature of the cells makes experimental investigation challenging and the processes occurring in the melt are difficult to study *in situ*. There is a certain inconsistency between published studies in terms of the experimental arrangement and little is known about the mechanism of fuel oxidation within the liquid metal media.

Being a comparatively novel system, LMA SOFC technology is some way from being fully commercialized due to technical challenges unique to this class of fuel cell. The objective of this review paper is therefore to provide a summary of emerging LMA SOFC technologies, discuss the advantages of the approach and set out the research and technological challenges that need to be addressed and overcome.

Background to fuel cells

Fuel cells are electrochemical energy conversion devices that convert chemical energy in fuel directly into electricity (and heat) without involving the process of combustion. The technology is highly efficient, can be applied to a range of fuels (depending upon the type of fuel cell), quiet in operation (the fuel cell itself has no moving parts) and scalable from mW to MW. As such, fuel cells are considered as one of the most promising technological solutions for sustainable power generation. Fuel cells can be used in a broad range of applications, including: transportation, residential combined heat and power (CHP), large scale distributed power generation and battery replacement uses.

Fuel cells come in a range of architectures and material sets, but central to all is the electrolyte onto which there is attached the anode and cathode electrodes on each side. Figure 1 illustrates the simplistic operation of a fuel cell (using the SOFC as an example). The cell is constructed from two porous electrodes separated by an electrolyte. Oxygen or air is supplied to the cathode (also referred to as the 'air electrode') where electrochemical reduction of the oxygen takes place to form oxide ions (O^{2-}) that migrate through the electrolyte, to the anode (also referred to as the 'fuel electrode'), and oxidize the fuel (hydrogen in this case) releasing water, heat and electrons that travel around an external circuit to the cathode and do useful work.

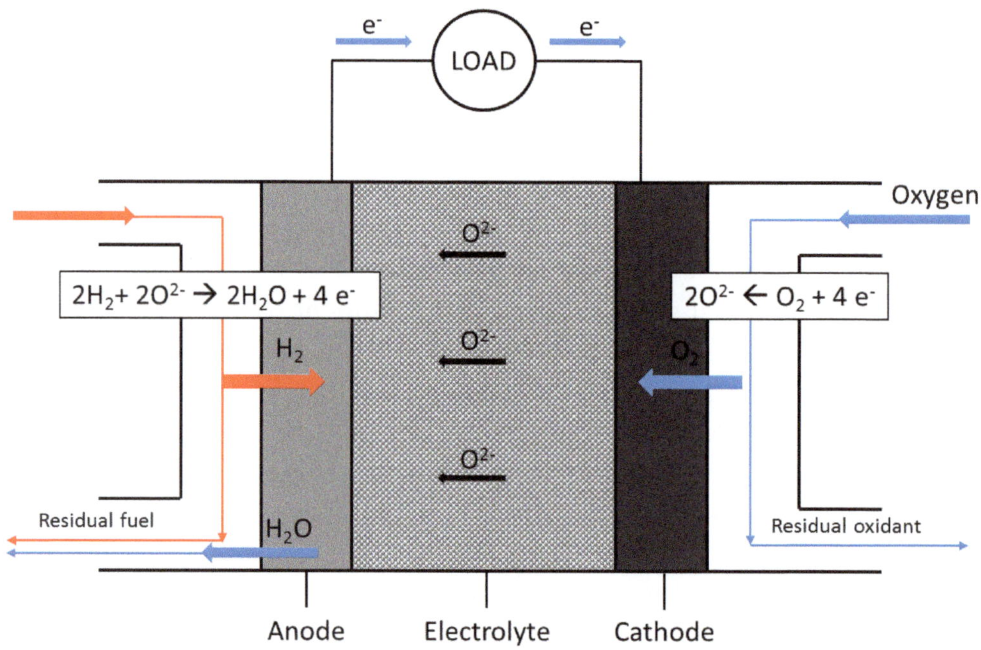

Figure 1. *Schematic operation of an individual fuel cell (taking a solid oxide fuel cell as an example).*

Different types of fuel cells

Fuel cell types vary from one to another in operating parameters and technical characteristics (e.g. power density, efficiency, *etc.*). However, the fundamental feature of the fuel cell which is different, and indeed names the fuel cell type, is the electrolyte. There are at least five main fuel cell types: Phosphoric Acid Fuel Cell (PAFC), Proton Exchange Membrane Fuel Cell (PEMFC), Alkaline Fuel Cell (AFC), Molten Carbonate Fuel Cell (MCFC) and Solid Oxide Fuel Cell (SOFC). In the first two types, hydrogen ions (protons) flow through the electrolyte from anode to cathode to react with oxygen, and produce water; whereas in the last three types (Alkaline, Molten and Solid Oxide fuel cells) anions (OH^-, CO_3^{2-}, O^{2-}) migrate through the electrolyte from cathode to anode to react with fuel and similarly produce water (Figure 2). Fuel cell electrolytes are electronically insulating but ionically conducting, allowing certain types of ions to transport through them.

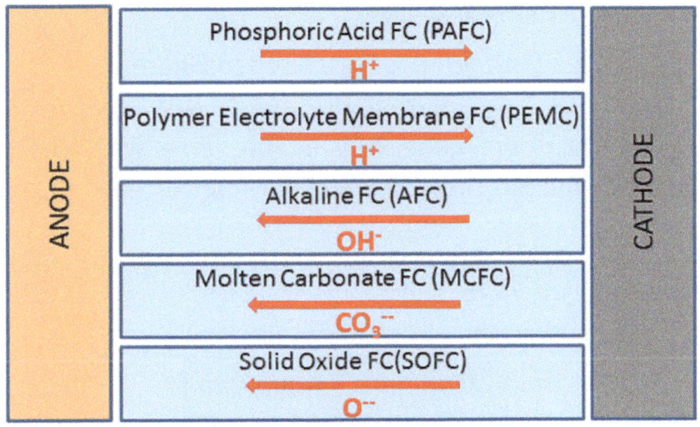

Figure 2. *Flow of ions through different fuel cell electrolyte types.*

Fuel cell applications

The most common types of fuel cell systems and their main characteristics, including fuel efficiency, operating temperature, lifetime, *etc.* are summarised in Table 1. The type of fuel cell and the range of operating temperatures are primarily related to the electrolyte material. Hence, the first three types (AFC, PAFC, and PEMFC) are considered as low temperature fuel cells, and the other two types (MCFC and SOFC) require high temperature. Low temperature fuel cells have the advantage of rapid start-up time but necessitate precious metal electrocatalysts and high purity hydrogen as fuel.

Table 1. *Technical characteristics of the five main fuel cell types (1 - [14]; 2 - [15]; 3 -[16])*

Fuel Cell Type	Operating temperature [1,2], C°	Power density[1,2], mW/cm²	Typical stack size[3]	Fuel Efficiency[2], %	CO tolerance[1]	Lifetime[2] × 1000, hr	Applications[2,3]; Capital Cost[2], $/kW	Advantages[3]	Disadvantages[3]
PAFC	150-200	150-300	400 kW 100 kW module	55	Poison (<1%)	> 40	Distributed Power; 3000 $/kW	- Higher temperature enables CHP; -Increased tolerance to fuel impurities	- Pt catalyst; - Long start up time; - Low current and power;
PEMFC	50-100	300-1000 (350[2])	< 1-100kW	45-60	Poison (<50ppm)	> 40	- Backup power; - Portable power; - Distributed generation; - Transportation; - Specialty vehicles; >200 $/kW	- Solid electrolyte reduces corrosion & electrolyte management problems; - Low temperature; - Quick start-up	- Expensive catalysts; - Sensitive to fuel impurities; - Low temperature waste heat
AFC	90-100	150-400	10-100 kW	40-60	Poison (<50ppm)	> 10	- Military; - Space; >200 $/kW	- Cathode reaction faster in alkaline electrolyte, leads to high performance; - Low cost components	- Sensitive to CO_2 in fuel and air; - Electrolyte management
MCFC	600-700	100-300	0.3 – 3 MW 300 kW module	60-65	Fuel	> 40	- Electric utility; - Distributed power generation; 1000 $/kW	- High efficiency; - Fuel flexibility; - Can use a variety of catalysts; - Suitable for CHP;	- High temperature corrosion and breakdown of cell components; - Long start-up time; - Low power density
SOFC	800-1000	250-350	1kW-2MW	55-65	Fuel	> 40	- Baseload power generation; - Auxiliary power; - Electric utility; - Distributed generation; 1500 $/kW	- High efficiency; - Fuel flexibility; - Can use a variety of catalysts; - Solid electrolyte; - Suitable for CHP&CHHP; - Hybrid/GT cycle	- High temperature corrosion and breakdown of cell components; - High temperature operation requires long start up time

Liquid Metal Electrode Electrochemical Systems

The generic construction of a LMA SOFC is illustrated in Figure 3. The arrangement is ostensibly identical to a conventional SOFC except that the anode is a molten metal. Fuel is fed to the electrode and oxidized products are generated in the same way. The difference is that both gaseous and solid fuel can be fed directly and effectively converted with no pre-treatment stage.

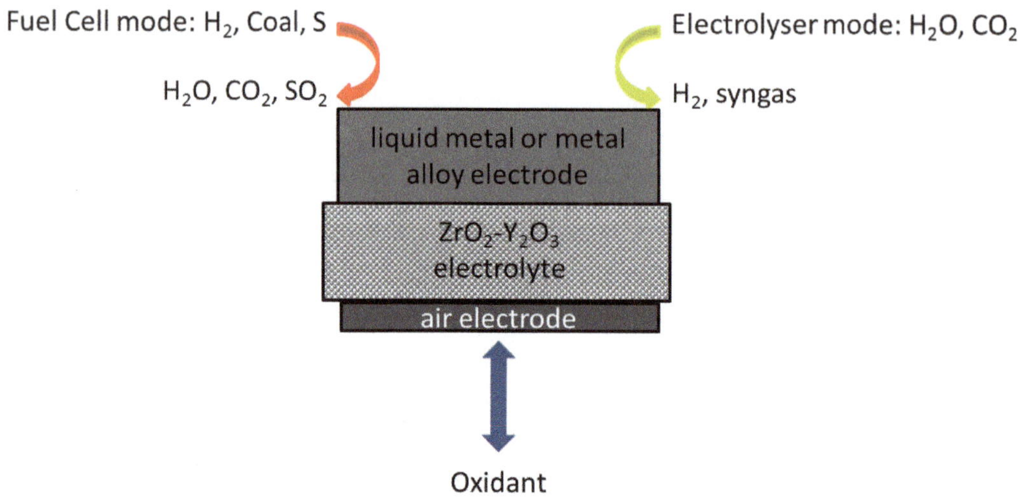

Figure 3. *Operation of liquid metal electrode reactor in fuel cell and electrolyser mode.*

Figure 3 also shows the potential for operation in electrolyser mode. This concept has the cell fed with CO_2 and steam (H_2O) that are electrochemically converted to synthesis gas (a mixture of hydrogen and carbon monoxide), that can be used to make liquid fuel (*i.e.* via the Fischer-Tropsch process) or used in a fuel cell directly [17].

Liquid Metal Anode SOFCs

The liquid metal electrode resides in a layer between the fuel chamber (gaseous or solid fuel) and the solid electrolyte (Figure 4). The oxygen reduction reaction occurs at the cathode/electrolyte interface resulting in oxygen ions, which then migrate though the solid electrolyte, typically represented by yttria-stabilised zirconia (YSZ), to the liquid metal anode. The O^{2-} ions react electrochemically with the liquid metal, generating metal oxide which is the active species for the oxidation of the fuel (gaseous or solid - carbon and hydrogen in the diagram), producing carbon dioxide or water respectively. However, the exact mechanism occurring and the species involved in the liquid metal anode media are not well defined and depend upon the metal used.

The molten metal blocks direct contact of electrolyte with gaseous impurities and hence reduces electrolyte degradation by inhibiting the reaction between contaminants and electrolyte [4]. Moreover, the fuel contaminants can become a fuel source themselves as they undergo electrochemical oxidation. All of the electrolyte surface in contact with the anode is available to supply a flux of oxide ions for reaction - there is no need to engineer a complex ceramic metal (cermet) electrode with optimized triple phase boundary (TPB) as is necessary in conventional SOFCs [18,19].

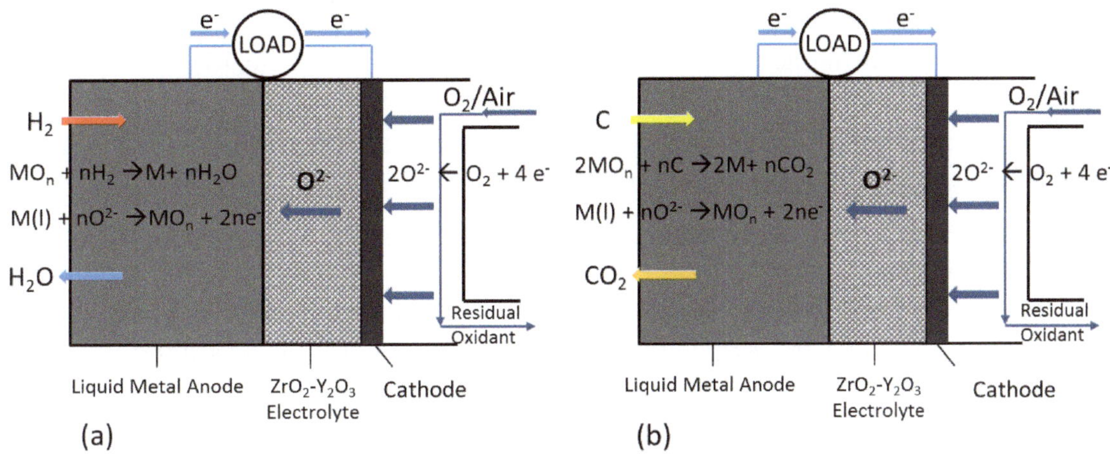

Figure 4. *Schematic showing operation of LMA SOFC with* ***(a)*** *hydrogen and* ***(b)*** *solid carbon, as the fuel.*

Operation of Liquid Metal Anode SOFCs

The liquid metal anode acts as an intermediary for the oxidation of fuel [4]. Taking a molten metal (M) and hydrogen as a fuel, the reactions at the LMA-SOFC anode follows the generic mechanism:

$M + xO^{2-} \rightarrow M[O_x]_M + 2xe^-$ Reaction (1)

$xH_2 + M[O_x]_M \rightarrow M + xH_2O$ Reaction (2)

Reaction (1) proceeds at the liquid anode-electrolyte interface, whereas reaction (2) occurs at the fuel-anode interface. The oxide ions (O^{2-}) are delivered from the cathode via the electrolyte. $[O_x]_M$ is a state of oxygen within the liquid metal anode. The state of oxygen in the metal varies with operating conditions, and hence 'x' is used to represent this uncertainty. When using tin, oxygen may form oxide or suboxide or remain dissolved in the melt [20]. The state of oxygen may vary, though it has no particular effect on the overall reaction, which is the sum of Reactions (1) and (2):

$H_2 + O^{2-} \rightarrow H_2O + 2e^-$ Reaction (3)

Reaction (3) is applicable to any liquid metal anode SOFC system including Sn or Cu [21].

Thermodynamics of LMA SOFC operation

The maximum electrical voltage generated by an electrochemical system can be determined via the standard equilibrium electrode potential (E_{eq}), expressed by the Nernst Equation (1) as follows:

$$E_{eq} = E^0 + \frac{RT}{nF}\sum_i v_i \ln a_i \qquad \text{Equation (1)}$$

Where a_i and v_i are the activities (fugacities for gases) and stoichiometric factors of the species i (the stoichiometric factors v_i taking positive and negative values for oxidised and reduced species respectively), R is the molar gas constant, 8.314 J mol-1 K-1, T is absolute temperature and E^o is the standard electrode (equilibrium) potential of the reaction that can be defined from the equation (2):

$$E^0 = \frac{-\Delta G_r^0 RT}{nF} \qquad \text{Equation (2)}$$

where ΔG_r^0 is the Gibbs free energy change of electrochemical reaction, n – number of electrons transferred during electrochemical reaction and F is Faraday's constant, 96,485 C mol^{-1}.

Figure 5 shows the standard equilibrium potentials of several metal-metal oxide couples as a function of temperature. The data for the figure was generated using the chemical reaction and equilibrium software HSC Chemistry 6.1 (Outotec, Finland).

It can be seen that the cell voltage is expected to decrease with increasing temperature, as a result of the negative entropy change associated with the reactions. However, the voltage obtained under operating conditions (with a net current flowing) is difficult to predict. In addition to the usual loss mechanisms associated with reaction kinetics, ohmic and mass transport limitations, cell voltage will depend on the actual reaction at the anode: whether it is directly with dissolved fuel or the oxidation of the metal to oxide (electrochemical reaction) followed by oxidation of the hydrogen by the oxide (chemical reaction).

If the reaction is directly with the fuel then conventional fuel cell reaction OCVs will hold, which are generally higher than those for the reaction with the metal. Therefore, from a thermodynamic perspective, it may be preferable for the metal to have a low affinity for oxygen and a high solubility of fuel and oxygen.

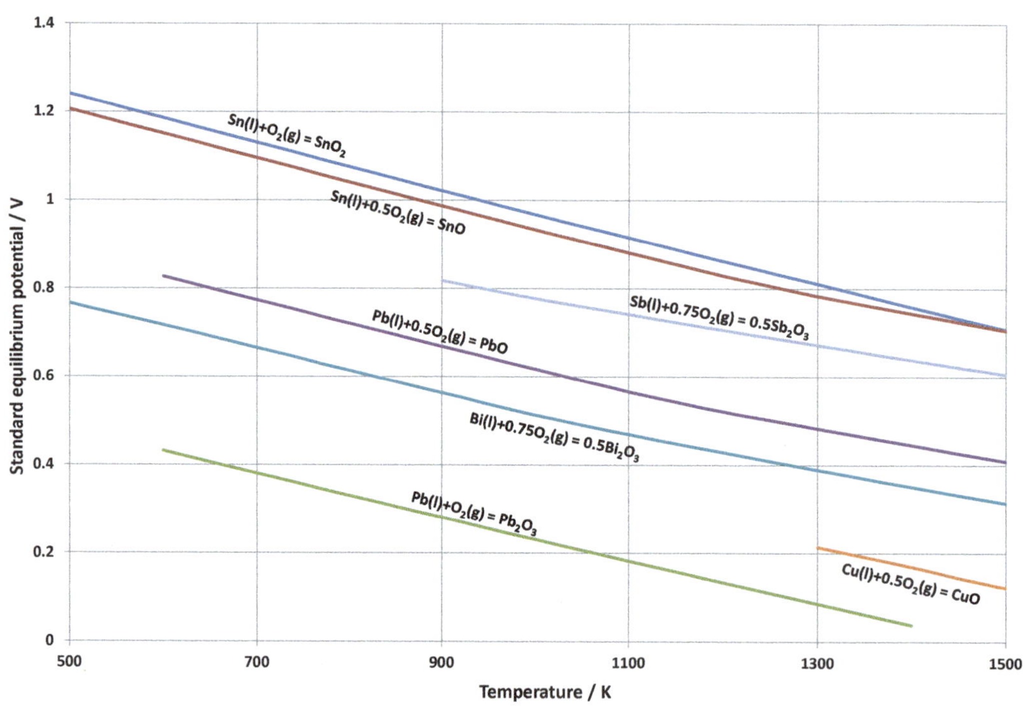

Figure 5.*Effect of temperature on standard equilibrium cell potentials for various metal oxidation reactions.*

Battery Effect

The use of LMAs can have an additional positive impact on fuel cell performance by creating the so-called 'battery' buffering effect. It is well known that SOFCs (similar to other fuel cells) can be detrimentally affected by power demand surges during operation, since such surges can cause fuel starvation, which in turn may cause irreversible structural changes at the electrodes. The presence of a certain amount of LMA acts as a metal-oxygen battery, where the metal itself is oxidised to provide sufficient load [4,5,7]. The oxidised metal can be reduced again by the fuel fed into the anode when normal load is resumed.

Liquid Tin Anode SOFC

Liquid tin anode SOFCs (LTA SOFC) have been demonstrated with a variety of fuels including gaseous, solid and liquid carbonaceous fuels that did not require any processing, reforming or pretreatment to remove sulfur. The power density reported was >160 mW cm^{-2} with hydrogen, and 120 mW cm^{-2} with JP-8 fuel[20]. Operated at 900 °C, a liquid tin layer of 6 mm was exposed to humidified H_2 gas (3 % H_2O). From the analysis of transient characteristics time, the oxygen (oxide) diffusion coefficient of 1×10^{-3} cm^2 s^{-1} was derived from test results. The value differs significantly from previously measured literature values [20]. Oxide diffusivity is one of the most critical fundamental parameter that determines cell performance. Inconsistencies in literature reported values call for further work to develop robust methods for the determination of this parameter.

Physical and chemical properties of tin

Abernathy *et al.* make the case for tin as an ideal metal for LMA SOFCs [4]. A comparison table (Table 2) summarises basic physical properties and abundances of various metals that can be used as a liquid anode. The metal should have relatively low melting point (simplifying operational conditions) and high boiling point. Tin has a low vapour pressure (9.87×10^{-4} Pa at 1000 °C) and hence, low volatility [12], *i.e.* the anode is not vaporised into the fuel exit stream during operation. Being technically easier to operate with, Sn remains the most commonly used anode material for such LA SOFC systems, though some researchers are seeking better anode materials and suggest the use of metal alloys based upon tin.

Such factors as metal abundances in the Earth's crust, annual production and associated prices also have to be taken into consideration. Research studies so far have shown results working with tin, copper and bismuth. Other metals may not be suitable due to hazards (Sb), high price (Ag) or low abundance (In).

***Table 2**. Common metals properties and abundances, prices (adapted from [4])*

Metal	T_m / °C	T_b /, °C	Price*, $/kg	Abundance in Earth's crust, ppm	Reported experimental studies using metal and/or metal alloys
Aluminum	660	2520	1.7	83,000	-
Antimony	631	1587	5.0	0.20	[22], [23]
Bismuth	271	1564	16.3	0.063	[21], [24–27]
Cadmium	321	767	2.7	0.10	-
Copper	1085	2563	5.2	79	[21]
Indium	157	2073	390	0.05	[23]
Lead	328	1750	1.7	7.9	Pb - [23];Sn-Pb alloy-[25]; Sn-Pb-Bi alloy-[25],
Tin	232	2603	13.9	2.5	[4–6], [11–13], [20], [24], [26–28]
Silver	962	2163	471	0.079	-
Zinc	420	907	1.7	79	-

**The metal prices are taken from US Geological Survey Minerals Information Team, Mineral Commodity Summaries 2010*

The following table (Table 3) shows the change of thermophysical properties of tin at a range of temperatures, including the solubility of oxygen, hydrogen and sulfur. Electrical resistivity of tin at 1000 °C has a reasonable value of 67.1 μΩ cm compared to the resistivity of nickel (53.1 μΩ cm) used in conventional ceramic metal (cermet) SOFC anodes [12].

Table 3. *Thermophysical properties of tin (adapted from [4])*

Property	Temperature, °C	Value
Surface tension	232	5.44 mN cm^{-1}
Viscosity	232	1.85 mN s m^{-2}
Expansion on melting	232	2.3 %
Density	800	6580 kg m^{-3}
	1000	6480 kg m^{-3}
Resistivity	800	62.1 μΩ cm
	1000	67.1 μΩ cm
Gas solubility-oxygen	536	0.000018 wt %
	750	0.0049 wt %
Gas solubility- hydrogen	1000	0.04 cm^3 H_2 / 100 g Sn
	1300	0.36 cm^3 H_2 / 100 g Sn
Gas solubility - sulfur	800	4.5 wt%
	1000	8.0 wt%

The density of tin and nickel at room temperature are 7.28 g cm^{-3} and 8.90 g cm^{-3} respectively. This makes tin an even more preferable metal on the basis of weight [12]. High solubility of sulfur in the melt is an attractive property of tin, allowing 'dirty' fuels to be used without sulfur removal or fuel reforming.

Fuels Investigated

Various fuels have been studied for LTA SOFCs including biodiesel [5], military fuels, such as JP-8 [6,11,13], gasoline, methane, *etc.* All these studies confirm the LTA SOFC fuel flexibility. Figure 6 summarises measured efficiency of LTA SOFC operated on different fuels [6].

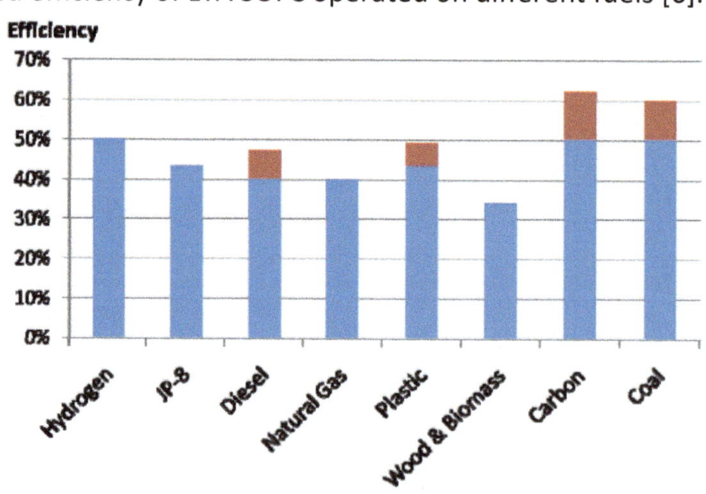

Figure 6. *Measured efficiency of LTA SOFC over the range of fuels [6]*

Electrochemical Characterisation of LTA SOFCs

The first studies on LTA SOFC developed by Cell Tech Power LLC were reported in 1998 [12] and subsequently mainly focused on applying LTA SOFC for distributed power generation using natural gas. The next step was to test alternative fuels such as waste plastics and military logistic fuels (*e.g.* JP-8) applicable for portable applications, followed by studies where biomass and coal are considered to be directly converted in LTA SOFC. Tao *et al.* [12] initially developed the Gen 3.0 cell and stack and demonstrated a power density of 40 mW cm^{-2} with JP-8 fuel. Cell performance was significantly improved in the next generation cell - Gen 3.1, which was lighter by a factor of four. The operational power density was 120 mW cm^{-2} with JP-8 fuel that contained 1400 ppm of sulphur [12]. This study confirmed the operation of LTA SOFC without fuel desulfurization or preprocessing. To overcome existing drawbacks of the previous cell design (Gen 3.0) an alumina porous separator was introduced to increase fuel mass transport to liquid tin. Prior to testing with JP-8, Gen 3.1 was tested with hydrogen to ensure proper cell operation (Figure 7 (left)).

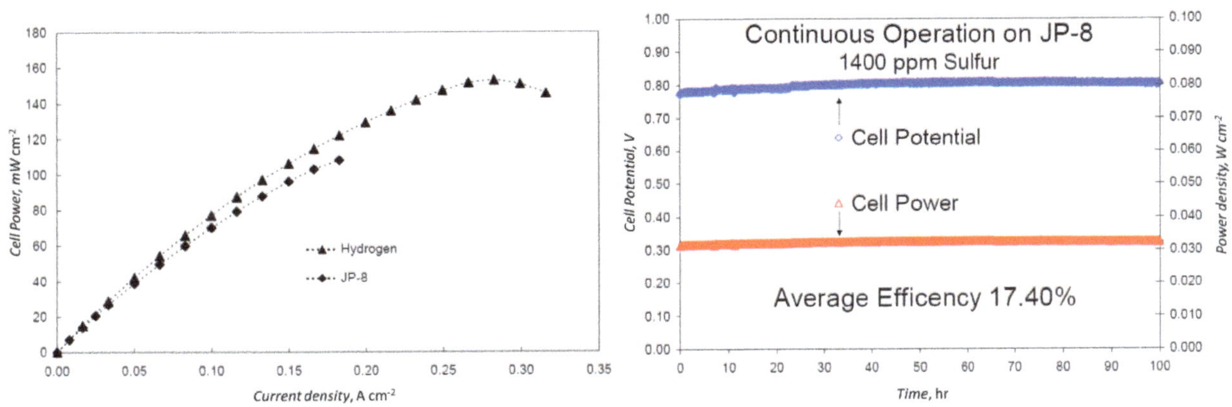

Figure 7. *Power curves for tin based cells operating on hydrogen and JP-8* ***(left)****; durability test on JP-8 (V, W)* ***(right)*** *[12]*

At equal operating conditions the cell fed with JP-8 showed 70-80% of the cell performance with hydrogen: the maximum current density for hydrogen was 316 mA cm^{-2} and 220 mA cm^{-2} for JP-8, corresponding to a maximum power density of 153 mW cm^{-2} and 120 mW cm^{-2} respectively. In terms of durability, Gen 3.1 with JP-8 demonstrated an average efficiency of 41.3% at 76 mW cm^{-2} for 1 hour. The whole durability test is illustrated in Figure 7 (right). The next generation of cells aims to reach >200 mW cm^{-2} power density with the military fuels. To achieve this a more detailed evaluation of separator properties such as porosity and stability for long-term operation is still needed [12]. It has been pointed out that interactions between the liquid tin anode and fuel have not been systematically studied and that better modelling and a fundamental study is needed.

Direct conversion of biodiesel has been demonstrated using the LTA SOFC [5]. In an approach similar to that adopted for JP-8, a tubular LTA SOFC cell was tested for 4.5 h with biodiesel (B100) prepared from virgin and waste cooking oils [5]. Peak power and current density values of 117 mW cm^{-2} and 217 mA cm^{-2} and overall cell efficiency of 40% were reported. The results of the study also verify the fuel flexibility of the LTA SOCF system.

An interesting approach of using a separate electrochemical reactor coupled to an external chemical reactor with liquid tin anode SOFC was recently proposed by Colet Lagrille *et al.* [26]. The system oxidizes carbon directly in fuel cell mode and produces H_2 in the electrolyser mode. The schematic of two separate reactors is illustrated in Figure 8. The approach has the advantage of allowing the 'combustion' (fuel oxidation) process and the electrochemical oxidation of the melt to be optimized in two separate reactors; and it also improves mass transport losses. The disadvantage is that pumping of liquid metal between the reactors is necessary.

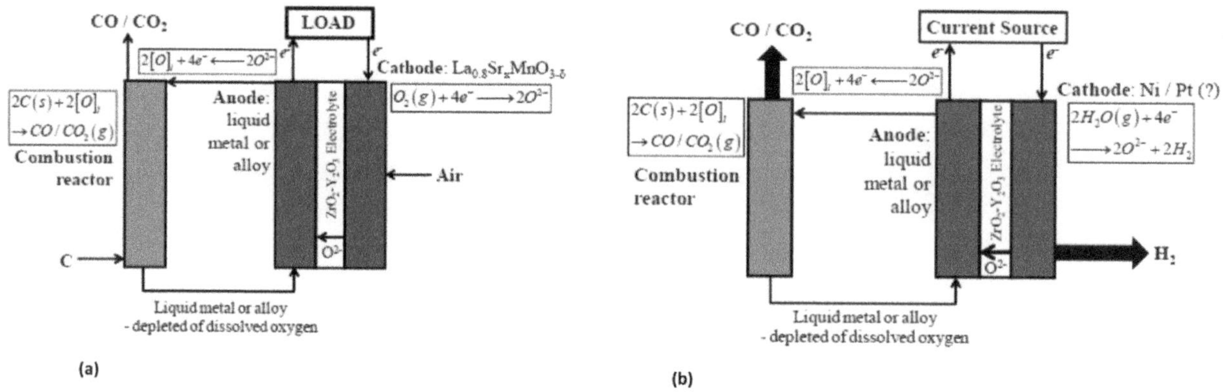

Figure 8. ***(a)*** *Schematic of a LMA SOFC coupled to an external combustion reactor,* ***(b)*** *Schematic of a LMA Solid Oxide Electrolyser (SOE) coupled to an external combustion reactor [26]*

This study also reports a 2D model of a micro-tubular hollow fiber fuel cell to predict distribution of oxygen concentration within the liquid tin anode [26]. The diffusion coefficient of oxygen in tin was assumed, the authors highlighted the need for accurate determination of this parameter for reliable model prediction.

Alternative metals and metal-alloy anodes for LME SOFC

Other metals and metal alloys have been demonstrated as interesting anode materials capable of direct carbon oxidation in novel SOFCs. As summarised earlier in Table 2, the most common liquid metal anodes are made of: tin, bismuth, antimony, lead, copper, as well as indium and silver.

Liquid Copper Anode SOFCs

Successful performance of LME direct carbon fuel cell (DCFC) has been demonstrated using a molten copper anode operating at 1373 K [21]. The overall conversion efficiency measured at maximum power density was 62.5 %.

Design of this system was different to those discussed earlier in the way that the cathode-supported YSZ tube was immersed into liquid copper media. Figure 9(a) shows the schematic presentation of the system [Pt, Cr Cermet, Cu-O (l) +C // (Y_2O_3) ZrO_2//$La_{0.9}Sr_{0.1}$ (Mn, Fe, Co)$O_{3-\delta}$ O_2 (air), Pt]. The fuel (coke particles) was added via an alumina inlet tube and stirred thoroughly with a stirrer; the entire cell was suspended in a vertical furnace. The OCV of the system was measured to be 1.2 V, and generated a maximum power density of 1.7 W cm^{-2} (2.25 A cm^{-2} at 0.75 V) as shown in Figure 9(b).

Despite good overall performance, liquid copper SOFCs are still difficult to operate. The relatively high vapour pressure of copper, compared to tin and other metals, is a challenge for practical operation. In addition, copper anodes necessitate very high operating temperatures (m.p. = 1357.8 K), which inevitably complicates operation of the system.

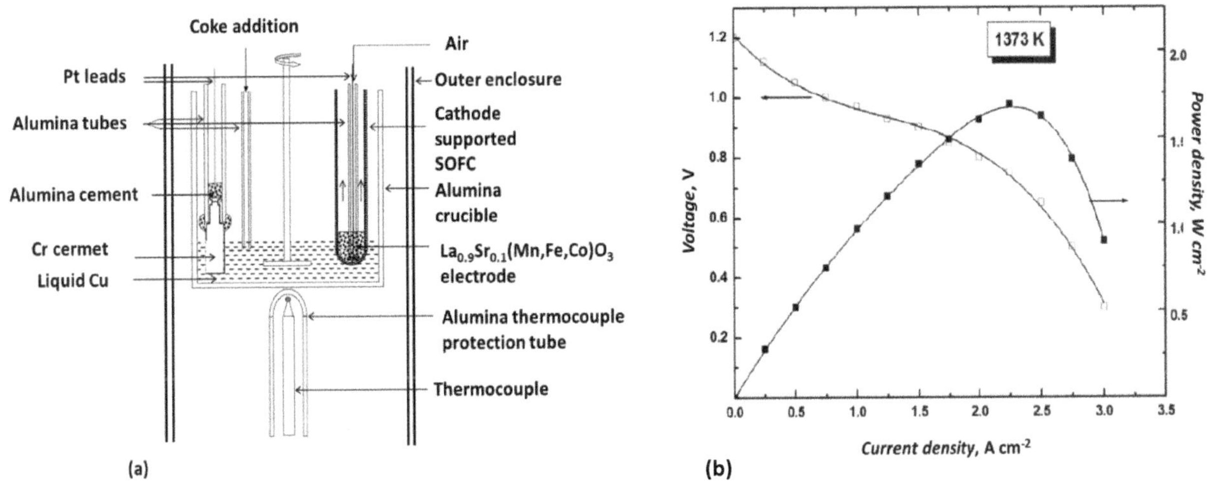

Figure 9: (a) *Schematic of liquid copper anode SOFC,*
(b) *Polarisation and power density curves for liquid copper anode SOFC[21]*

Liquid antimony

Liquid antimony has been assessed as an anode for DCFCs [29]. Operating at 973 K and employing a Sc-stabilized zirconia (ScSZ) electrolyte, the system, incorporating an air electrode but with no input of fuel, has demonstrated an open-circuit voltage (OCV) of 0.75 V for the redox couple Sb - Sb_2O_3. Having an anode electrode resistance of 0.06 Ω cm^2, the operated power density was 350 mW cm^{-2}. As expected, the addition of sugar char immediately increased the cell voltage as the Sb_2O_3 was reduced with fuel, preventing formation of a metal oxide layer on the electrolyte surface. Scaling up of such liquid antimony DCFC technology may require an increase in the rate of reaction between fuels and antimony oxide, as some carbonaceous fuels (carbon black, rice starch) have not demonstrated any interaction with liquid Sb [29].

Liquid bismuth

Liquid bismuth has been reported to have an inferior performance to tin anodes [25]. The OCV was reported to be slightly lower than theoretical (440 mV): 424 mV with hydrogen supplied and 408 mV without [25]. Being ion-conductive, bismuth oxide may solve the problem with oxide transport blockage through YSZ-anode media [26]. However, operating at lower power densities, liquid bismuth SOFCs require in-depth investigation and are not considered to be a feasible alternative to SOFCs with a tin anode [21,22] .

Other metal anodes for LME SOFCs

Indium (In), lead (Pb), antimony (Sb) [23] and silver (Ag) [30] have also been examined in battery and fuel cell modes (if feasible) to test their anodic efficiency in LME SOFCs. Similar to tin, the performance of indium was adversely affected by formation of metal oxide at the electrolyte interface when higher current was drawn. This was not the case for Pb and Sb, where the melting points of their metal oxides were lower than the operating temperature. Direct oxidation of coal has also been proposed to operate in a liquid iron anode SOFC [31].

The application of a liquid silver anode for DCFCs has been experimentally explored by Javadekar [30] who achieved an OCV of 1.12 V with charcoal. Liquid silver has lower melting point and higher oxide solubility compared to other metals (*e.g.* copper) and therefore is advantageous. Another approach was to use a Ag-Sb alloy, where the high solubility of oxide in silver might be combined with the low melting point of Sb, to improve overall anode performance [30].

Metal alloy anodes

In an effort to optimize anode properties, various alloy systems have been investigated. The work involving the pure metal anodes (bismuth) mentioned above was developed further to explore anodes of liquid tin-lead and liquid tin-lead-bismuth alloys in battery mode followed by direct oxidation of coal [25]. For the tin-lead alloy, performance degradation (from 11.7 mW cm^{-2} to 10.1 mW cm^{-2}) occurred within 13 hours, with an increase in performance to 14 mW cm^{-2} when coal was introduced. The addition of bismuth to tin-lead alloy decreased the overall anode effectiveness.

Uncertainties and technological challenges of LME SOFC

Prior to commercialization of liquid metal anode solid oxide fuel cells, greater understanding of the electrochemical reactions, redox and transport processes within liquid metal electrodes is needed [20]. The following research challenges are particularly highlighted: the nature of oxidised species in liquid metal electrode media; transport of oxygen or oxidized species in LME SOFC; reaction kinetics between oxide and fuel within liquid anode; kinetics of oxygen transfer at liquid metal anode - electrolyte interface; oxygen (oxide) diffusivity / transport through liquid metal layer and solubility of oxide and fuels in the melt.

As argued by Abernathy *et al.* [4], improvement of oxygen transport within liquid metal media can be achieved by increasing oxygen solubility through alloying. Wetting characteristics of liquid metal with the electrolyte and fuel oxidation kinetics need also to be considered.

Most of the work so far has been focused on the testing of single cells, which limits its further application. In this regard, it would be beneficial to model the performance of LME SOFCs stacks and prototype based on derived design guidelines.

In order to be able to model and later design practical LME SOFC systems the information on oxygen solubility and transport in liquid metals is essential as it is a central process that may limit the performance of the fuel cell. Fundamental investigation of these interactions within a liquid anode, as well as analysis of other physical metal properties with temperature (*e.g.* vapour pressure, melting point, surface tension, contact angle, *etc.*) will be beneficial in designing and building durable and stable fuel cells.

Other important factors requiring investigation include: aspects concerned with the introduction of solid fuel into the melt; optimization of thickness of the melt; prevention of unfavorable formation of metal oxide tending to result in blocking of the electrolyte; fundamental studies concerned with transportation of the oxide from the generation zone at the interface with the electrolyte to the reaction zone at the interface with the fuel (relevant aspects include melt thickness, diffusion of oxidised species and convection in the melt); effect and optimization (which may include surface structuring) of contact area between the liquid anode and electrolyte.

Engineering considerations include: matching of electrolyte composition to temperature of operation; choice of electrolyte thickness based upon optimization between mechanical properties and ionic conductivity; agitation of the melt; generation of suitable electrolyte surface structures; cell design; scale-up of laboratory cells to pilot and then to commercial plant.

Conclusions

LMA SOFCs are a highly efficient option for direct electrochemical conversion of solid fuels into electrical energy. A range of possible metal anodes have been reviewed with tin the material on which most development work has been performed. Despite numerous studies, the understanding of the reaction kinetics at liquid metal-fuel and liquid metal-electrolyte interfaces, as well as oxygen diffusivity through the liquid metal layer needs further systematic investigation.

Acknowledgements: *The authors wish to thank Nazarbayev University and the Government of Republic of Kazakhstan for the BOLASHAK International Scholarship for Aliya Toleuova and the EPSRC Supergen Fuel Cells programme (EP/G030995/1) for supporting Brett's research.*

References

[1] International Energy Agency, World Energy Outlook 2010.
[2] C. Gough, S. Shackley, UK Technical report **47** (2005) 1–256.
[3] D. Cao, Y. Sun, G. Wang, *J. Power Sources* **167** (2007) 250–257.
[4] H. Abernathy, R. Gemmen, K. Gerdes, M. T. Koslowske, T. Tao, *J. Power Sources* **196** (2011) 4564–4572.
[5] W. A. McPhee, M. Boucher, J. Stuart, R. S. Parnas, M. T. Koslowske, T. Tao, B. A. Wilhite, *Energy Fuels* **23** (2009) 5036–5041.
[6] T. Tao, M. T. Koslowske, W. A. McPhee, L. S. Bateman, M. J. Slaney, J. Bentley, Coal Based Fuel Cell Technology: Status, Needs and Future Applications, Liquid Tin Anode Solid Oxide Fuel Cell for Direct Conversion of Carbon Based Fuels Including Coal, Morgantown WV, USA, 2007, p. 93–103.
[7] T. Tao, M.T. Koslowske, J. Bentley, 219th ECS Meetings, Liquid Tin Anode SOFC for Direct Fuel Conversion - Cell, Stack and System Development, Montreal, Canada, 2011, abstract No.820.
[8] S. Singhal, K. Kendall, High Temperature Solid Oxide Fuel Cells Fundamentals, Design and Applications, Elsevier, Oxford, UK, 2003, p. 89–92.
[9] T. S. Li, W. G. Wang, *J. Power Sources* **196** (2011) 2066–2069.
[10] D. J. Brett, M. N. Manage, E. Agante, N. P. Brandon, E. R. Brightman, J. C. Brown, I. Staffell, in *Functional materials for sustainable energy applications*, J. A. Kilner, S. J. Skinner, S. J. Irvine, P. P. Edwards, Woodhead Publishing Limited, Cambridge, UK, 2012, p. 249–274.
[11] T. Tao, L.S. Bateman, J. Bentley, M. J. Slaney, *ECS Trans.* **5** (2007) 463–472.
[12] T. Tao, W. A. McPhee, M. T. Koslowske, L. S. Bateman, M. J. Slaney, J. Bentley, *ECS Trans.* **12** (2008) 681–690.
[13] T. Tao, M. J. Slaney, L. S. Bateman, J. Bentley, *ECS Trans.* **7** (2007) 1389–1397.
[14] R. O'Hayre, S. Cha, W. Colella, F. B. Prinz, *Fuel Cell Fundamentals,* John Wiley & Sons, New Jersey,USA, 2006, p. 10–12.
[15] S. Kakac, A. Pramuanjaroenkij, X. Y. Zhou, *Int. J. Hydrogen Energy* **32** (2007) 761–786.
[16] Energy Efficiency & Renewable Energy, http://www1.eere.energy.gov/hydrogenandfuel-cells/fuelcells/pdfs/fc_comparison_chart.pdf (accessed on 1.12.2012).
[17] M. N. Manage, D. Hodgson, N. Milligan, S. J. Simons, D. J. Brett, *Int. J. Hydrogen Energy* **36** (2011) 5782–5796.
[18] N. P. Brandon, D. J. Brett, *Philos. Trans. R. Soc. London, Ser. A* **364** (2006) 147–159.
[19] P. R. Shearing, D. J. Brett, N. P. Brandon, *Int. Mater. Rev.* **55** (2010) 347–363.
[20] R. Gemmen, H. Abernathy, K. Gerdes, M. T. Koslowske, W. A. McPhee, T. Tao, in *Advances in Solid Oxide Fuel Cells V,* N. P. Bansal, P. Singh, John Wiley & Sons, New Jersey, USA, 2009, p. 37–52.
[21] K. T. Jacob, *ECS Trans.* **35** (2011) 573–582.
[22] A. Javadekar, A. Jayakumar, R. J. Gorte, J. M. Vohs, D. J. Buttrey, *J. Electrochem. Soc.* **159** (2012) A386–A389.
[23] A. Jayakumar, J. M. Vohs, R. J. Gorte, *Ind. Eng.Chem. Res.* **49** (2010) 10237–10241.

[24] A. Jayakumar, S. Lee, A. Hornés, J. M. Vohs, R. J. Gorte, *J. Electrochem. Soc.* **157** (2010) B365–B369.

[25] M. LaBarbera, M. Fedkin, S. Lvov, *ECS Trans.* **35** (2011) 2725–2734.

[26] M. Colet Lagrille, U. Doraswami, G. H. Kelsall, *ECS Trans.* **42** (2012) 137–148.

[27] T. Tao, M. T. Koslowske, J. Bentley, J. Brodie, C. MacKean, *ECS Trans.* **41** (2012) 125–135.

[28] S. C. Rayman, M. T. Koslowske, L. S. Bateman, T. Tao, R. E. White, *ECS Trans.* **33** (2011) 93–121.

[29] A. Jayakumar, R. Küngas, S. Roy, A. Javadekar, D. J. Buttrey, J. M. Vohs, R. J. Gorte, *Energy Environ. Sci.* **4** (2011) 4133–4137.

[30] A. D. Javadekar, Ph.D. Thesis ,University of Delaware (2012).

[31] I. V. Yentekakis, P. G. Debenedetti, B. Costa, M. Konsolakis, V. Kiousis, *Ionics*, **5** (1999) 460–471.

14

Development of an electrochemical sensor for the determination of the total antioxidant capacity in berries based on boron doped diamond

BRUNA PEKEC*,***, BIRGIT FEKETEFÖLDI**, VOLKER RIBITSCH**, ASTRID ORTNER*** and KURT KALCHER*✉

**Institute of Chemistry, Department of Analytical Chemistry, Karl-Franzens University, Graz, Austria*

***Joanneum Research Forschungsgesellschaft mbH - Materials, Weiz, Austria*

****Institute of Pharmaceutical Science, Department of Pharmaceutical Chemistry, Karl-Franzens University, Graz, Austria*

✉Corresponding Author

Abstract

Many antioxidants can be electrochemically oxidized using graphite-based electrodes; nevertheless problems arise due to the strong adsorption of redox species at the sensing area. We have demonstrated that boron doped diamond (BDD) electrodes do not show this property, which can be exploited for the design of a new amperometric sensor for the quantification of antioxidants as "total antioxidant capacity" (AOC). As reference substances hydroquinone (HQ) and 6-hydroxy-2,5,7,8-tetramethylchromane-2-carboxylic acid (Trolox) were studied in more detail. The supporting electrolyte was a phosphate buffer solution (PBS, 0.1 mol/L, pH 7.0). The limits of detection (LOD) were 1.5 mg/L and 2.5 mg/L for HQ and Trolox, respectively. The repeatability was 3 % RSD for concentration of 200 mg/L HQ. The method could be applied for the determination of AOC in different berry samples, such as strawberry, blueberry, grape and bramble. A comparison with a standard photometric assay showed good correlation between both methods. The BDD sensor features good reproducibility without fatiguing over at least two months of operation.

Keywords

Boron doped diamond electrode; total antioxidant capacity; amperometric sensor; berry extracts.

Introduction

The term "antioxidant" identifies a huge number of substances, in majority of cases phenolic or polyphenolic substances that inhibit the oxidation of molecules. They are strongly reducing agents that neutralize reactive oxygen species (ROS). These ROS are oxidizing radicals or molecules that are formed during the respiratory chain and cause damage on cell own material. Antioxidants neutralize these species and therefore they are of enormous scientific interest.

The free scavenger effect of antioxidants has been in use for many years to prolong the stability of lipid foods especially of edible oils. Reports go back to 1947 where butylated hydroxyanisole was used as the first antioxidant for the prevention of the deterioration of fat-containing food [1-2]. Nowadays antioxidants are also in the medicinal focus because of their tissue protecting effects by neutralization of ROS. Therapeutic progress has been made with many diseases that show a high degeneration rate, such are diabetes, coronary heart failure or Alzheimer disease [3-5]. Ever since in 2003 Sinclair *et al.* found out that resveratrol, an antioxidant occurring in grape, extends the lifespan of yeast cells and causes a decrease of typical aging effects, antioxidants became also interesting for cosmetic branches and for anti-aging research [6-7].

Antioxidants are mostly secondary plant metabolites. Especially berries, such as grape, red or black currant, bramble or raspberries, are known to form high amounts of flavonoids. Their synthesis is a response of the plant to stress attacks like strong light exposure or fungal or microbial infection. In the latter case, the plant uses them as phytoalexines for tissue protection. Unfortunately, isolated antioxidants are highly sensitive to light, temperature and oxygen. Although natural antioxidants are an interesting tool to prevent lipid products, e.g. oily food or cosmetics, to become rancid, their stabilization is still a problem to be solved.

Responding to a growing interest some analytical methods, mostly photometric assays, were developed. Standardized techniques are the ORAC-assay (oxygen radical absorbance capacity) involving hydrogen transfer reactions, the Folin-Ciocalteu-assay and the TEAC-assay (Trolox equivalent antioxidant capacity) that base on electron transfer reactions [8]. Although these methods have been well approved they are rather complex, time-consuming and use high priced chemicals or test kits.

Most antioxidants show electrochemical activity and could be determined using voltammetric or amperometric methods. Unfortunately, during the electrochemical process reduced and oxidized species adsorb very strongly on the surface of many electrode types, such as graphite or carbon paste electrodes (CPE).

Diamond electrodes offer an alternative approach. Ever since in 1987 Pleskov *et al.* [9] commenced studies on the electrochemistry of boron doped diamond (BDD) their development and fundamental research gained rising tendency. In contrast to graphite electrodes the C-sp^3 hybridization and tetrahedral bonding of boron doped diamond electrodes (BDDE) leads to properties like chemical inertness, hardness, thermal conductivity and low fatigue [10]. Their main field of application is the treatment of wastewater [11]. The potential window of BDDEs is much wider (-0.75 to +2.35 V) than of graphite electrodes (-0.5 to +1 V) because neither oxygen nor hydrogen evolution interfere in the analysis. Because commercial BDDEs are usually synthesized by chemical vapor deposition (CVD) they contain non-diamond and metallic impurities [10]. Also, surface oxygen can cause disturbances and it needs to be removed. For surface cleaning and pretreatment of BDDEs a few suggestions, like acid washing or anodic polarization, have been offered [12,13].

In this work the pretreatment of boron doped diamond electrodes by polarization in Na_2SO_4 at +1 V, and the use of this electrode for the electrochemical determination of antioxidants in ethanolic berry extract, was investigated.

Experimental

Chemicals and materials

The investigated target analytes were phenolic and polyphenolic ingredients of different berries (grape, bramble, blueberry, strawberry) that are known to show antioxidative potential [14]. Therefore ferulic acid, vanillic acid, syringic acid, coumaric acid, sinapic acid, caffeic acid and *p*-hydroxybenzoic acid, all purchased from Sigma-Aldrich GmbH (Vienna, Austria), were selected. Trolox [(±)-6-Hydroxy-2,5,7,8-tetramethylchromane-2-carboxylic acid] and hydroquinone (HQ) were chosen as reference substances and were purchased, like all other standards, from Sigma-Aldrich GmbH (Vienna, Austria). To prepare the standard solutions the appropriate amount of substance was dissolved in deionized water. The aqueous solutions were prepared freshly before use, kept cool and protected from light. Phosphate buffer (Na_2HPO_4/NaH_2PO_4 0.1 mol/L, pH 7) was selected as electrolyte medium.

Apparatus

All electrochemical experiments were performed using potentiostat (Metrohm AUTOLAB® PGstat302N) in a three electrode arrangement controlled by the corresponding software (NOVA 1.6). The BDDEs purchased from Fraunhofer (Michigan, USA), plated on titanium substrate with gold as a contact, a saturated calomel electrode (SCE) and a platinum wire were used as working, reference and auxiliary electrode respectively. All potential values in this work are given against SCE.

Cyclic voltammetry and hydrodynamic amperometry were chosen as techniques to obtain the electrochemical data. In the case of cyclic voltammetry the explored potential window was between -1 V and +1.2 V. The other operating parameters were: scan rate: 50.0 mV/s, step potential: 2.0 mV and time interval: 0.050 s. For hydrodynamic amperometry measurements the characteristic potential of the oxidation current peak was selected.

Measuring procedure

The BDDE was conditioned at +1 V for 1 h in 0.1 mol/L Na_2SO_4 solution prior to the start and also between the measurements to remove impurities and for the baseline correction.

All experiments were carried out in 20 mL phosphate buffer (0.1 M, pH 7) after degassing with argon. Aliquots of the standard solutions were added to the cell and cyclic voltammograms and hydrodynamic amperograms, respectively, were recorded by applying the aforementioned parameters.

Photometric analysis

The photometric Trolox Equivalent Antioxidant Capacity assay (TEAC-Assay) was selected as a reference method. According to the report of Li *et al.* [15] the redox indicator 2,2′-Azinobis-(3-ethylbenzthiazolin-6-sulfonic acid) (ABTS) forms a radical in contact with potassium persulfate. The ABTS radical cation is of green color which decreases in intensity upon addition of substances with antioxidative potential. The solution was diluted using phosphate buffer saline (pH = 7) until it showed an absorption of 0.700 ± 0.050 at 734 nm. To 3 mL of that solution 100 μL of the sample were added and the absorption decrease was measured. For the calibration Trolox and HQ solutions of 0.1, 0.2, 0.3 and 0.4 mol/L were used and the AOC of the berry samples was calculated as Trolox- or hydroquinone-equivalent (TE, HQE).

Preparation of the samples

The berry marcs (donated from Grünewald Fruchtsaft GmbH) were dried at 50 °C. For the recovery of the phenolic compounds the milled raw material was extracted with a mixture of absolute ethanol and water (75:25 v/v) at a 8:1 solvent-to-sample ratio (volume to mass of oven-dried marc). The extraction was performed using sonication at a temperature of 50 °C for 15 min. After centrifugation (7000 rpm, 20 min) the extract was then filtered through a glass fiber filter (type AP2029325, Millipore) and stored at 4 °C.

Results and Discussion

Cyclic voltammetry

Cyclic voltammetry (CV) was basically used to determine the oxidation potential of the standard berry antioxidants with CPEs and BDDEs (Table 1). The quoted potentials correspond to the maximum of the oxidation current registered in the anodic scan. Typical CVs of hydroquinone and Trolox are shown in Fig. 1. As can be seen the electrochemical behavior of both substances on both electrodes is rather similar with the exception of carbon paste electrode where the currents are significantly higher. This behavior is due to the adsorption of the analyte at CPE and therefore due to the strong π electron interaction of the aromatic ring with the graphite structure. Both substances were used as reference compounds, by converting the response of other antioxidants to Trolox- (TE) and hydroquinone-equivalents (HQE), respectively.

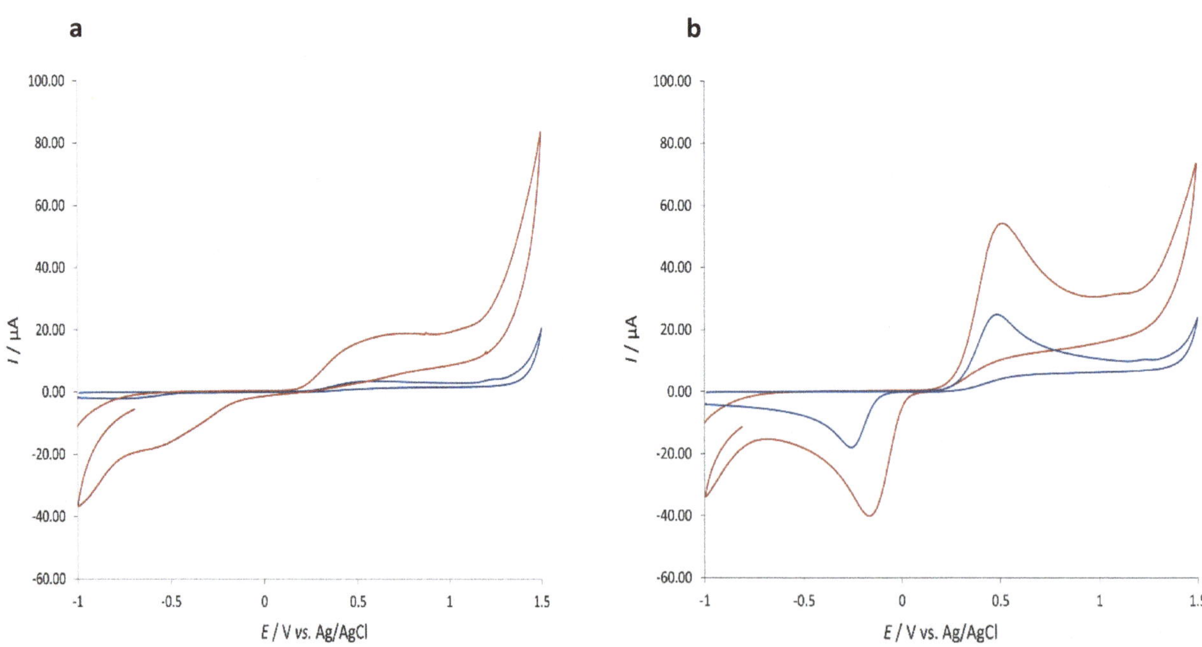

Figure 1. *Cyclic voltammograms of 30 mg/L **(a)** Trolox and **(b)** HQ measured in PBS (0.1 mol/L, pH 7) on • BDDE and • CPE.*

The investigated antioxidants (Table 1) produce typical peak currents in CV at BDDEs in the range of 4 to 8 mA mol^{-1} with the exception of HQ which is about four times higher; probably because HQ ideally interacts with the graphite surface finally yielding benzoquinone, whereas other oxidants are reduced to the corresponding R-O· radical with ensuing follow-up effects. Again it is evident that carbon paste electrode favors adsorption of the oxidation product which exerts a positive effect for the first cycle in cyclic voltammetry but it is deteriorating the signals in further scans and in amperometry (see below) by blocking the active surface area.

Since the sensors based on boron-doped diamond should be used for alcoholic extracts of berries, the influence of ethanol on the signal was investigated. Concentrations from 0.5 to 20 % ethanol were added to the supporting electrolyte; no influence on peak position or peak height was observed.

Table 1. *Oxidation potentials, E_{pa}, and peak currents, i_{pa}, from cyclic voltammetry of typical antioxidants found in berries. Concentration: 30, respectively 90 mg/L, supporting electrolyte: PBS (0.1 mol/L, pH 7); potential window: -1.5 to +1.5 V, scan rate: 50 mV/s.*

Substance	**Formula**	**CPE**		**BDDE**	
		E_{pa} / V	i_{pa} / mA mol^{-1}	E_{pa} / V	i_{pa} / mA mol^{-1}
Syringic acid	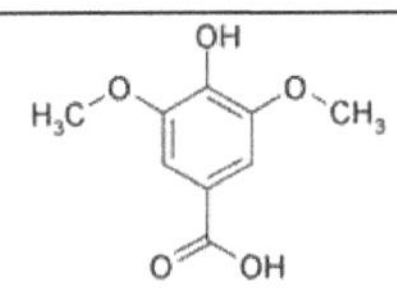	0.95	23.3	0.80	8.2
Sinapic acid	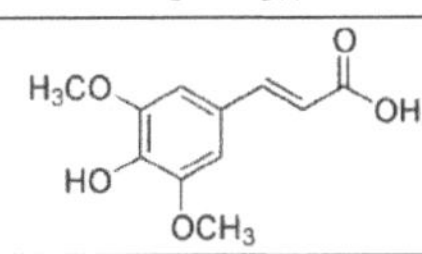	0.70	15.2	0.65	4.3
Caffeic acid	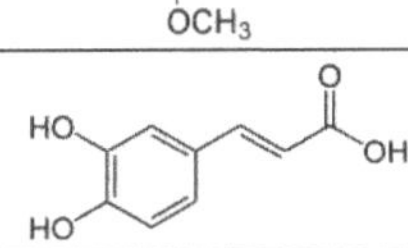	0.48	10.2	0.50	7.5
Vanillic acid	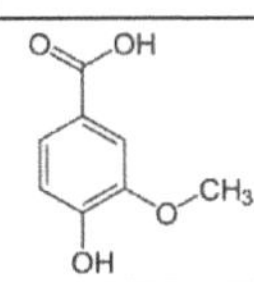	0.88	78.6	0.88	4.9
Ferulic acid	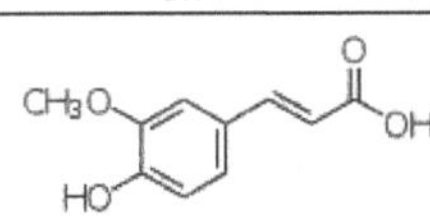	0.50	19.2	0.50	7.9
Coumaric acid	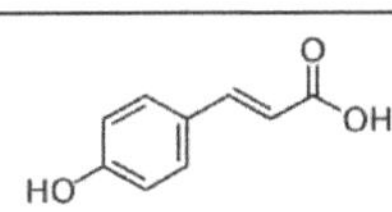	0.90	37.5	0.75	5.1
p-Hydroxybenzoic acid	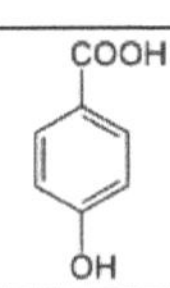	0.90	148.5	1.0	6.9
Trolox	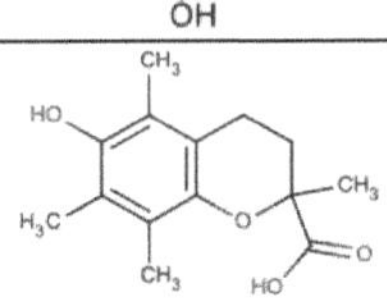	0.50	14.2	0.50	6.2
Ascorbic acid	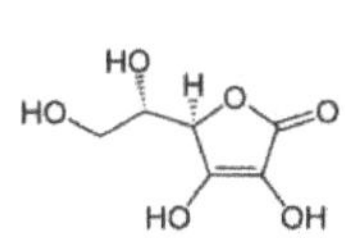	0.61	56.4	0.54	36.4
Hydroquinone	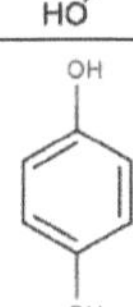	0.50	44.0	0.50	26.7

Hydrodynamic amperometry

Basic studies

Hydrodynamic amperometry was applied to the antioxidants listed in Table 1 applying a working potential which corresponds to the maximum of the oxidation current in CV. After attaining a stable baseline the analytes were added to a stirred solution.

Figure 2 shows a comparison of the electrochemical behavior of HQ between BDDE (a) and CPE (b) electrodes under identical conditions. As can be seen a strong adsorption of the quinone on the CPE leads to a fast decrease of the signal due to blocking of the active electrode surface area with a very small current steps after ensuing additions. BDDE does not show any obvious adsorption therefore, after addition of the electrochemically active analyte, the signal remains constant. Repetitive additions lead to clear step-like increases of current. For this reason BDDE seems to be an ideal electrode material for the determination of phenolic antioxidants due to the absent tendency for adsorption.

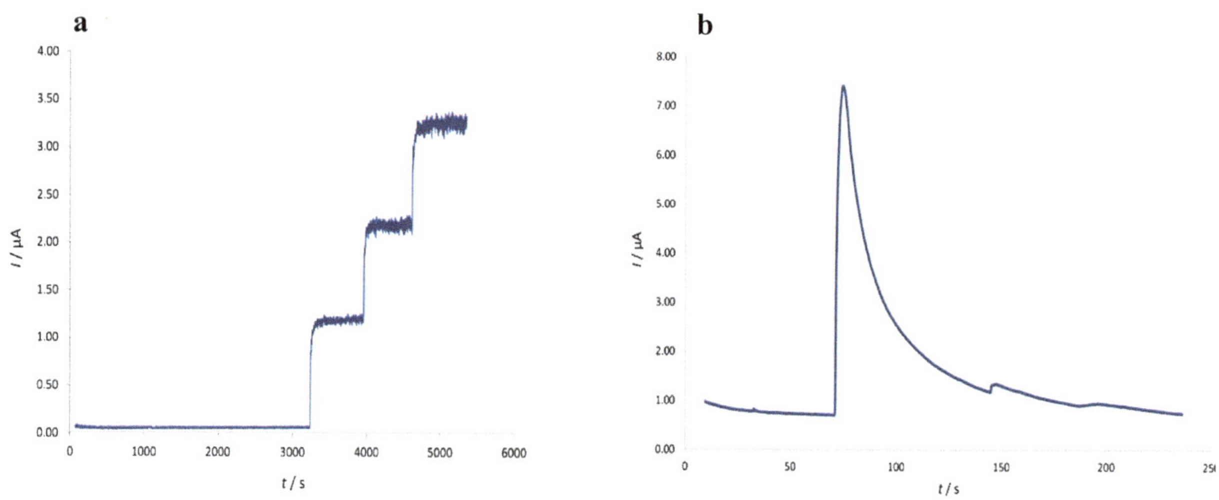

Figure 2. *Hydrodynamic amperograms of subsequent additions of 30 mg/L HQ in phosphate buffer (0.1 M, pH 7) using **(a)** BDDE and **(b)** CPE. Working potential: +1V.*

Validation of analytical parameters

The analytical method based on hydrodynamic amperometry was evaluated with hydroquinone and Trolox as standards. Under *optimized* conditions the linearity range of the BDD-sensor was estimated to be between 10 and 400 mg/L for Trolox and 10 to 300 mg/L for HQ (Fig. 3). The calibration curves can be described by the following regression equations:

$i_{\mathrm{HYDROQUINONE}}$ / μA = 43.04 × γ / (mg/L) + 0.387 and

i_{TROLOX} / μA= 22.09 × γ / (mg/L) + 0.609,

both with a mean correlation coefficient of $R = 0.999$. For the concentrations above 400 mg/L (Trolox) and 300 mg/L (HQ) the calibration graph levels off but still remains linear to concentrations above 1 g/L, in fact with a very small sensitivity. The limit of detection (LOD) and the limit of quantification (LOQ estimated as the three-fold of LOD) were determined as 1.5 mg/L and 4.5 mg/L for HQ, and 2.5 mg/L and 7.5 mg/L for Trolox according to the International Conference on Harmonisation, ICH, Quality Guidelines for Analytical Evaluation [Q2(R1)] guidelines for visual evaluation [16].

The precision experiments (100, 200, 375, 625, 750 and 875 mg/L; n = 4) showed a relative standard deviation (RSD) of 3-8 % and a RSD of the slope of 3 %. The intercept of the calibration curves ranged from around 0.4 (HQ) to 0.6 μA (Trolox) with a RSD of about 5 % in different calibrations (n = 4).

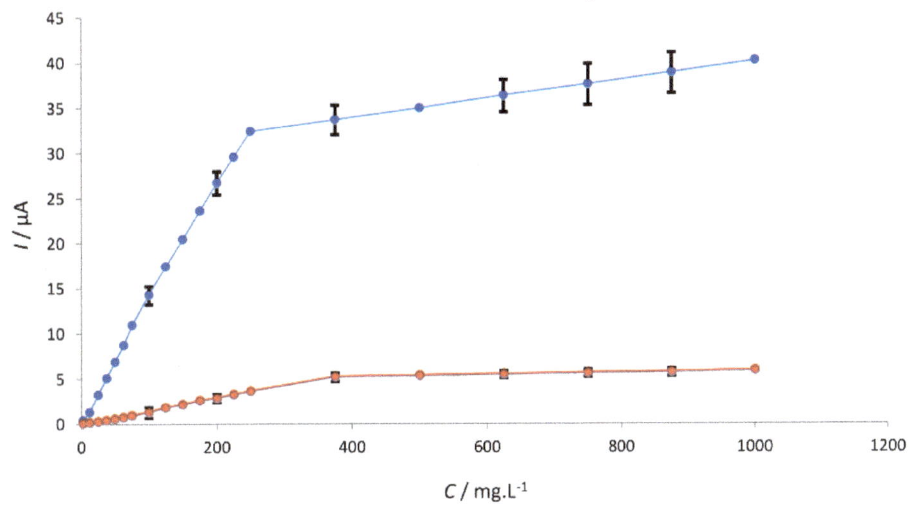

Figure 3. *Calibration curve of the BDD-sensor 10-1200 mg/L concentration range of • HQ and • Trolox.*

Application of the sensor in samples

After it was demonstrated that the BDDEs showed only insignificant adsorption effects to the redox products of the investigated antioxidants they were applied to chosen samples. Thus, ethanolic extracts were prepared using red grape, bramble, blueberry and strawberry. The extracts were injected directly to the analytical cell without further dissolution or preparation. 20 mL of phosphate bufferwas transferred into the electrochemical cell and after the baseline stabilization, 100 μL of the extract was injected. The extracts were analyzed using hydrodynamic amperometry at +1 V and the AOC was estimated as Trolox equivalent (TE) or hydroquinone equivalent (HQE) using the corresponding calibration curves. The capacities of the berry samples were determined in the range of 8 and 23 mM TE/HQE (Table 2) and are in a good agreement with the results of the photometric reference method.

Table 2. *Comparison of the electrochemical and photometric total antioxidant capacity measurements of berry extracts.*

	Trolox equivalent, mmol/L		Hydroquinone equivalent, mmol/L	
	TEAC Assay	Electrochemical BDD sensor	TEAC Assay	Electrochemical BDD sensor
Grape	23.4 ± 1.2	23.9 ± 1.2	22.6 ± 1.1	23.3 ± 1.2
Bramble	12.4 ± 0.6	12.4 ± 0.6	12.1 ± 0.6	13.4 ± 0.7
Blueberry	18.5 ± 0.9	18.1 ± 0.9	18.4 ± 0.9	17.9 ± 0.8
Strawberry	8.9 ± 0.4	8.9 ± 0.4	8.8 ± 0.4	9.1 ± 0.5

Figure 4 shows a comparison between electrochemical signals determined using the BDD sensor and the decrease of absorption (ΔA) measured using the TEAC-Assay. It can be seen clearly that there is a directly proportional relation between the absorption signal decrease and amperometric currents. The method employing electrochemical sensors seems well applicable to detect the antioxidant capacity of ethanolic berry extracts.

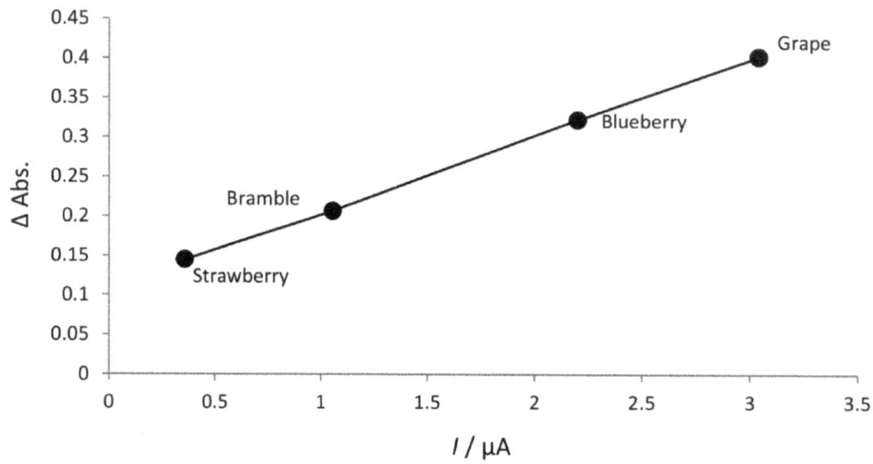

Figure 4. *Comparison of electrochemical data recorded in current intensity and photometric results charted as decrease of absorption.*

Our studies show that the evaluated Trolox and hydroquinone equivalents from amperometric measurements with BDDEs are correlating well with the optical reference method (Table 2). This was also verified with individual antioxidants (Table 1) which produce the same antioxidant equivalent capacities with both procedures, amperometric and optical.

Maybe, in complex samples, substances which can be oxidized at an operational potential of 1.0 V could be present and do not contribute to the antioxidant capacity. In fact, we could not, so far, observe such a behavior with ethanolic extracts of berries.

Conclusions

The investigations show that boron doped diamond used as working electrode in electrochemical analytics is an attractive tool because of its many advantages. Due to its low feature for the adsorption of redox products it can be used for quantification of antioxidants even in complex material. The designed sensor is inexpensive because it can be used for repetitive measurements, easy to prepare, and with high reproducibility .

Acknowledgements: *The financial support of the Österreichische Forschungsförderungsgesellschaft (FFG) for project "Antiflavo" (No.827334) is highly acknowledged.*

References

[1] E.R. Sherwin, *J. Am. Oil Chem. Soc.* **55** (1978) 809-814
[2] E.N. Frankel, *Food Chem.* **57** (1996) 51-55
[3] H. Kaneto, Y. Miyagawa, T. Matsuoka, Y. Fujitani, Y. Umayahara, T. Hanafusa, Y. Matsuzawa, Y. Yamasaki, M. Hori, *Diabetes* **48** (1999) 2398-2406
[4] K.L. Hamilton, J.L. Staib, T. Phillips, S.L. Lennon, S.K. Powers, *Free Radical Biol. Med.* **34** (2003) 800-809
[5] M.J. Engelhart, M.I. Geerings, A. Ruitenberg, J.C. van Swieten, A. Hofman, J.C.M. Wittermann, M.M.B. Breteler, *J. Amer. Med. Assoc.* **287** (2002) 3223-3229
[6] K.T. Howitz, K.J. Bittermann, H.Y. Cohen, D.W. Lemming, S. Lavu, J.G. Wood, R.E. Zipkin, P. Chung, Kisielewski, L.L. Zhang, B. Scherer, D.A. Sinclaire, *Nature* **425** (2003) 191-196

[7] D. Valenzano, E. Terzibasi, T. Genade, A. Cattaneo, L. Domenici, A. Cellerino, *Curr. Biol.* **16** (2006) 296-300
[8] R.L. Prior, X. Wu, K. Schaich, *J. Agr. Food Chem.* **53** (2005) 4290-4302
[9] Y.V. Pleskov, A.Y. Sakharova, M.D. Krotova, L.L. Bouilov, B.V. Spitsyn, *J. Electroanal. Chem.* **228** (1987) 19-27
[10] J.H.T. Luong, K.B. Male, J.D. Glennon, *Analyst* **134** (2009) 1965-1979
[11] M.A.Q. Alfaro, S. Ferro, C.A. Martinez-Huitle, Y.M. Vong, *J. Brazil. Chem. Soc.* **17** (2006) 227-236
[12] M.C. Granger, M. Witek, J. Xu, J. Wang, M. Hupert, A. Hanks, M.D. Koppang, J.E. Butler, G. Lucazeou, M. Mermoux, J.W. Strojek, G.M. Swain, *Anal. Chem.* **72** (2000) 3793-3804
[13] H.B. Suffredini, V.A. Pedrosa, L. Codognoto, S.A.S. Machado, R.C. Rocha-Filho, L.A. Avaca, *Electrochim. Acta* **49** (2004) 4021-4026
[14] M. Monagas, B. Bartolomé, C. Gómez-Cordovés, *Food Sci. Nutr.* **45** (2005) 85-118
[15] H.B. Li, K.W. Cheng, C.C. Wong, K.W. Fan, F.Chen, Y. Jiang, *Food Chem.* **102** (2007) 771-776
[16] Validation of analytical procedures – Methodology, *Federal Register* **62** (1997) 27463

15

A multichannel frequency response analyser for impedance spectroscopy on power sources

QUENTIN MEYER, SIMON BARASS, OLIVER CURNICK*, TOBIAS REISCH*, DANIEL J. L. BRETT✉

Electrochemical Innovation Laboratory, Centre for CO_2 Technology, Department of Chemical Engineering, University College London, UK
**Intelligent Energy Ltd, Loughborough, Leicestershire, UK*
✉Corresponding Author

Abstract
A low-cost multi-channel frequency response analyser (FRA) has been developed based on a DAQ (data acquisition)/LabVIEW interface. The system has been tested for electric and electrochemical impedance measurements. This novel association of hardware and software demonstrated performance comparable to a commercial potentiostat / FRA for passive electric circuits. The software has multichannel capabilities with minimal phase shift for 5 channels when operated below 3 kHz. When applied in active (galvanostatic) mode in conjunction with a commercial electronic load (by discharging a lead acid battery at 1.5 A) the performance was fit for purpose, providing electrochemical information to characterize the performance of the power source.

Keywords
LabVIEW, multi-channel impedance, electrical circuit, fuel cell, lead acid battery.

Introduction

Electrochemical impedance spectroscopy (EIS) is a powerful diagnostic technique that has been the subject of significant technical development over the last fifty years [1–3]. It has proven to be particularly useful in the fields of electrochemistry, corrosion [4], primary and secondary batteries [5,6] and fuel cells [7–9]. Advances in electronics and reduction in the cost of hardware has made the technique increasingly popular. However, EIS is still mainly used as a high-end lab-based diagnostic that requires relatively expensive hardware. This has hindered its uptake as a diagnostic tool used in routine industrial analysis or for on-line measurements of electrochemical power systems. For applications that would benefit from multichannel input (on different cells in a fuel cell stack, for example) the cost and complexity multiplies. Therefore, there is a need for a low-cost, modular approach for electrochemical impedance spectroscopy analysis.

Alternative methods for performing multichannel EIS analysis have been attempted previously using Matlab GUI Builder [10] and a built-in 10 channels frequency analyser. Single channel EIS using commercially available DAQ (data acquisition) hardware and a LabVIEW interface has been attempted for fuel cell applications [11], but multichannel metrology has yet to be reported.

This paper details the development of UCL-FRA, a low-cost EIS system based on the use of a commercially available DAQ interface and LabVIEW software (DAQ/LabVIEW interface). Issues relating to the successful application of the technique are discussed, results of its operation in galvanostatic mode applied to a battery are presented and operation in multi-channel mode described.

Experimental

Software

The software used for data collection and processing was developed using LabVIEW 2011 (National Instruments) and was based on the use of VIs (virtual instruments) in the Sound and Vibration Toolkit library for generating and analyzing AC waveforms. The coding methodology required special consideration to account for low frequency detection, simultaneous sampling and multichannel responses. The multichannel response was achieved by outputting single discrete sine waves sequentially across the frequency range of choice, and measuring on 5 response channels simultaneously. The software is currently capable of 5 response channels, but can be modified to a higher number of channels to suit the application, provided that the performance of the hardware is adequate to ensure robust data collection. The processed signals are displayed as Nyquist and Bode plots in real time as the data is collected; the time for the software to process the signal does not impact on the rate at which the measurement can be made. An entire measurement from 10 kHz to 0.1 Hz with 10 points per decade, takes about 400 s, which is comparable to other commercial devices.

The signal sampling frequency is limited by the hardware used. High frequency sampling improves the accuracy of the measurement when analyzing high frequency signals. However, when sampling low frequencies, where the minimum sampling time is the inverse of the signal frequency, the data buffer becomes saturated. It is also necessary to respect the Nyquist-Shannon principle for robust data sampling [12]; *e.g.* for a maximum bandwidth of 10 kHz, the sampling rate cannot be smaller than 20 kS / s. In practice, to avoid aliasing, a clock speed 10 times greater than the maximum speed of the measurement was used for each channel (100 kS/s). Therefore, the DAQ hardware suitable for this application needs at least 500 kS/s, for 5 response channels.

Hardware

A USB data acquisition card (USB 6363, National Instrument) was used for data collection and signal generation. The DAQ has 16 bit resolution, with 32 input channels (16 differential), 4 analog outputs, a sampling rate of 1 MS/s and an output rate of 2.86 MS / s. It was used in the ±5 V range (or lower), with a resolution of 153 μV (*i.e.* $10\ V / 2^{16}$) . All the response signals were recorded in differential mode.

The AC perturbation was generated as a single sine wave, switching between discrete levels from high to low frequency. The AC+DC stimulus and response signals are both recorded by the DAQ and processed using a Fast Fourier Transform (FFT) to convert the signals from the time domain to the frequency domain.

Figure 1 shows the set-up used in this work. Electrical impedance measurements on 'dummy cells' were performed in potentiostatic mode, where the stimulus voltage is imposed to the system and the current response is recorded using the 22 Ω series resistor as a shunt resistor. The electrochemical impedance measurements on the battery were performed in galvanostatic mode; the stimulus current is controlled using an analogue control port on the load unit and recorded using a current transducer.

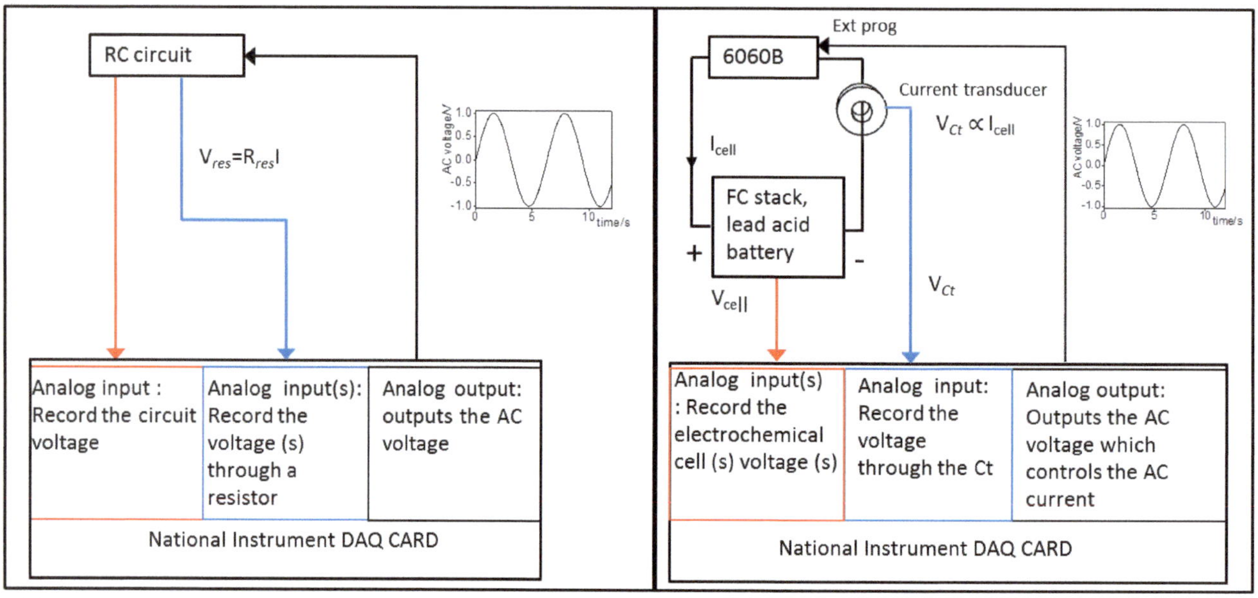

Figure 1. *Electrical connection schematic for (a) electrical and (b) electrochemical impedance spectroscopy. Black line represents the voltage stimulus from the DAQ; blue the voltage as a proxy for the current and red the actual voltage from the cell.*

Measurement procedure for passive components in potentiostatic mode

An electric circuit (dummy cell) was used for electrical impedance testing. This was composed of a 22 Ω resistor R_1 in series with an RC parallel combination (4.7 μF capacitor C_1 and 200 Ω resistor R_2). This circuit was connected to the DAQ card to enable single and multiple channel response. The series resistor R_1 is used as a shunt to derive the current through the circuit.

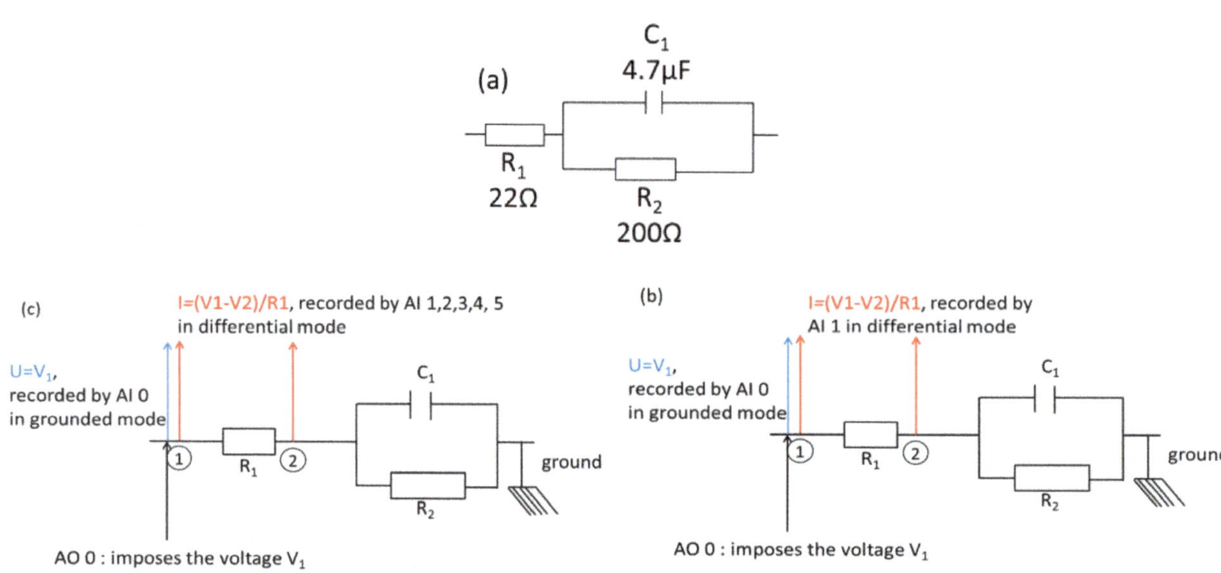

Figure 2. *(a) Electrical circuit comprising the 'dummy cell'; (b) connections for measuring voltage across the shunt resistor in single and (c) multiple-input mode. AO: analog output. AI: analog input.*

Measurement procedure for electrochemical impedance in galvanostatic mode

In order to trial the system on an electrochemical power source, a commercially available lead acid battery (6 V, 10 Ah, Yuasa) was tested. An electronic load (6060B Agilent) with a bandwidth of 10 kHz was used to control discharge rate. The load is equipped with an analogue input capable of remotely controlling the current that is passed [13], a 0-10 V signal from the DAQ can be converted into a 0-60 A current by the load unit. A current transducer, with a frequency bandwidth of 150 kHz (ITS-ULTRASTAB-60, LEM) capable of zero flux detection was used to record the AC and DC current, and convert it to a voltage across a resistor. The voltage of the battery and the voltage across the burden resistor are then recorded at the same time; consequently the transfer function can be extracted, as described in Figure 1(b). The load unit has a bandwidth of 10 kHz, and therefore some attenuation could be expected for frequencies higher than 2 kHz. For comparison with a commercially available high current potentiostat an IviumSTAT A11700 (10 V, 10 A, Ivium, Netherlands) was used. For both systems the AC perturbation was equivalent to 75 mA imposed across a frequency range of 10 kHz – 1 Hz.

Results and Discussion

Assessment of DAQ hardware as a function generation and frequency response analyser.

It is important to ensure that there are no artifacts or significant phase shifts introduced by the hardware/software combination. Figure 3(a) shows that when the input and output of the DAQ are connected together there is agreement between the two in terms of the frequency generated and measured. Figure 3(b) shows a Bode plot of magnitude and phase. It can be seen that the magnitude of the signal is constant across the range and that there is only a small phase shift introduced above 1 kHz (<0.03 degrees at frequencies up to 10 kHz).

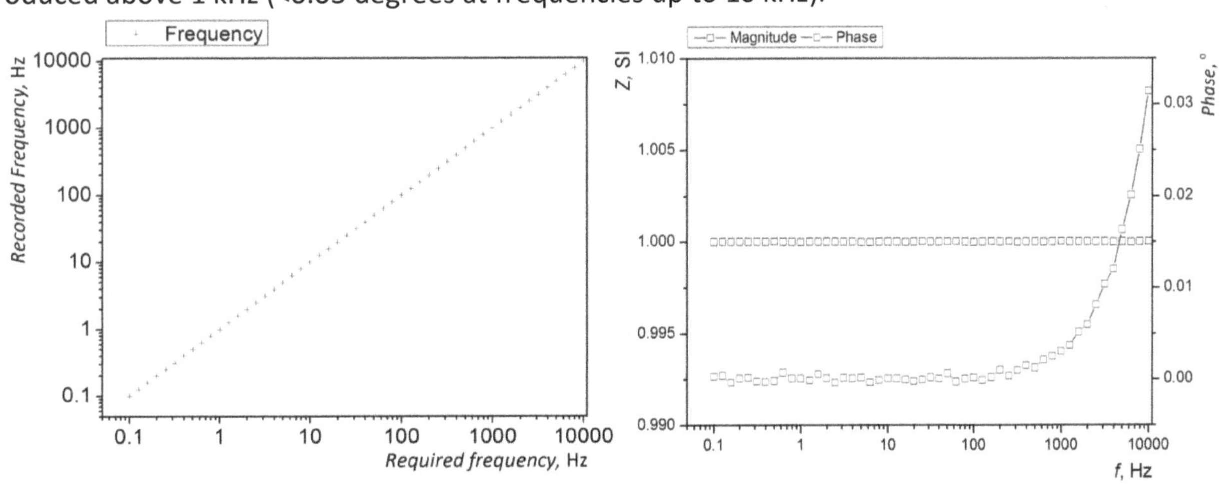

Figure 3. *(a) Frequency generated versus recorded frequency sweep, from 10 kHz to 0.5 Hz. (b) Phase and Magnitude for the impedance between two channels recording the same signal.*

Electrical impedance: single channel mode

The EIS system was tested in single channel mode on the circuit shown in **Figure 2**(b). Figure 4 shows Nyquist and Bode plots comparing the response obtained from the UCL system, the commercial (Ivium) frequency response analyzer and the model simulation based on the components used in the circuit. It can be seen that excellent agreement is obtained between each, over the entire frequency range (10 kHz to 0.1 Hz). This demonstrates that the system is suitable for performing EIS on passive (non power source) systems.

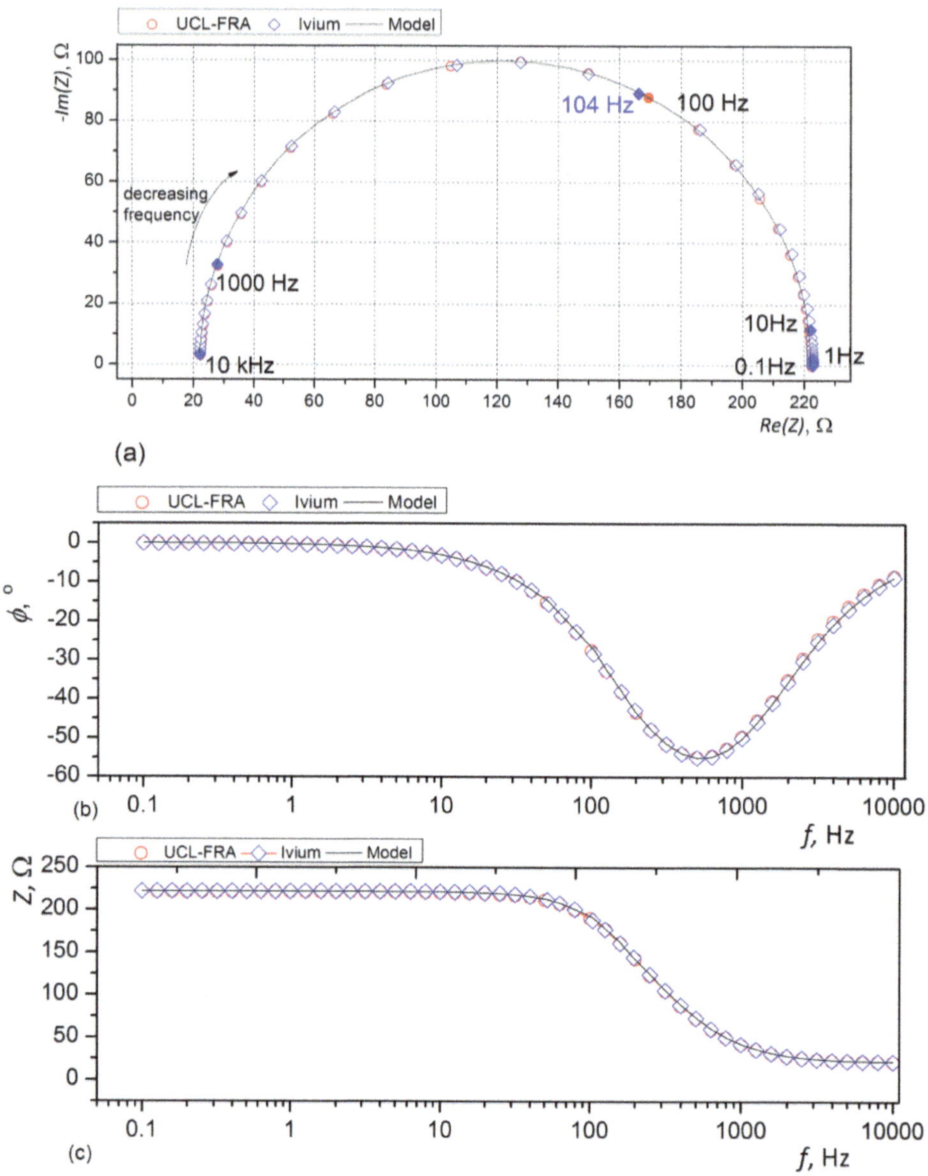

Figure 4. *(a) Nyquist and (b-c) Bode plots for a single response channel using UCL-FRA and commercial Ivium systems compared with the modeled transfer function. Frequency sweep between 10 kHz and 0.1 Hz, AC amplitude 0.1 V for a DC signal of 2 V with 10 frequencies per decade.*

Electrical impedance in multi-channel mode

Operation in multi-channel mode imposes more of a challenge on the DAQ system. Lower cost DAQ hardware uses a single clock that triggers each of the channels sequentially. A phase shift between each of the channels is therefore possible at high frequencies and care should be taken when introducing additional channels to the measurement. Higher end hardware, with dedicated clocks on each channel, avoids this potential artefact. Figure 5 shows the response from 5 channels on the DAQ system (Z1 – Z5) connected to the circuit described in **Figure 2**(c). In the Nyquist projection the system shows little difference between each channel across the full frequency range, so allowing an accurate measurement of the real resistances in the circuit. In the Bode projection it can be seen that some phase difference is introduced at high frequency. However, below 3 kHz this difference is less than 3 degrees.

Overall, these results suggest that this DAQ/LabVIEW interface system can be used for multi-channel EIS data acquisition over the frequency range of 3 kHz to 0.1 Hz.

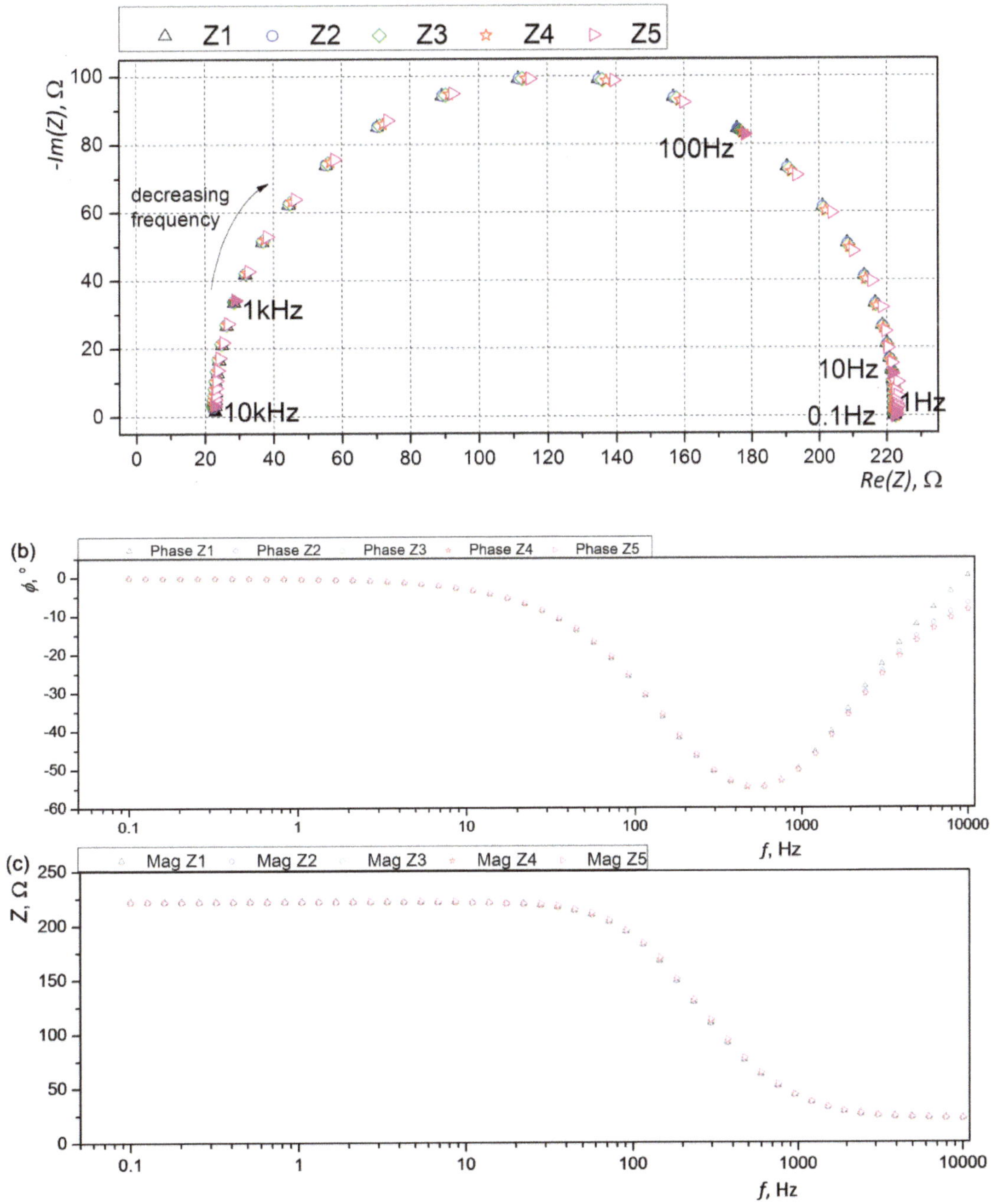

Figure 5. *(a) Nyquist plots and (b-c) Bode plots showing the response from 5 channels, sampled concurrently, using the circuit described in Figure 2(c) with UCL-FRA. Frequency sweep between 10 kHz and 0.1 Hz, AC amplitude 0.1 V for a DC signal of 2 V with 10 frequencies per decade.*

Electrochemical impedance spectroscopy on power sources

Power sources are typically very low impedance devices and notoriously challenging for FRAs. A lead acid battery was tested (in discharge) to assess the performance of the system in active mode when incorporating an electronic load. Figure 6 shows the Nyquist and Bode (phase and magnitude) responses for the UCL and commercial (Ivium) systems. Discharge from the battery was performed in galvanostatic mode at 1.5 A using the set-up of Figure 1 (b). The measurement was achieved with a lowest frequency of 1 Hz, to avoid discharging the battery too much due to the longer measurement time at low frequencies.

The Nyquist plot shows that there is a difference (shift) in the real impedance between the DAQ/LabVIEW system and that from the Ivium analyser. However, the form of the Nyquist plot is very similar for each, exhibiting the shape of two depressed semi-circular arcs, representing two processes with different time constants (RC combinations). This is typical of an EIS response for a battery of this kind [14]. The systematic difference in real resistance between the two (<10 mΩ) is attributed to the different leads and connectors used for each system causing different ohmic resistance.

The Bode plots in Figure 6(b-c) show that the two magnitudes are very similar across the range, although the one recorded with the UCL-FRA exhibits a constant lower resistance associated with the different lead configuration. The two phases remain in close agreement up to a frequency of ~3 kHz. Above this frequency, the bandwidth limitation of the load starts to become apparent.

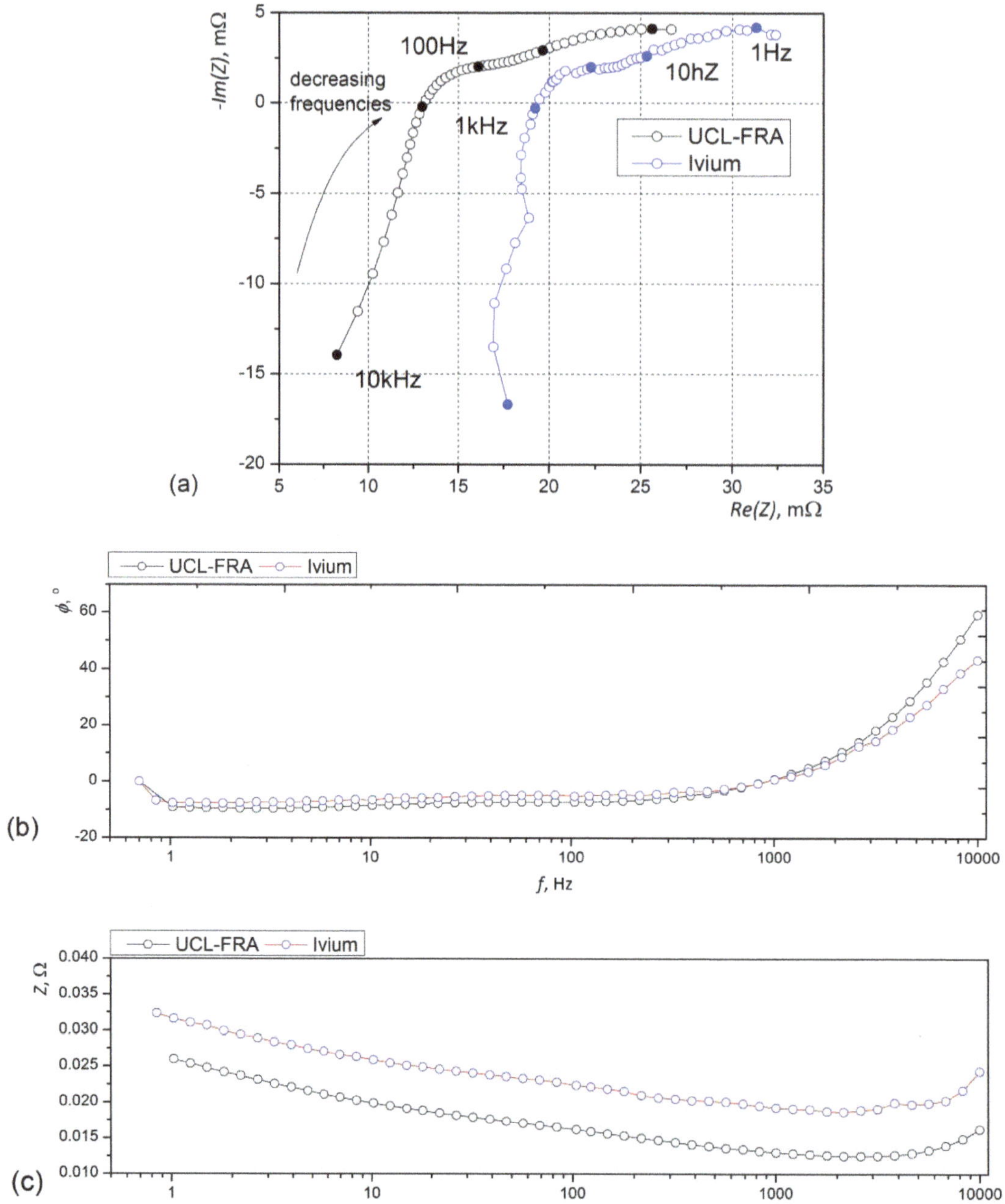

Figure 6. *Electrochemical impedance response for UCL-FRA and commercial FRA (Ivium) in the form of: (a) Nyquist, and Bode, (b) phase and (c) magnitude for a lead acid battery being discharged at 1.5 A. Frequency sweep between 10 kHz and 1 Hz, AC amplitude 75 mA for a DC signal of 1.5 A with 10 frequencies per decade.*

Conclusion

A DAQ/LabVIEW interface has been developed capable of performing frequency response analysis for electrochemical impedance spectroscopy of low-impedance power sources. The system was tested using electrical circuits to demonstrate that multi-channel data acquisition is possible (here using 5 channels) without significant phase shift between the channels up to 3 kHz.

EIS on a battery discharging at 1.5 A compared favorably with a commercial FRA; the main difference being a systematic shift in real impedance and high frequency inductance, both of with can be attributed to differences in the cabling used for each arrangement.

This method is attractive for applications where a low cost solution is required for measuring electrochemical power systems and a multichannel input is needed. For example, performing individual cell EIS on fuel cell stacks or on-board diagnostics.

Acknowledgements: *The authors would like to acknowledge University College London and Intelligent Energy Ltd. for supporting the work of Meyer, the EPSRC Supergen Fuel Cells programme (EP/G030995/1) and EPSRC Flexible Fuel Cell project (EP/G04483X/1) for supporting Brett's research.*

References

[1] V. Sense, M. Cell, *Scribner Associate-Tutorial*, 1-5.
[2] M. E. Orazem, B. Tribollet, *Electrochemical Impedance Spectroscopy*, Wiley, Gainsville, USA 2008, p. 91-95.
[3] J.-P. Diard, B. Le Gorrec, C. Montella, *Cinetique electrochimique*. Hermann, 1996, p. 231-241.
[4] F. L. Floyd, S. Avudaiappan, J, Gibson, *Prog. Org. Coat.* **66** (2009) 8-34
[5] S. Rodrigues, N. Munichandraiah, A. K. Shukla, *J. Power Sources* **87** (2000) 12-20
[6] E. Karden, S. Buller, R. W. D. Doncker, *J. Power Sources* **85** (2000) 72-78
[7] X. Yuan, H. Wang, J. Colinsun, J. Zhang, *Int. J. Hydrogen Energ.* **32** (2007) 4365-4380
[8] Q. Huang, R. Hui, B. Wang, J. Zhang, *Electrochim. Acta* **52** (2007) 8144-8164.
[9] X.-Z. Yuan, *Electrochemical Impedance Spectroscopy in PEM fuel Cells*, Springer 2010, p 193-313.
[10] I. A. Schneider, H. Kuhn, A. Wokaun, G. G. Scherer, *J. Electrochem. Soc.* **152** (2005) A2092-A2103.
[11] S. J. Andreasen, J. L. Jespersen, and S. Kaer, *Fuel Cells* **4** (2009) 463-473
[12] C. E. Shannon, *Proceedings of IEEE* **86** (1998) 447-457.
[13] Agilent Technologies, *Agilent Sales and Support Office*, **5951-2826** (2004) 31.
[14] H. Blanke, O. Bohlen, S.Buller, *J. Power Sources*, **144** (2005) 418-425

16

Electrochemical Engineering - its appearance, evolution and present status. Approaching an anniversary

VELIZAR STANKOVIĆ

Department of Metallurgical Engineering, Technical Faculty Bor, University of Belgrade, Serbia

E-mail: vstankovic@tf.bor.ac.rs

The goal of electrochemical engineering is to achieve
in a quantitative way an optimal cell design
N. Ibl

Abstract

Through this story an attempt has been made to present a chronology of electrochemical engineering - from its appearance as an individual science, via its growth throughout the five decades of its existence as an individual science, until today. The collaboration and linkage of electrochemical engineering with other disciplines has also been discussed in this essay. The role and duties of electrochemical engineering in the 21st *century have been touched upon to the extent that it was possible.*

Keywords

Electrochemical engineering, Electrochemical reactors, Cell productivity, Education in electrochemical engineering

---

1. Circumstances of the emergence of electrochemical engineering as a new scientific discipline

The term “electrochemical engineering“ first appeared in the literature as a novel title of the fourth edition of C.L. Mantel’s book Industrial Electrochemistry [1], published in 1960 (the first edition was published in 1931 and later editions in 1940 and 1950). Just looking through its contents, one can easily realize that this book mainly describes the electrochemical technologies for production of some particular electrochemical products. Although there was an attempt to group and systematise the exposed technologies in separate chapters, such as: basic electrochemical relationships and laws; molten salt electrochemistry, metal electrowinning and electrorefining; gases production; corrosion and electroplating, a big gap still remained between the title and the contents. This book, along with others relating to industrial electrochemistry from around that time, demonstrated that, at the end of the '50s, the number of electrochemistry technologies had expanded to the extent that it was not possible even to describe each of them in a textbook without a lot of repetition in describing each technology. This made books more and more voluminous, making it troublesome for students, who had to learn each technology for the production of every particular electrochemical product. This was one of the reasons, maybe not the most important but still a very serious one, for starting to consider establishing a new discipline in a similar manner to what had been done when chemical engineering appeared at the end of the 19th and the very beginning of the 20th century - firstly through unit operations and afterwards through transport phenomena and reaction engineering. Hence, at the end of the '50s and in the early '60s, chemical engineering had already developed its capabilities to treat unit operations and processes in a way that interconnects physical and physico-chemical phenomena with chemical processes to chemical technology.

At that time (the end of the '50s), it became evident that similar transformations to those that had occurred in chemical technology had to be implemented in industrial electrochemistry to get more engineering and a more scientific approach to improving the existing technologies and establishing new ones. By then, almost all that had been achieved in improvements to any electrochemical technology was due to the art of engineering rather than to an applied science.

2. Interconnection between chemical engineering and electrochemistry - pioneering works linking the two sciences

During the '50s a close interconnection between electrochemistry and chemical engineering was established, on a theoretical level, through the penetration of electrochemistry into chemical engineering and *vice versa*. During that period, firstly C.S. Lin with his co-workers [2], and shortly afterwards Tobias, Eisenberg and Wilke [3] considered, both experimentally and theoretically, the mass transfer phenomenon of ions coming from the bulk electrolyte to the electrode surface by molecular diffusion, migration and convection. Their considerations resulted in the well-known equation connecting limiting current density with the other variables affecting it:

$$i_{\mathrm{L}} = k_{\mathrm{L}} z F c_{\mathrm{b}} \qquad (1)$$

Where: i_{L} - is limiting current density; k_{L} - mass transfer coefficient; z - number of exchanged electrons in electrochemical reaction; F – Faraday's constant; c_{b} – bulk concentration of reacting ions, all given in arbitrary units.

The published papers in which an electrochemical method of limiting current measurement was used to evaluate the mass transfer coefficient at convective mass transport forged a strong connection between electrochemistry and chemical engineering. The few early works [2,3] relating to local limiting current measurements had opened new opportunities – employing electrochemical methods for evaluation either local or overall, as well as instantaneous or time-averaged transport characteristics of flowing liquid systems, *e.g.* to quantify mass, heat and momentum transfer features. In this way, methods of adsorption, dissolution, or sublimation used till then for evaluation of the mass transfer coefficient, were replaced by a much faster and more reliable electrochemical method of the limiting current measurement and computation of a mass transfer coefficient from experimentally recorded polarization curves, using Eq. (1). This method was widely exploited during the '60s and particularly the '70s, until today, for studying the mass transport rate in different electrochemical model-systems. The electrochemical method of limiting current measurement, in its early or advanced form, contributed greatly towards providing evidence about the hydrodynamic image of flowing systems. Thanks to the Chilton-Colburn analogy widely used at that time in transport phenomena, data obtained for local or overall mass transfer coefficients were then employed to estimate heat transfer as well as momentum transfer in similar flowing systems. Indeed, these works confirmed the great influence of flowing conditions in electrochemical reaction rate *via* mass transport in electrochemical reactors, opening up a new research area for creating new, or improving the existing electrochemical cells.

In that period (the middle of the '60s), the method of conductivity measurement was also introduced in chemical engineering for local or overall porosity measurement in liquid fixed or fluidised beds and later for a gas hold-up quantification in gas-liquid systems, regardless of whether gas is dispersed in a liquid phase, or generated therein due to a chemical or electrochemical reaction.

This close collaboration between electrochemistry and chemical engineering also had a reverse direction – bringing benefits for electrochemistry. The marker pulse technique, a method widely used in chemical engineering for axial and radial dispersion measurement of a tracer in a reaction chamber, was adopted and modified at the end of the '60s, to be used for investigation of ionic species distribution, generated in a flowing electrolyte in electrochemical reaction systems. At that time, terms like "plug flow reactor" or "perfectly mixed reactor" were reserved for chemical and/or catalytic reactors, but not yet used for electrochemical cells.

Mass transfer in a cell when changing the hydrodynamic conditions was investigated at that time as a method for mass transfer enhancement when a diffusion-controlled electrochemical reaction takes place. The use of dimensional analysis and similarity theory adapted for and applied in chemical engineering, was at that time introduced in electrochemical studies in order to minimize the possible combinations of operating variables that have to be tested in the research and development of new electrochemical processes, new electrochemical reactors, or new technologies. This was very important in the scaling of laboratory results to the pilot-plant and then from pilot data to the industrial scale. The experimental data obtained were then expressed in the form of dimensionless relationships, either as:

$$\mathrm{Sh} = f(\mathrm{Re}, \mathrm{Gr}, \mathrm{Sc}, \mathrm{Fo}, \Gamma_1, \ldots \Gamma_n) \quad (2)$$

or, based on the Chilton-Colburn analogy, as:

$$f_{\mathrm{fr}}/2 = j_{\mathrm{h}} = j_{\mathrm{D}} = f(\mathrm{Re}), \quad (3)$$

describing mass, heat and momentum transfer. Equations like these had been introduced as a tool in the consideration of diffusion-controlled electrochemical reaction systems and published at that time in the literature. In equations (2) and (3) Sh is Sherwood number; Re – Reynolds number; Gr – Grashof number; Sc – Schmidt number; Fo – Fourier number; Γ_{i} – dimensionless groups dictated by geometric similarity; f_{fr} – Fanning friction factor; j_{D}, j_{h} – mass and heat transfer factor, respectively.

Moreover, new similarity criteria, like the Wagner number (Wa), relating particularly to the nature of electrochemical reaction systems, were derived and introduced in dimensionless expressions similar to equations (2) and (3).

Obviously, both sciences – electrochemistry and chemical engineering - profited from their mutual collaboration during the '50s and '60s.

The main benefit from this interrelation between electrochemistry and chemical engineering was the appearance of a new discipline – named electrochemical engineering.

3. Establishing electrochemical engineering on the scientific scene

In the early '60s, as a result of the interrelation between chemical engineering and electrochemistry, numerous qualitatively new papers appeared in the relevant literature (mainly in chemical engineering rather than in electrochemical journals), founding the conditions for the appearance of a new multi-disciplinary science – electrochemical engineering.

C. Wagner, V.G. Levich, N. Ibl, C. Tobias, P. Le Goff, and J. Newman with their collaborators, as well as some others not mentioned here due to limited space, contributed a lot to the establishment and early development of the new discipline. These scientists may be considered as the founders of electrochemical engineering - a separate scientific discipline born from the mutual influence of chemical engineering and electrochemistry. Wagner first introduced the name to this discipline, publishing a paper in the second volume of Advances in Electrochemistry and Electrochemical Engineering in 1962, entitled *"The Scope of Electrochemical Engineering"* [4]. In that article, Wagner makes a clear distinction between electrochemical engineering science and electrochemical engineering technology.

In the middle of the 20th century there were not so many electrochemical technologies. From this time-distance, we might suppose that improvements of these technologies, starting from their appearance, occurred little by little through decades, more by intuition or by trial and error than by a scientific approach to basic phenomena affecting features of these processes. The contribution of electrochemical engineering, in its early stage (at the beginning of the '60s), to the design of industrial electrochemical cells and processes was very modest - almost negligible. Mass transfer considerations in electrochemical reactors had an academic rather than an engineering approach. This was mainly because there was no mass transfer limitation when the process operated with electrolytes having high concentrations of reacting ions, as was mostly the case with industrial electrochemical processes from that period. Current density distribution was not of major importance in those cases and for the side by side electrode configuration that was mainly used in electrochemical cells from that time. The operating current density was not limited by diffusion of reacting ions, but more by the cell voltage and consequently the specific energy consumption than by mass transport limitations or irregular current distribution.

At that time, electrochemical reaction engineering was a term of no significance for industrial practice. It was enough to find empirically the cell volume and inter-electrode distance needed to estimate the highest degree of conversion and the highest current efficiency at the lowest cell voltage. Not much attention was paid to enhancing the cell productivity by changing either the hydrodynamics conditions in a cell or the cell voltage in order to get optimal operating energy conditions.

A great step towards new cell design and its optimisation was made when the pioneering work of Jottrand and Grunhard [5] appeared in 1962, concerning the mass transfer enhancement in a tubular cell with a fluidised bed of inert particles in the inter-electrode space. Very soon after, numerous articles on the same subject but with different aims were published in the literature, providing evidence on how many times various inert turbulence promoters (ITP), placed in inter-electrode space in a cell, enhance the limiting current density due to an increase in the mass transfer coefficient. Besides higher cell productivity, a smoother electrode surface was obtained in the case of either metal deposition or dissolution, so that these types of cells were further developed with the aim of employing them in electroplating, electro-polishing, electroforming and other similar processes of electrochemical surface treatment, but also for metal removal from various sources. Several years after this series of articles, a new type of cell with expanded mesh electrodes, in which mass transport is facilitated by a fluidised bed of glass beads, was developed, and later marketed by BEWT engineers, UK, as a "Chemelec cell" for the recovery of metals from wastewaters and industrial effluents.

At that period (the second half of the '60s), a new electrochemical reactor also appeared, named the three-dimensional electrode cell (TDE-cell). The first articles on this subject were published at the end of the '60s and early '70s by English scientists from Newcastle and Southampton Universities and shortly afterwards by French scientists from the University of Nancy, as well as by German scientists from DECHEMA Institute.

Two revolutionary things, with a strong influence on further electrochemical engineering development and electrochemistry technology, also happened in the field of materials engineering during the '60s:

- dimensionally stable anodes (DSA®) were introduced in 1968 in industrial practice, providing many advantages over the existing electrodes at that time, first of all a lower cell voltage, thus making a great contribution in energy saving, and better anti-corrosive features, thus having a longer lifetime;
- a co-polymer, having ionic features (ionomer) was synthesised and patented in 1966, called NAFION and, due to its ability to conduct protons, it was considered and applied first as a proton exchange membrane (PEM) in fuel cells and later in many other electrochemical technology areas (chlor-alkali electrolysis, water electrolysis, metal electrowinning, electroplating, electrodialysis, sensors and battery fabrication, ...), mainly for dividing the anode from the cathode chamber in various electrochemical reactors. These two discoveries had a great influence on the philosophy of thinking of scientists and engineers, opening up new frontiers in the design and development of electrochemical cells and processes.

4. Growth (expansion) of electrochemical engineering as a science and art from the '70s till the end of the '90s

Through these three decades of electrochemical engineering growth, several directions were profiled and developed, mainly simultaneously. These directions are:

- Electrochemical reactors – development, modelling and optimisation;
- Electrochemical processes – improvements of the existing ones, and research and development of new ones;
- Interconnection of electrochemical engineering with other sciences, in some cases forming new scientific sub-disciplines;
- Publishing affairs – education of new generations in the electrochemical engineering community.

4.1. Electrochemical reactors research and development

During the '70s and '80s electrochemical engineering had a very fruitful and progressive period contributing to electrochemical reactor engineering development, which resulted in new electrochemical reactors, as well as new electrochemical processes that appeared and found their place in electrochemical technology. One can say that the '70s and '80s were the peak of the electrochemical reactor research and development, when a lot of new cells appeared. Some of them were developed on an industrial scale and marketed in that period, as:

- "Chemelec" cells (already mentioned) for metal removal from rinse and wastewaters;

- Filter press cells for multipurpose use, either in mono- or bipolar mode (applied in water electrolysis, in electrowinning, in electro-organic synthesis, known as ICI Cell, or SU Cell);
- Swiss-roll cells employed in the electro-organic synthesis of vitamin C in the period when anodic oxidation had still to play a role;
- Pump cells;
- Capillary-gap mono- or bipolar cells, also for multipurpose usage;
- Rotating cylinder and rotating drum cells, both used for silver removal from spent photo-chemicals;
- TDE cells either as a fixed bed cell, such as an en-Viro cell, or as a fluidised bed cell for heavy metal ions removal from various industrial effluents;
- Cylindrical cells with coaxial electrodes configuration and TDE anodes used in the NALCO process;
- Bipolar trickle bed cells for electro-organic synthesis, as well as some others that can be found in the relevant literature that have found their use for all the aforementioned purposes [6].

Particular attention in that period was paid to the TDE cells with fluidised bed cathode and en-Viro cells, considered as a new very promising and very powerful electrochemical reactor. Very quickly after the first papers were published on the subject of TDE cells, a pilot-plant for the fluidised bed cell was constructed by CJB Developments UK in 1972, and afterwards the first industrial plant was built in 1978 in Germany by AKZO Zout Chemie Netherlands, only to be closed a few years later, having exhibited certain disadvantages [7]. Later, during the '90s, thanks to the appearance of advanced conductive materials with a very developed surface area and high porosity, a new generation of TDE cells, including the Poro-cell, Reno-cell, and Retech-cell, was developed and marketed. Although these cells were originally investigated and developed mainly for metal recuperation from various industrial waste solutions containing a low level of metal ions, sometimes a few p.p.m. only, they were later modified and used for electro-generation of different oxidants used for oxidation of different organic or inorganic compounds.

The theoretical knowledge and laboratory experience in electrochemical reactor engineering accumulated in that period contributed significantly to producing an optimal version of some of these cells on an industrial scale.

In fact, scientists working in electrochemical reactor engineering were faced with the problem of how to explore in the best way a simple relationship connecting the cell productivity with other variables affecting it. Cell productivity per unit of installed volume and unit of time (termed also as space-time yield [8]) is defined by the basic, well-known equation:

$$\frac{1}{V}\frac{\mathrm{d}m}{\mathrm{d}t}=\eta_{\mathrm{e}}\alpha k_{\mathrm{L}}aMc_{\mathrm{b}}(t) \tag{4}$$

where: η_e is the current efficiency; $\alpha = i/i_L$ – ratio between operating and limiting current density; $a = A/V$ – specific surface area (working electrode surface per unit of cell volume); k_L - mass transfer coefficient; M – molar mass; $c_b(t)$ – concentration of metal ions in the bulk.

It follows from Eq. (4) that, for a given and constant bulk concentration $c_b(t)$, as is usually the case in effluent and wastewater treatments, there are three possibilities for an increase in the limiting current density as a measure of the maximum cell productivity:

A. to increase the mass transfer coefficient;
B. to increase the specific surface area of the working electrode; or
C. to increase both the mass transfer and specific electrode surface area in the same cell.

A: Mass transfer around an electrode can be enhanced by making and keeping forced convection in a cell. This can be carried out by:

1. changing the electrolyte flow conditions around working electrodes by either pumping or stirring, or using an ultrasound source; rotating, shaking or vibrating a working electrode, or in some other way;
2. introducing different ITPs in the inter-electrode space that will locally change electrolyte flow close to the electrode surface;
3. bubbling gas around the electrodes.

All three methods of mass transfer enhancement were extensively investigated and the contribution of each of them was quantified by scientists through the '60s and particularly the '70s.

B: Specific electrode surface (*a*) area can be increased by:

1. decreasing the inter-electrode distance;
2. using either dispersive or porous conductive material connected with a current feeder as a working electrode (TDE).

C: A cell with a TDE fulfils both requirements – it has a high specific surface area and, at the same time, it has good mass transport properties independently, whether electrolyte flows by, or through, the TDE. This aspect of enhancing the productivity of a TDE cell particularly occupied the attention of scientists at that time, resulting in many papers relating to this issue.

Table 1 presents an approximate specific surface area and corresponding maximum space-time yield (productivity) for different cells as an illustration supporting the statements made in the previous text, but also to demonstrate an appreciable progression in designing a cell of high performances.

Table 1 *Specific surface area and maximum productivity of some particular cells compared under similar operating conditions [7]*

Type of cell	Specific surface area, m^{-1}	Productivity at α =1, mol m^{-3} h^{-1}
Conventional cell	7.5	0.14
Filter press cell	30 – 170	0.56 – 3.17
Capillary gap cell	100 – 500	1.9 – 9.33
Fixed bed cell	1000 – 10 000	18.65 – 186.5
Fluidized bed cell	1000 – 10 000	37.3 – 186.5
Rotating drum cell	50 – 5000	9.32 – 93.3

It is important to emphasise that, when the bulk concentration in Eq. (4) is time-dependent (metal removal, for example), the current efficiency - η_e is also a complex and time-dependent variable, depending, among other things, on the galvanostatic or potentiostatic mode of operation. Numerous papers on the modelling and optimization of electrochemical reactors were published at that time and later, in order to achieve the best working conditions, *i.e.*, the highest cell productivity at optimum energy consumption for not only the cells considered in Table 1 but also for others. These papers made a great contribution to the development of electrochemical reactor theory and engineering practice.

4.2. Establishing new and improving the existing electrochemical processes

Thanks to the achievements in developing the new electrochemical reactors that are the core of every electrochemical process, great success was attained in developing new processes and technologies and in improving the existing ones.

i. A particularly significant contribution was made in electro-organic synthesis, where various organic compounds were produced electrochemically on an industrial scale, such as: tetraethyl lead, sorbitol/mannitol, glyoxylic acid, methoxy-benzylalcohol, adiponitrile, methoxy-benzaldehyde, methylethylketone, propiolic acid, anthraquinone, naphthoquinone and others. Many other organic products are under research and development on a pilot plant scale and many more are still on a bench scale and will be commercialized soon.

ii. In electrochemical metal surface treatment, comprising electro-coating, electro-polishing, electro-sinking and grinding, many new electrochemical tools and new electrolytes were developed and commercialised. Great success was achieved in electroforming, where numerous micro-products appeared, and various micro-fabrication methods were developed and marketed. Micro-reactor technology was one of the more propulsive areas, and very promising for the electrochemical production of some special chemicals in very small amounts. In that period – the early '70s – the so-called LIGA process had already been developed and employed, spreading its role into many areas of present day techniques and technologies: in micro-mechanics, micro-acoustics, micro-optics, in information technologies, in biotechnologies, biomedicine, and many other areas.

iii. The principle of the fuel cell (FC) was first described by W. R. Grove in 1839. Since then, many efforts have been made and many types of FCs have been considered for producing energy at low or high temperatures using mainly hydrogen as a fuel but also natural gas and coal gas, in a mixture with hydrogen or alone. The greatest improvements in development of proton exchange membrane (PEM) FCs capacity was achieved during the '90s. An almost 25-fold increase in output power was achieved in that period, starting from 70 kW m^{-3} in 1989 and reaching 1.8 MW m^{-3} in 1997. This leap was a consequence of the increased interest of the automotive industry, backed with adequate financial support, and was due also to new catalytic materials and cells development.

iv. In parallel with FC developments, electrochemical engineering contributed a lot to developing new and improving the existing water electrolysis and hydrogen production technologies. The improvements of alkaline water electrolysis technology consist mainly in improving the electrolytic cells to have an increased specific surface area; using advanced materials as separators has led to a reduced inter-electrode gap, thus reducing specific cell resistance. This, coupled with an increased operating temperature, as well as the introduction of new catalytic materials, makes the whole process more energy/economy viable. Besides these improvements, new technologies were developed in that period, based on Nafion membranes offering qualitatively new approach to hydrogen/oxygen production as well as water steam splitting on oxide ceramic.

v. Several substantial improvements were made at that time in chlor-alkaline electrolysis in that period (from the '70s till the end of the '90). The most important was substitution of graphite with DSA, thus prolonging the anode life-time many times and preserving lower chlorine evolution overpotential. The other improvements were similar to those applied in water electrolysis, and relate to decreasing the inter-electrode space of electrolysers; mono-

or bi-polar mode of operation; and substitution of amalgam process by diaphragm and membrane processes, thus making the whole technology more environmentally sustainable.

4.3. Electrochemical engineering and other disciplines

The period between the '70s and the end of the past century was also characterized by the penetration of electrochemical engineering into or interconnection with other disciplines. Here we will discuss only those that I consider to be the most important. The others will only be touched upon.

Extractive metallurgy

Electrochemical engineering has particularly contributed a great deal to the development of hydrometallurgy, which also made its own very substantial progress during the '60s and on, of the past century. Now, electrowinning is usually the last link in the technology chain for production of various metals from ores by leaching followed by solvent extraction/ion exchange and an electrowinning pathway. As a response to a significant growth in the theoretical knowledge and its experimental confirmation, many attempts were also made in that period to improve the existing or establish new purely electrochemical or metallurgical processes, as well as processes linked with other adjoining disciplines. We will mention here only those processes and technologies relating mainly to metallurgical engineering and environmental processes - the fields I am more familiar with:

- Copper electrowinning from pregnant leach solutions, namely, the leaching-solvent extraction-electrowinning (L-SX-EW) process of copper production, as a qualitatively new technology was introduced on an industrial scale at the end of the '60s. As the major cost component of this technology is the electrowinning step, which requires 8 to 10 times more power than electrorefining, shortly afterwards many attempts were made with the aim of making electrowinning less expensive and more efficient. One approach was an attempt to replace the anodic reaction of O_2 evolution by SO_2 oxidation at copper or zinc electrowinning, which would be a great step in making the process more economically acceptable than it is now. So far, it is still at a development level. Another interesting attempt was to carry out the electrowinning from an electrolyte containing cuprous instead of cupric ions, thus cutting significantly the energy costs. Another interesting process is electrolytic production of metals from corresponding sulphide mattes cast and served as anodes in electrorefining plants. So far, only in nickel metallurgy has such an approach been successfully implemented.
- The improvements made in extractive metallurgy of basic heavy metals were implemented in some other technologies, in which electrowinning is the final step;
- Replacing the cementation processes by direct electrowinning in zinc electrolyte purification from other metal ions using advanced cells based on TDE, prior to the zinc electrowinning;
- Recuperation of metals from spent electroplating baths, from spent solutions in electronics, from spent pickling solutions in the metal working industry, from mother solutions in salt production and from some other similar effluents that are produced in metallurgy or the inorganic chemical industry;
- Electrolytic metal powder production is nowadays an important segment, providing starting materials for sinter metallurgy, as well as in electronics, in metallic pastes production and for many other products;

- Applying different electrochemical coating and surface finishing methods leads to minimising the corrosion effects in wastage of materials, to which electrochemical engineering contributed greatly in the past and still currently does.
- Organic chemistry and organic chemical technology (electrochemical reactor design for massive production; micro-reactors design and implementation for production of very hazardous chemicals).

Environmental protection

In that period ('70s to '90s), particular attention was given to environmental pollution and environmental protection, due to an increasing demand for environmentally sustainable chemical processes. As an answer to this demand, electrochemical engineering opened a new direction in the treatment of different industrial effluents. Coming into this important and broad area, electrochemical engineering started to play a key role in different aspects of pollution abatement, in both aqueous and gaseous phase. Electrochemical methods and devices were developed and employed for wastewater purification, either for removal of hazardous heavy metal ions from them or for cyanide or chromate destruction; for direct anodic destruction of organic species that must be destroyed prior to releasing such wastewaters into the recipient; for "in situ" water disinfection for human usage or for special purposes.

Nowadays, also thanks to the development of advanced materials, boron-doped diamond (BDD) anode is an industrial reality in drinking and wastewater treatment processes, and particularly useful for anodic destruction of those organic pollutants from wastewaters that can only be decomposed into less hazardous or even inert compounds at higher overpotentials with no water electrolysis [9].

Electrochemical engineering has found its place also in off-gas cleaning technologies for either direct electrochemical or indirect (using a proper electrochemically generated red/ox mediator) conversion of harmful gaseous off-gas stream constituents such as: SO_2, H_2S, NO_x, ammonia, Cl_2, amines, hydrocarbons and other organic constituents, previously absorbed by a proper absorbing solution, from off-gas streams. So, for example, different variants were considered and developed for SO_2 removal from flue gases and its oxidation by an electro-generated oxidant (Ispra-Mark 13), using a proper redox pair. A similar concept was also considered for NO_x and SO_2 removal from gaseous mixture *via* absorption and oxidation of absorbed species by Ce^{4+} ions electrochemically generated in a separate electrochemical reactor. An interesting and promising approach to direct gas purification from organic and nitrogen compounds is the electrochemical catalyst promotion, known also as NEMCA. By polarising a catalyst layer deposited onto solid electrolyte, its catalytic activity increases 10 to 100 times. So far this method is still at the laboratory and bench scale.

Soil remediation is another area where electrochemical engineering has found itself contributing significantly to resolving the problem of the purification of soil polluted by previous industrial activities.

Recycling technologies

Following the concept of "zero discharge effluent technology", electrochemical engineering found its place in different recycling technologies, as an ultimate stage, for reclaiming and recovery of useful or harmful metals from different waste devices such as: spent batteries, printed boards and other electronic scrap –the amount of which grows ever greater; for metals removal from spent electroplating baths, as already discussed; from spent pickling solutions; from mother liquors; from spent photo-chemicals, different rinsing waters and other similar liquid sources.

Significant results were achieved in the recycling of acids and alkalis by splitting corresponding salts from large volume solutions produced in chemical industry or in metallurgy by means of electrodialysis or a combination of membrane and ion-exchange resin techniques assisted by electrochemical regeneration of loaded resin.

Wastewater purification and recycling in a process is in many cases more economically viable than to treat it to a purity level tolerable to be discharged into a recipient. Besides removing metal ions or destroying organic molecules, new electrochemically assisted techniques were developed and included in technologies for removal of fine suspended particles from rinse waters such as: electro-coagulation and electro-flotation, both efficiently employed on an industrial scale.

Some of the above-mentioned processes or technological innovations were developed and marketed, while others remained only as an attempt on a laboratory or pilot-plant scale waiting for further research, or were completely abandoned.

Other areas

Electrochemical engineering also penetrated the medical sciences, contributing to developing new medical tools, new devices, new indicators, new drugs, new compatible materials for orthopaedic purposes, for example, and many others.

In biochemistry, thanks to close collaboration with electrochemistry and electrochemical engineering, a new sub-discipline has appeared recently, named bio-electrochemistry, where a lot of new projects have been driven in developing new alternative sources for green energy production, such as bio-fuel cells; for promoting or inhibiting bacterial activity and its proliferation; for bioelectrocatalysis and for similar purposes using small currents to manage the bacterial life and features.

Obviously the three decades of the past century we have considered were very productive for electrochemical engineering science, as well as for electrochemical engineering technology.

4.5. Education of new generations of electrochemical engineers

Due to a significant increase of papers relating to electrochemical engineering subjects, a strong need appeared, at the end of the '60s, to establish a new, specialized journal for publishing such articles. So, the first issue of the Journal of Applied Electrochemistry was published in 1970 [10].

In 1974, based on many theoretical and experimental contributions in electrochemical engineering, **Electrochemical Reactor Design**, written by D.J. Picket, was published as the first book on this subject. Later, several books with the keyword ***electrochemical engineering*** in their titles appeared, with the aim of spreading the basic principles of the new science among younger scientists and students studying chemical and metallurgical engineering. Some of the writers of these textbooks, who significantly contributed to the development of electrochemical engineering, are listed here chronologically as their books appeared: F. Coeuret and A. Stork (1984), I. Roushar (1986), E. Heitz and G. Kreysa (1987), T. Fahidy (1988), K. Scott (1991), F. C. Walsh (1993), H. Wendt and G. Kreysa (1999).

At that time (the early '80s) the first scientific symposium entitled Electrochemical Engineering was held at Loughborough University UK, as a response to the increased interest in presenting the research results amongst a specialised group of scientists. After this first attempt, the next symposia were held every three years, always at Loughborough. These events were a good indicator for identifying major areas of research interest and key directions which electrochemical engineering had moved towards, in the period between two events.

In the meantime, the EEWP (Electrochemical Engineering Working Party) was constituted as a branch of the EFCE (European Federation of Chemical Engineers).

Following a successful tradition of organizing electrochemical engineering symposia, the EEWP took over this duty from Loughborough University, and the 1st European Symposium on Electrochemical Engineering was organized soon after in Bad-Soden, Germany, in 1990. The 9th was held in 2011 in Patras, Greece. Besides this symposium, held every three years and having a European character, similar symposia are organized in the frame of world meetings such as ISE Annual meetings and CHISA congresses, allowing the gathering of scientists from other parts of the world to present their results.

Also, taking care of the young scientific population interested in coming into electrochemical engineering science, the European Summer School on Electrochemical Engineering was held for the first time in Toulouse, France in summer 1995 and thereafter in Ferrara, Italy; Patras, Greece; then in Palić, Serbia and most recently in Almagro, Spain, in 2009.

5. Electrochemical engineering at the end of the first decade of 21st century

Electrochemical engineering has extended far beyond the limits which were drawn by the early works in the '60s when the main theme was the mass transport of ions to or from an electrode. One can still occasionally find in the literature papers concerning this subject, but this is rather a reflection on the works done during the '60s to '80s. Such papers are mainly directed towards the corrosion process induced by mono- or two-phase flows. The published relationships at that time on mass transfer in electrochemical systems can today cover any particular technical problem. Also, current density distribution is something that can be routinely solved employing modern computational techniques and reliable numerical methods. With these tools, modelling of any electrochemical reaction system with mono- or two-phase flow of electrolyte and a dispersed gas- or solid phase in a cell with or without separator is now possible and something that can be successfully used in everyday practice. In these circumstances, what does electrochemical engineering at the end of the first decade of the new century mean? What is the role of the electrochemical engineer in following the established principles and in finding challenging new subjects and directions for further upgrading electrochemical engineering science? It may sound too declarative, but the following text could be an answer to the above questions:

> *"It is the responsibility of the electrochemical engineer in industry to simultaneously manage electrical consumption and chemical production. He or she must apply relevant scientific and engineering principles to design, construct, and operate a process in an economical, safe, and environmentally conscious manner. Improved understanding of scientific principles and the application of new materials can lead to more efficient cell designs and processes. The constant evolution of technology provides challenging and rewarding careers to engineers and scientists in a range of disciplines. Among the considerations that influence cell efficiency and lower consumption are temperature, spacing between electrodes, electrode material, electrolyte consumption, cell size, source of raw material, and production rate. Clearly, engineering skill is required for understanding these effects and for achieving optimum production conditions".*

From "Industrial electrolysis and electrochemical engineering"
by M. Grotheer, R. Alkire and R. Varian [10].

6. New challenges and frontiers in electrochemical engineering [11]

According to the last sentence from the cited text, there will be a great period for further development of electrochemical engineering.

Indeed, one should expect the further influence of electrochemical engineering on improvements of existing electrochemical technologies, and development and establishment of new ones that are still on the laboratory scale. Some of them have already been mentioned in the previous sections.

In the past, electrochemical engineering faced much more towards electrochemistry and less towards chemical engineering. Closer collaboration with the second science has to be established in the future, particularly in solving problems arising in cell scaling and optimisation, and particularly those with three-dimensional electrodes (their implementation in different separation/purification processes is not as expected), as well as those with membrane processes. Knowledge about transport processes across membranes, for example, is not complete yet, in spite of the fact that many industrial membrane processes exist.

How to design and construct more efficient cells and large processes for electrowinning of copper to work at lower specific energy consumption? Few new hydrometallurgical processes are at present under development. The question is when they will be marketed. Each of them has metal electrowinning as the last stage in the whole technology chain. The contribution of electrochemical engineering to research and development of these processes could be extremely valuable.

How to design cells and the process for electrorefining of metals to improve their purity? In this view, fundamental principles of thermodynamics, kinetics, hydrodynamics, mass transport and potential and current distribution have to be transformed into engineering concepts in order to achieve economical operation and high quality products, lessening impurities into deposit as much as is permitted.

The contribution of electrochemical engineering to the research and development of these processes could be very valuable.

How to make fuel cells less expensive for wider applications? This is mainly connected with the kind and amount of materials to be built in, but also with fuel cells design and their scaling, with fuel production and storage and with many other things, which electrochemical engineering may contribute towards. How to make fuel cells capable of operating and surviving in extreme conditions?

What has electrochemical engineering still to do in the pollution abatement area? How to achieve a zero discharge of pollutants in both gaseous and liquid exit streams? Beside improvement of the existing technologies, a problem with the existing relationships, describing kinetics and thermodynamics, and with their validity at concentrations approaching zero, needs to be reconsidered. This is a great challenge on the theoretical level. Many opportunities exist in areas a few of which are listed below:

- Wider application of BDD electrode in wastewater treatment can be expected, which will allow a successful anodic destruction of hazardous organics from wastes.
- Electrokinetic soil remediation is an emerging technology that has attracted increased interest among scientists and officials because of its great potential.
- It can be expected that electrochemical promotion of catalyst (NEMCA effect) should emerge from the laboratory scale, turning a new page in exhaust gas purification.

- The appearance of room temperature ionic liquids (RTIL) opens new opportunities where electrochemical engineering has to find its place and role in establishing new processes based on a new approach.

How can electrochemical engineering contribute in materials recovery, waste minimisation and recycling? This problem is an area where electrochemical engineering is already involved, but it is also an area where there are a lot of still unsolved problems. In particular, this relates to the recycling of electronic devices and scrap and the recovery of valuable materials.

What could bioelectrochemistry and bioelectrochemical engineering, as sub-disciplines undergoing very fast development, do to establish new electrochemical processes, products and cells for different purposes?

What more can electrochemical engineering do in and for nano-technologies?

What should be next? Where then is the limit? I do not know. Do you?

In the end, what does electrochemical engineering mean nowadays compared to the definition given below the title, provided almost half a century ago? Do we have to redefine it according to the current state?

Acknowledgement *The author would like to express his gratitude to F. C. Walsh, G. Kelsall and F. Lapicque for their suggestions, help and useful critiques in composing this paper.*

References

[1] C.L. Mantell, *Electrochemical Engineering,* McGraw Hill, New York, USA, 1960
[2] C.S. Lin, E.B. Denton, N.S. Gaskil, G.L. Putnam, *Ind. Eng. Chem.,* **43 (9)** (1951) 2136–2143
[3] C.W. Tobias, M. Eisenberg, and C.R. Wilke, *J. Electrochem. Soc.* **99(12)** (1952) 359C-365C
[4] C. Wagner, in *Advances in Electrochemistry and Electrochemical Engineering*, Vol 2, P. Delahey and C.W. Tobias, Eds., Wiley Interscience, New York, USA, 1962
[5] R. Jotrand, F. Grunchard, *3rd Congr. Eur. Fed. Chem. Eng., Proceedings,* London, UK ,1962, p. 211
[6] H. Wendt, G. Kreysa, *Electrochemical Engineering*, Springer-Verlag, Berlin, Germany, 1999
[7] V. Stankovic, *Chem. Biochem. Eng. Q.* **21(1)** (2007) 33-45
[8] K. Yuettner, U. Galla, B. Schmieder, *Electrochim. Acta* **45** (2000) 2575-2594
[9] C. Comninellis, G. Chem; *Electrochemistry for the Environment,* Springer, New York - London, 2010
[10] D. Inman; Editorial Statement, *J. App. Electrochem.*; **1(1)** (1971) 1
[11] M. Grotheer, R. Alkire, R. Varian; *The Electrochemical Society's Interface* **15(1)** (2006) 52-54
[12] F. Lapicque; *Chemical Engineering Research and Design*, **82(A12)** (2004) 1571-1574

17

Synthesis and electrochemical performance of hierarchical Sb_2S_3 nanorod-bundles for lithium-ion batteries

XIAOZHONG ZHOU✉, SHUHUA HUA, LIANHUA BAI and DONG YU

Key Laboratory of Eco-Environment-Related Polymer Materials of Ministry of Education, Key Laboratory of Polymer Materials of Gansu Province, College of Chemistry and Chemical Engineering, Northwest Normal University, Lanzhou 730070, Gansu Province, P. R. China

✉Corresponding Author

Abstract

Uniform hierarchical Sb_2S_3 nanorod-bundles were synthesised successfully by L-cysteine hydrochloride-assisted solvothermal treatment, and were then characterised by X-ray diffraction, field emission scanning electron microscopy, and high-resolution transmission electron microscopy, respectively. The electrochemical performance of the synthesised Sb_2S_3 nanorod-bundles was investigated by cyclic voltammetry and galvanostatic charge–discharge technique, respectively. This material was found to exhibit a high initial charge specific capacity of 803 mA h g^{-1} at a rate of 100 mA g^{-1}, a good cyclability of 614 mA h g^{-1} at a rate of 100 mA g^{-1} after 30 cycles, and a good rate capability of 400 mA h g^{-1} at a rate of 500 mA g^{-1} when evaluated as an electrode candidate material for lithium-ion batteries.

Keywords

Sb_2S_3 nanorod-bundles, electrochemical performance, lithium-ion batteries

Introduction

Lithium-ion batteries (LIBs) are currently the most advanced rechargeable batteries for powering portable electronic devices such as laptop computers and cellular phones in view of their high energy density and benign design flexibility [1]. Nowadays, graphitic materials are extensively spread as commercial anode materials because of their low operating potential close to that of Li^+/Li and a good structural stability during cycling [2]. Unfortunately, the small theoretical specific capacity of the graphite anode ($Li_{1/6}C$, 372 mA h g^{-1}) restricts its future applications for powering electric vehicles (EVs). As a result, a lot of research efforts have been made to investigate various alternative anode materials with improved performance over the last decade [3]. Among alternative anode materials, metal sulfides have attracted particular attention because of their unique structures and high specific capacity [4,5].

As a typical metal sulfide, Sb_2S_3 is an important V-VI semiconductor. Owing to their prominent optical, photoelectronic and electrochemical properties, Sb_2S_3 nanomaterials can be potentially applied in photosensors [6], near-infrared optical devices [7], photoelectronic devices [8,9], lithium-ion batteries [10,11], etc. Zheng et al. synthesised Sb_2S_3 nanostructures with various dimensional nanostructures by a hydrothermal method and found that the reversible capacity of column-like superstructures, nanorods, and sheaf-like superstructures Sb_2S_3 electrodes are all around 700 mA h g^{-1} [10].

Recently, biomolecule-assisted synthetic routes have become a promising strategy in the preparation of various nanostructured materials because they are green chemistry approaches without toxic reagents and solvents and have obtained unique structures [12]. As an available biomolecule, L-cysteine has attracted considerable attention because of its special structure, which contains multifunctional groups ($-NH_2$, –SH, and –COO–) [13]. According to previous reports, L-cysteine can form a polymeric network structure under solution-phase reaction because its multifunctional groups can help to form interactions between L-cysteine molecules [14]. Besides, the presence of L-cysteine is critical to the formation of the final product, which showed excellent cycle stability with a high specific capacity and outstanding rate capability when used as a material for the anodes in lithium-ion batteries [15]. Therefore, it would be interesting to develop L-cysteine-assisted methods to prepare metal sulfide-based composites with excellent properties.

In this paper, we introduce a simple process for the fabrication of hierarchical Sb_2S_3 nanorod-bundles on the basis of a hydrothermal method assisted by L-cysteine hydrochloride (L-Cys·HCl). In this method, L-Cys·HCl can serve as a template, structure-directing agent and environmentally friendly sulfur source, while $SbCl_3$ serves as the antimony source. In addition, the formation mechanism of the Sb_2S_3 nanorod-bundles is also discussed, and it has been found that the hierarchical Sb_2S_3 nanorod-bundles exhibited high reversible capacity with good cyclic stability and high-rate capability.

Experimental

Synthesis

All chemicals were of analytical grade and were adopted without further purification. A typical solvothermal experiment for synthesising Sb_2S_3 material was conducted as follows. At first, 0.03 mol of L-cysteine hydrochloride with the formula $HSCH_2CH(NH_2)COOH{\cdot}HCl$ and 0.01 mol of $SbCl_3$ were successively dissolved in 50 mL of anhydrous ethanol under magnetically stirring for 0.5 h in air. Next, the above-prepared solution was transferred to a 100 mL stainless steel Teflon-lined autoclave, followed by being sealed and then heated inside a conventional oven at 180 °C for 8 h. At last, after the autoclave was naturally cooled to room temperature, a black precipitate was collected by centrifugation and was then washed thoroughly successively with anhydrous ethanol and deionised water. The collected precipitate was dried in vacuum at 80 °C overnight for later uses.

Material characterisations

The structure, composition, and morphology of the as-prepared Sb_2S_3 material were characterised by powder X-ray diffraction (XRD, Rigaku D/max 2400, operating with Cu Kα radiation of $\lambda = 0.15416$ nm), field emission scanning electron microscopy (FESEM, JEOL JSM-6701F, operating at 5 kV), and transmission electron microscopy (TEM, FEI TECNAI TF20, operating at 200 kV), respectively.

Electrochemical measurements

To evaluate the electrochemical performance of the as-prepared Sb_2S_3 material, galvanostatic charge-discharge (GSCD) and cyclic voltammetry (CV) techniques were employed. The GSCD technique was performed in the voltage range of 0.001-2.5 V *vs.* Li^+/Li on a battery testing system (LAND CT2001A, Wuhan Jinnuo Electronics Co., Ltd., China) at room temperatures. The CV technique was performed in the voltage range of 0.001-2.5 V *vs.* Li^+/Li at a scan rate of 0.2 $mV \cdot s^{-1}$ on an electrochemical workstation (Autolab PGSTAT128N, Metrohm, Switzerland) at room temperatures.

The testing electrodes were prepared by coating a copper foil substrate with a slurry comprising 80 wt. % active material of Sb_2S_3, 10 wt. % conducting additive of carbon black, and 10 wt. % binder of polyvinylidene fluoride (PVDF). This composition of the active material, the conducting additive, and the binder was found to be optimal for Sb_2S_3, The electrodes were cut into discs with a diameter of 10 mm, and then were assembled in a CR2032-type coin cell with a lithium foil as the counter electrode, a Celgard 2400 polypropylene foil as the separator, and a liquid solution of 1M $LiPF_6$ in ethylene carbonate (EC)/ethylmethyl carbonate (EMC)/dimethyl carbonate (DMC) (1:1:1 by volume, Shenzhen Capchem Technology Co., Ltd., China) as the electrolyte in an Ar-filled glove box. For clarification, discharging here refers to intercalation of Li into Sb_2S_3, whereas charging here refers to deintercalation of Li from Sb_2S_3.

Results and Discussion

Fig. 1 gives the XRD patterns of the as-prepared Sb_2S_3 and the standard peaks of Sb_2S_3 (JCPDS No. 42-1393). We can see that all of the obtained diffraction peaks could be well indexed to the orthorhombic Sb_2S_3 phase, and no obvious peaks of other crystalline phases were detected. This means that a well-crystallised form of Sb_2S_3 phase has been produced by our one-pot solvothermal treatment.

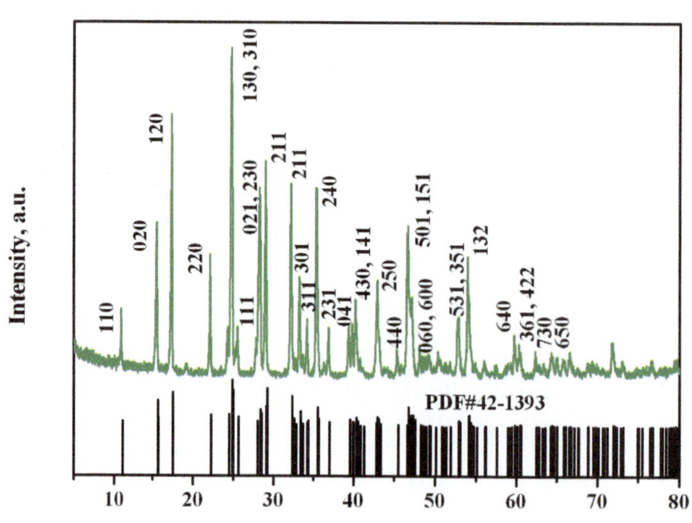

Figure 1. *XRD patterns of the as-prepared Sb_2S_3 material and standard peaks of Sb_2S_3 (JCPDS No. 42-1393)*

The morphology and structure of the as-prepared Sb_2S_3 material characterised by FESEM and TEM are presented in Fig. 2. It shows that the as-prepared Sb_2S_3 samples consist of cylindrical rods with a diameter of about 100 nm put into bundles. Fig. 2b shows the TEM and the relevant selected-area electron diffraction (SAED) patterns, which are composed of regular sharp diffraction spots characteristic of a single crystal of Sb_2S_3. This clearly implies that the nanorod grows preferentially along the [010] direction in a single crystalline form.

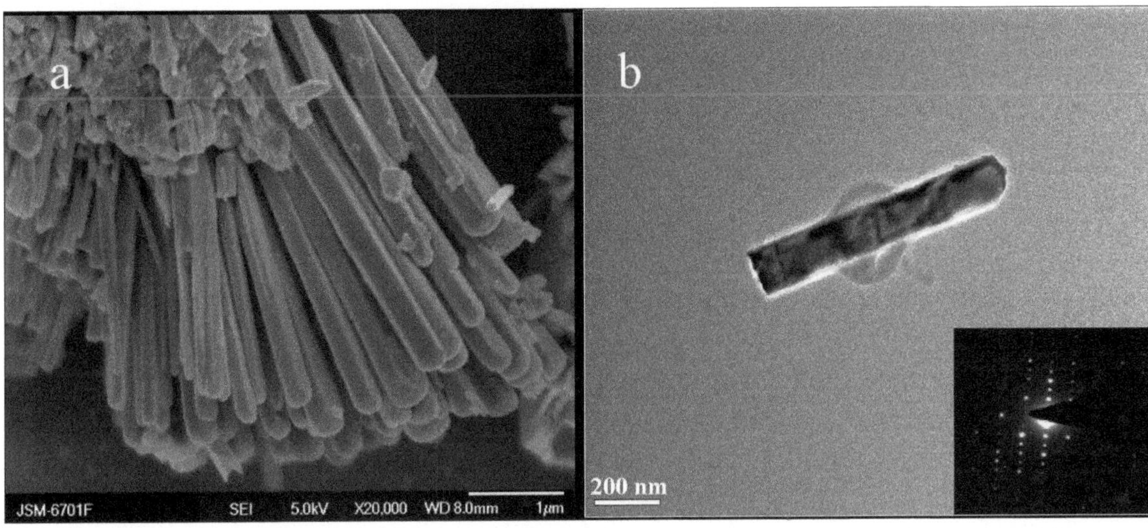

Figure 2. *Morphology and structure of the as-prepared* Sb_2S_3 *material characterised by* ***a*** *- FESEM and* ***b*** *- TEM*

Based on the previous results, it is reasonable to conclude that uniform hierarchical Sb_2S_3 nanorod-bundles are synthesised successfully by an L-Cys·HCl-assisted solvothermal treatment. L-Cys·HCl contains some multifunctional groups (-SH, $-NH_2$ and $-COO^-$) [16,17], which can be used for the conjugation of metallic ions or other functional groups [18]. When heated, L-Cys·HCl can release H_2S, which acts a sulfide source as well as a reducing agent, resulting in the formation of metal sulfide nanoparticles. The proposed growth process for the formation of hierarchical Sb_2S_3 nanorod-bundles is similar to the CoS nanowires [13], as shown in Fig. 3. And the reaction routes for the synthesis of Sb_2S_3 by L-Cys·HCl could be expressed as follows [12,19]:

$$HSCH_2CH(NH_2)COOH \cdot HCl + CH_3CH_2OH \rightarrow CH_3CH_2CH_2COCOOH + NH_4Cl + H_2S + H_2O \quad (1)$$

$$Sb^{3+} + H_2S + CH_3CH_2CH_2COCOOH \rightarrow Sb_2S_3 + SO_4^{2-} + CH_3COCOO^- + H_2O \quad (2)$$

The electrochemical properties of the as-prepared Sb_2S_3 material are first evaluated by a cyclic voltammetry (CV) test. Fig. 4 presents the CV curves of Sb_2S_3 sample for the initial three cycles in the voltage range of 2.5 to 0.001 V at a scanning rate of 0.2 mV s^{-1}. For the first cycle, there are two broad reduction peaks at around 1.0 and 0.4 V, respectively. The peak around 1.0 V can be attributed to the lithiation decomposition of the pristine Sb_2S_3 nanorods directly to fresh Sb and Li_2S. The peak around 0.4 V corresponds to the formation of Li_3Sb. Corresponding to the reduction peak, the oxidation peak at approximately 1.1 V was observed, which can be attributed to the dealloying process of Li_3Sb. Besides, three small and broad oxidation peaks at approximately 1.4 V, 1.9 V and 2.1 V may originate from the reconstruction of fresh Sb_2S_3 [10,11,20] and the transition of Li_2S into S, respectively, just as already found in CoSbS [21] and sulfide-graphene composite anodes [22].

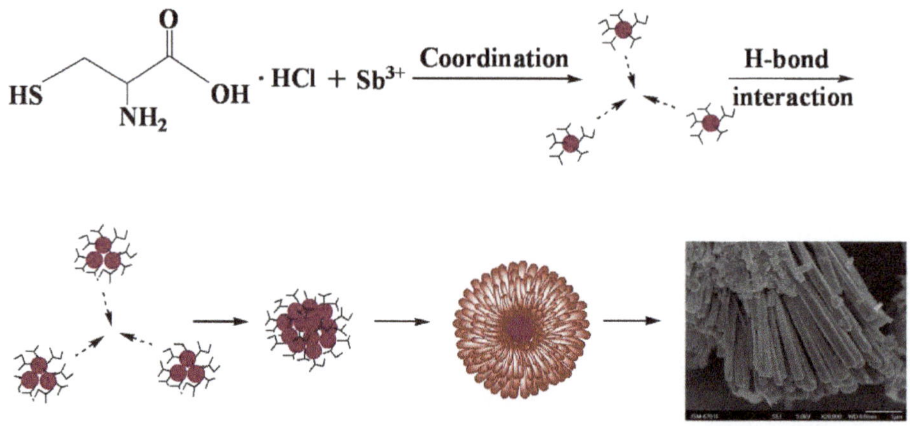

Figure 3. *The proposed growth process for formation of hierarchical Sb_2S_3 nanorod-bundles*

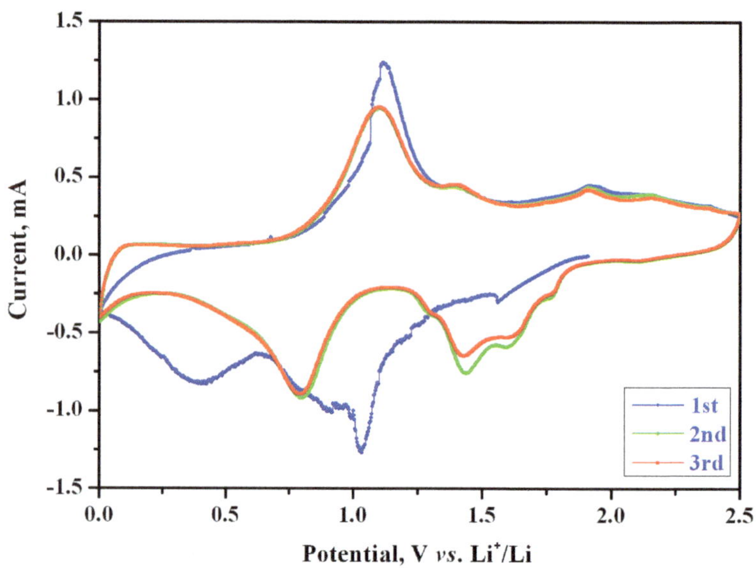

Figure 4. *CV curves of the as-prepared Sb2S3 material in the first three cycles at a scanning rate of 0.2 mV s−1*

From the second cycle on, we can see that the reduction peaks are different than that during the first cycle, which indicates a different electrochemical reduction mechanism. As shown in Fig. 4, the reduction and oxidation peaks at approximately 0.8 and 1.1 V originate from the alloying and dealloying process of fresh Sb metallic particles, respectively. The others are attributed to the decomposition/reconstruction of fresh Sb_2S_3 [10,11,20] and the electrochemical conversion between S and Li_2S [21,22]. Apparently, all the reduction peaks have their corresponding counterparts of oxidation peaks; this result suggests that the nanosized fresh Sb_2S_3 phase could be reversibly reconstructed and re-decomposed during repeated charging and discharging processes.

The potential profiles of the as-prepared Sb_2S_3 sample in cycles 1, 2, 5, 10, 30 and 49 are shown in Fig. 5. We can see that the discharging potential plateaus in the first cycle differ evidently from the following cycles, implying a different lithiation mechanism in the first cycle. This is in good agreement with the above CV analysis. The first discharging and charging specific capacities are 1166 mA h g^{-1} and 803 mA h g^{-1}, respectively, and the initial capacity loss could come from the irreversible loss of lithium ions due to the formation of a solid electrolyte interphase (SEI) layer [23], and the partial irreversibility of the electrochemical reaction (3) [20]. The initial coulombic efficiency is 68.9 %, which is higher than that (50 %) of alloying-dealloying mechanism.

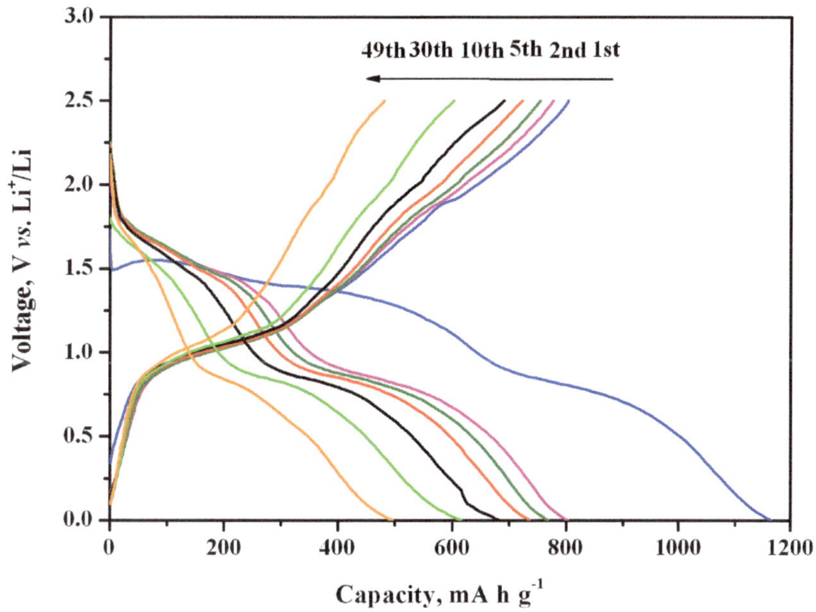

Figure 5. *Potential profiles of the as-prepared Sb_2S_3 sample in cycles 1, 2, 5, 10, 30 and 49 at a current density of 100 mA g^{-1}*

The cycle performance of the as-prepared Sb_2S_3 material at a current density of 100 mA g^{-1} is shown in Fig. 6. We can see that the Sb_2S_3 material displays a reversible capacity of 614 mA h g^{-1} even in the 30th cycle, the capacity retention after the 30th cycle is approximately 74.9 % with respect to the first charging specific capacity. During the cycling, the discharging specific capacity is slightly larger than the previous charging specific capacity, indicating that the electrode material becomes gradually activated, and this is beneficial to the capacity retention. These excellent electrochemical characteristics might be attributed to its special structure and smaller size of the nanorods. The sheaf-like Sb_2S_3 superstructures are composed of nanorods with a relatively smaller diameter; this special structure favors both the diffusion of the lithium ion and the electrolyte and Sb_2S_3.

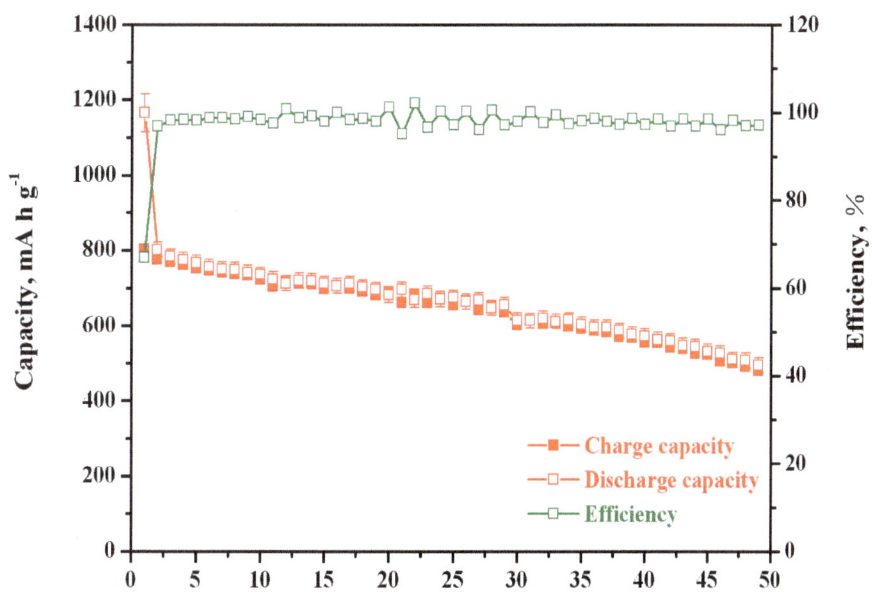

Fig. 6. *Cycle performance of the as-prepared Sb_2S_3 material at a current density of 100 mA·g^{-1}.*

The entire reaction in relation to Sb_2S_3 can be expressed as the following reaction [10,11,20-22].

$$Sb_2S_3 + 6Li^+ \leftrightarrow 2Sb + 3Li_2S \quad (3)$$

$$2Sb + 6Li^+ \leftrightarrow 2Li_3Sb \quad (4)$$

$$Li_2S \leftrightarrow Li^+ + S \quad (5)$$

The reversible specific capacity of the Sb_2S_3 is 473 mA h g^{-1} if we only consider the reversible reaction (3). However, if we consider both the reversible reactions (3) and (4), the reversible specific capacity of the Sb_2S_3 can be achieved as high as 946 mA h g^{-1}. In our work, the first reversible specific capacity of the Sb_2S_3 sample is as high as 803 mA h g^{-1}, and 68.9 % can be achieved for the initial coulombic efficiency. This suggests that our as-prepared Sb_2S_3 undergoes both the conversion reaction mechanism and alloying-dealloying lithiation mechanism, which contributes to the nanometer-sized effect. According to the nanometer-sized effect, the nanosized electrode material could have reactive activity. As a result, the reduction and oxidation of metal antimony could be observed, and the electrode exhibited larger capacity and better cycling performance than powder electrode [24].

The rate capability of the as-prepared Sb_2S_3 material is reflected in Fig. 7. We can see that the Sb_2S_3 material displays a good rate capability of 400 mA h g^{-1} at a current density of 500 mA g^{-1}. But when the charge/discharge current density changes from 1000 to 50 mA g^{-1}, the specific capacities of the Sb_2S_3 materials cannot return to the last values; the major reason for capacity fading is that Li_2S formed during the first discharge reaction and Li_2S_x (x>1, lithium polysulfide) are known to dissolve in the electrolyte [25]. A promising route to circumvent this drawback is by producing composite carbon materials whereby the carbon network provides good conductivity, prevents Li_2S and Li_2S_x from dissolving in the electrolyte and buffers the large volume changes induced by charging [26-28].

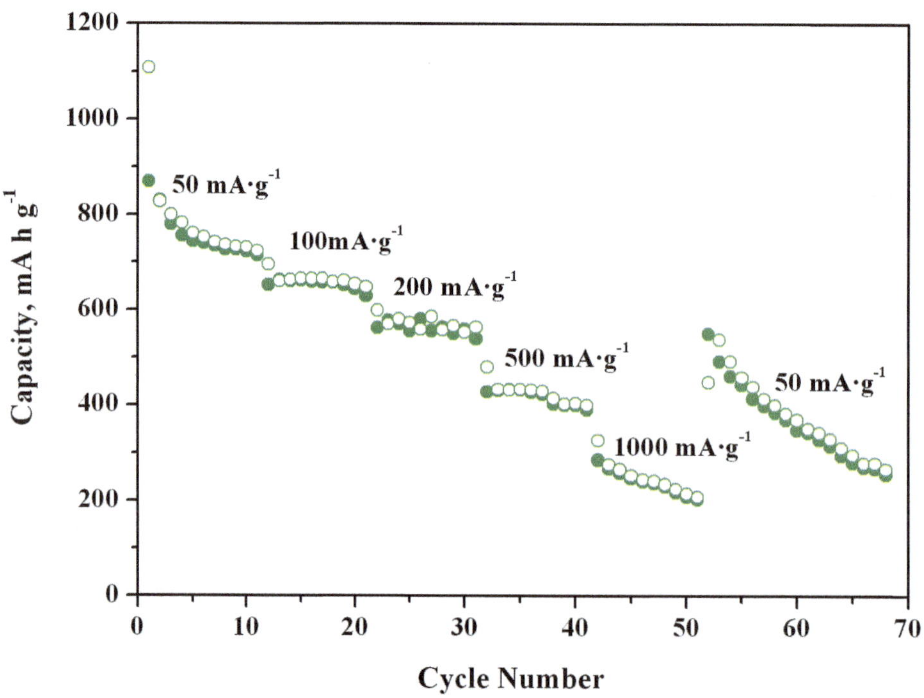

Fig. 7. Potential profiles of the as-prepared Sb_2S_3 material at varying current densities

Conclusions

In summary, uniform hierarchical Sb_2S_3 nanorod-bundles are synthesised successfully by a L-cysteine hydrochloride-assisted solvothermal treatment. The presence of L-cysteine is critical to the formation of the Sb_2S_3 material, which is found to display a high lithiation and delithiation specific capacity of 1166 and 803 mA h g^{-1} at a current density of 100 mA g^{-1}, a good cyclability of 614 mA h g^{-1} at a current density of 100 mA g^{-1} after 30 cycles, and a superior rate capability of 400 mA h g^{-1} at a current density of 500 mA g^{-1} when evaluated as an electrode candidate material for lithium-ion batteries. This good lithium storage performance can be ascribed to the nanosized structure. In addition, the preparative method could be a universal green chemistry approach to the synthesis of other metal sulfides.

***Acknowledgements**: Financial supports from the specialized research fund for the doctoral program of higher education of China under grant No. 20116203120005 and the natural science foundation of Gansu Province under grant No. 1107RJZA147 are acknowledged.*

References

[1] J. M. Tarascon and M. Armand, *Nature* **414** (2001) 359-367.
[2] M. Noel and V. Suryanarayanan, *J. Power Sources* **111** (2002) 193-209.
[3] W.-J. Zhang, *J. Power Sources* **196** (2011) 13-24.
[4] H. Hwang, H. Kim and J. Cho, *Nano Letters* **11** (2011) 4826-4830.
[5] J. Xiao, D. Choi, L. Cosimbescu, P. Koech, J. Liu and J. P. Lemmon, *Chem. Mater.* **22** (2010) 4522-4524.
[6] X. B. Cao, L. Gu, W. C. Wang, W. J. Gao, L. J. Zhuge and Y. H. Li, *J. Cryst. Growth* **286** (2006) 96-101.
[7] G. Y. Chen, W. X. Zhang and A. W. Xu, *Mater. Chem. Phys.* **123** (2010) 236-240.
[8] W. J. Lou, M. Chen, X. B. Wang and W. M. Liu, *Chem. Mater.* **19** (2007) 872-878.
[9] G. Y. Chen, B. Dneg, G. B. Cai, T. K. Zhang, W. F. Dong, W. X. Zhang and A. W. Xu, *J. Phys. Chem. C* **112** (2008) 672-679.
[10] J. M. Ma, X. C. Duan, J. B. Lian, T. Kim, P. Peng, X. D. Liu, Z. F. Liu, H. B. Li and W. J. Zheng, *Chem. Eur. J.* **16** (2010) 13210-13217.
[11] C. M. Park, Y. Hwa, N. E. Sung and H. J. Sohn, *J. Mater. Chem.* **20** (2010) 1097-1102.
[12] K. Chang and W. X. Chen, *ACS Nano* **5** (2011) 4720-4728.
[13] S.-J. Bao, C. M. Li, C.-X. Guo and Y. Qiao, *J. Power Sources* **180** (2008) 676-681.
[14] X. Y. Chen, H. L. Li, S. M. Wang, M. Yang and Y. X. Qi, *Mater. Lett.* **66** (2012) 22-24.
[15] S.-K. Park, S.-H. Yu, S. Woo, J. Ha, J. Shin, Y.-E. Sung and Y. Z. Piao, *Cryst.Eng. Comm* **14** (2012) 8323-8325.
[16] B. Zhang, X. Ye, W. Hou, Y. Zhao and Y. Xie, *J. Phys. Chem. B* **110** (2006) 8978-8985.
[17] S. Xiong, B. Xi, D. Xu, C. Wang, X. Feng, H. Zhou and Y. Qian, *J. Phys. Chem. C* **111** (2007) 16761-16767.
[18] J. Jiang, R. Yu, R. Yi, W. Qin, G. Qiu and X. Liu *J Alloy Compd* **493** (2010) 529-534.
[19] X.-L. Li and Y.-D. Li, *J. Phys. Chem. B* **108** (2004) 13893-13900.
[20] H. Yang, X. Su and A. Tang, *Mater. Res. Bull.* **42** (2007) 1357-1363.
[21] J.-O. Lee, J.-U. Seo, J.-H. Song, C.-M. Park and C.-K. Lee, *Electrochem. Commun.* **28** (2013) 71-74.
[22] M. Zhang, D. Lei, X. Z. Yu, L. B. Chen, O. H. Li, Y. G. Wang, T. H. Wang and G. Z Cao, *J. Mater. Chem.* **22** (2012) 23091-23097.
[23] J. Yan, B. J. Xia, Y. C. Su, X. Z. Zhou, J. Zhang and X. G. Zhang, *Electrochim. Acta* **53** (2008) 7069-7078.

[24] M.-Z. Xue and Z.-W. Fu, *Electrochem. Commun.* **8** (2006) 1250-1256.
[25] J.-S. Chung, H.-J. Sohn, *J. Power Sources* **108** (2002) 226-231.
[26] P. V. Prikhodchenko, J. Gun, S. Sladkevich, A. A. Mikhaylov, O. Lev, Y. Y. Tay, S. K. Batabya and D. Y. W. Yu, *Chem. Mater.* **24** (2012) 4750-4757.
[27] K. Chang, Z. Wang, G. Huang, H. Li, W. Chen and J. Y. Lee, *J. Power Sources* **201** (2012) 259-266.
[28] K.-J. Huang, L. Wang, J. Li and Y.-M. Liu, *Sens. Actuators, B* **178** (2013) 671-677.

18

Electrochemical combustion of indigo at ternary oxide coated titanium anodes

María I. León, Zaira G. Aguilar and José L. Nava[✉]

Departamento de Ingeniería Geomática e Hidráulica, Universidad de Guanajuato, Av. Juárez 77, Zona Centro, C.P. 36000, Guanajuato, Guanajuato, Mexico

[✉]Corresponding Author

Abstract

The film of iridium and tin dioxides doped with antimony (IrO_2-SnO_2–Sb_2O_5) deposited on a Ti substrate (mesh) obtained by Pechini method was used for the formation of •OH radicals by water discharge. Detection of •OH radicals was followed by the use of the N,N-dimethyl-p-nitrosoaniline (RNO) as a spin trap. The electrode surface morphology and composition was characterized by SEM-EDS. The ternary oxide coating was used for the electrochemical combustion of indigo textile dye as a model organic compound in chloride medium. Bulk electrolyses were then carried out at different volumetric flow rates under galvanostatic conditions using a filter–press flow cell. The galvanostatic tests using RNO confirmed that Ti/IrO_2-SnO_2-Sb_2O_5 favor the hydroxyl radical formation at current densities between 5 and 7 mA cm^{-2}, while at current density of 10 mA cm^{-2} the oxygen evolution reaction occurs. The indigo was totally decolorized and mineralized via reactive oxygen species, such as (•OH, H_2O_2, O_3 and active chlorine) formed in-situ at the Ti/IrO_2-SnO_2-Sb_2O_5 surface at volumetric flow rates between 0.1–0.4 L min^{-1} and at fixed current density of 7 mA cm^{-2}. The mineralization of indigo carried out at 0.2 L min^{-1} achieved values of 100 %, with current efficiencies of 80 % and energy consumption of 1.78 KWh m^{-3}.

Keywords

Dimensionally stable anodes; electrochemical degradation of organics; Pechini method; textile effluents; indigo textile dye

Introduction

Textile processing industries nowadays are widespread sectors in many countries. This industry is one of the most polluting industries in terms of the volume, color and complexity of its effluent discharge. Textile effluents include dyes that have a complex chemical structure, which most of the time are disposed on municipal sewers or into surface waters. Residual textile dyes tend to be transformed into toxic aromatic amines which cannot be degraded by sunlight and, once in the environment, they exhibit recalcitrant properties [1-3].

Electrochemical incineration [4-10] is a technique that has been found adequate for the treatment of colored wastewaters. It is important to point out that several color degradation studies mention systems with platinum electrodes [7] and dimensionally stable anodes (DSA) [6,8], which have shown mineralization of 50-70 %. Dogan and Turkdemir [7] consider that mineralization of indigo dye on Pt is induced by by-products of water and chloride discharge on the platinum surface; however, the indigo achieved mineralization of 60 %. Similar results in the degradation of acid red 29 [11], reactive blue 19 [8], mediated by active chlorine (given by the mixture of chlorine (Cl_2), hypochlorous acid (HOCl) and hypochlorite ion (OCl^-)), produced on DSA lead to mineralization of 56 % and 70 %, respectively. BDD electrodes exhibit a superior performance, since a large amount of hydroxyl radicals ($^\bullet OH$) are formed by water oxidation on the BDD surface [5-6,12-13], achieving 100 % efficiency in color removal and mineralization. The main problem encountered with BDD electrode is its high price limiting its industrial application.

For the above it is necessary to develop a DSA of metal oxides as an alternative to oxidize recalcitrant organic matter similar to a BDD electrode, in other words to produce DSA($^\bullet$OH) capable to oxidize recalcitrant organic matter. Comninellis and coworkers have developed a DSA electrode of SnO_2–Sb_2O_5 with an interlayer between supports (Ti) of IrO_2 by the spray pyrolysis technique, capable to produce hydroxyl radicals physisorbed on DSA (Eq. 1), by water discharge [14]. The interlayer of IrO_2 improves useful life of the electrode. These authors put on evidence that the physisorbed hydroxyl radical DSA($^\bullet$OH) cause predominantly the complete combustion of organics (*R*), Eq. (2); for example, these authors demonstrated that DSA($^\bullet$OH) reacts with *p*-clorophenol leading to complete combustion. On such electrode, IrO_2 acts as a catalyst, SnO_2 acts as a dispersing agent and Sb_2O_5 as a doping agent. Such ternary electrodes are among the best electrocatalysts for O_2 evolution, being able to produce physisorbed hydroxyl radicals on their surface from water discharge. The high catalytic activity of this ternary oxide electrode has been recently reported for the electrochemical oxidation of other organic compounds [15,16]. Another paper by Comninellis put on evidence the convenience of using Ti/SnO_2 to oxidize phenol matter via $^\bullet$OH radicals adsorbed onto Ti/SnO_2 [17]. However, the main problem encountered with the Ti/SnO_2 anode is its low stability under anodic polarization, which is not the case of the SnO_2–Sb_2O_5 coating having an IrO_2 interlayer between the Ti substrate [18].

$$\mathrm{DSA} + \mathrm{H_2O} \rightarrow \mathrm{DSA(^\bullet OH)} + \mathrm{H^+} + 1\mathrm{e^-} \quad (1)$$

$$\mathrm{R} + \mathrm{DSA(^\bullet OH)_z} \rightarrow \mathrm{CO_2} + z\mathrm{H^+} + z\mathrm{e^-} + \mathrm{DSA} \quad (2)$$

In a previous paper carried out by our group a film of iridium and tin dioxides doped with antimony oxide (IrO_2-SnO_2–Sb_2O_5) was deposited onto Ti substrate mesh and plate by the Pechini method [19]. The ternary oxide coating was used for the anodic decolorization of methyl orange (MO) azo dye via reactive oxygen species, such as ($^\bullet$OH, H_2O_2 and O_3) formed *in-situ* from water oxidation at the $Ti/IrO_2-SnO_2-Sb_2O_5$ surface. However, in that paper we did not follow the formation of $^{\bullet}OH$ at DSA surface and the electrochemical combustion of organic matter.

The indirect technique for the detection and identification of low concentration of $^{\bullet}OH$ radicals formed by water discharge at the oxide anodes involves trapping of the $^{\bullet}OH$ radical by an addition reaction (spin trap) to produce a more stable radical (spin adduct). A number of $^{\bullet}OH$ radical spin traps are available in the literature but *N,N*-dimethyl-*p*-nitrosoaniline (RNO) has demonstrated to be effective owing to the selective reaction of RNO with $^{\bullet}OH$ radicals, the high rate of the reaction with $^{\bullet}OH$ radicals (k = 1.2×10^{-4} M^{-1} s^{-1}) and the ease of application as one merely observes the bleaching of the sensitive absorption band at 440 nm [17, 20].

The goal of this manuscript is to prepare a film of iridium and tin dioxides doped with antimony ($IrO_2-SnO_2–Sb_2O_5$) onto titanium mesh (expanded metal) to produce $^{\bullet}OH$ radicals via water discharge for the electrochemical combustion of indigo textile dye (which resembles a denim laundry industrial wastewater). Bulk electrolyses were then carried out at different mean linear flow velocities and at constant current density using a filter–press flow cell. The integral current efficiency and the energy consumption of electrolysis were estimated. The detection of $^{\bullet}OH$ radicals formed by water discharge at the oxide anode using RNO as spin trap was also examined.

Experimental

Indigo dye solution was 1 mM indigo textile dye (536 ppm COD) in 0.05 M NaCl (which resembles a denim laundry industrial wastewater). The resulting solution exhibited a conductivity of 5.78 mS cm^{-1}, and a pH of 6.3 at 298 K. The solution was deoxygenated with nitrogen for about 10 minutes before each experiment. All the chemicals employed in this work were reactive grade.

Equipment

A potentiostat-galvanostat model SP-150 coupled to a booster model VMP-3 (20V-10A) both from Bio-Logic™ with EC-Lab® software were used for the electrolysis experiments. The potentials were measured versus a saturated calomel reference electrode (SCE), Bio-Logic model 002056RE-2B. All electrode potentials shown in this work are presented with regard to a standard hydrogen electrode (SHE).

COD analyses were performed using a dry-bath (Lab Line Model 2008), and a Genesys 20 spectrophotometer. Chloride volumetric titrations were confirmed by potentiometric measurements using a silver wire and a SCE, which was inserted in a glassy titration cell. The potential differences between silver wire and SCE were detected by a high impedance multimeter (Agilent-model-34401A). The colour removal was registered using a visible spectrophotometer (Genesys 20).

Microelectrolysis experiments

A 100-*mL* Pyrex electrochemical cell, with a three electrode system and nitrogen inlet was used for the construction of the anodic polarization curves. The working electrode was mesh-($IrO_2-SnO_2–Sb_2O_5$) with 1 cm^2 geometric area exposed to the electrolyte. The potentials were measured *vs.* SCE and the counter electrode was a glassy carbon. All the potential measurements shown in this work are presented with regard to standard hydrogen electrode (SHE).

A divided cell made of two compartment quartz cells of 3 mL capacity each one for the indirect detection of $^{\bullet}OH$ radicals was used. The anode was in the form of plate (1 cm^2) and the cathode was a vitreous carbon rod (1 cm^2). A home-made salt bridge to connect both semi-cells was employed; this was fabricated with vitreous Pyrex tube of 2 mm diameter sealed with Pt at the ends; this bridge was filled with phosphate buffer (pH 7.4). The quartz cell used as the anodic compartment was collocated into the UV-visible spectrophotometer (Perkin Elmer Lambda 35) to follow the bleaching (*in-situ*) of the yellow color of RNO during electrolysis.

Flow cell experiments

The flow cell FM01-LC that includes the turbulence promoter type D was used; the detailed description of this cell is depicted elsewhere [21]. In this work the spacer was 0.55 cm thick. DSA anode was a mesh-(IrO_2-SnO_2–Sb_2O_5), while platinum coated titanium flat sheet, was used as the cathode. DSA electrode was prepared by Pechini method described below. The platinum coated titanium was provided by De Nora. Details on the FM01-LC cell characteristics are given in Table 1.

***Table 1**. Mesh-(Ti/IrO_2-SnO_2-Sb_2O_5) electrode dimensions, experimental details of the FM01-LC electrolyzer.*

Electrode length, L	16 cm
Electrode height, B	4 cm
Electrode spacing, S	0.55 cm
Anode area, (Ti/IrO_2-SnO_2-Sb_2O_5)	112 cm^2
Cathode área, (Ti/platinized)	64 cm^2
Overall voidage, ε (Ti/IrO_2-SnO_2-Sb_2O_5)	0.93
Volumetric flow rate,	from 0.1 to 0.4 L min^{-1}

Overall voidage is the ratio of the free space in the channel to overall channel volume.

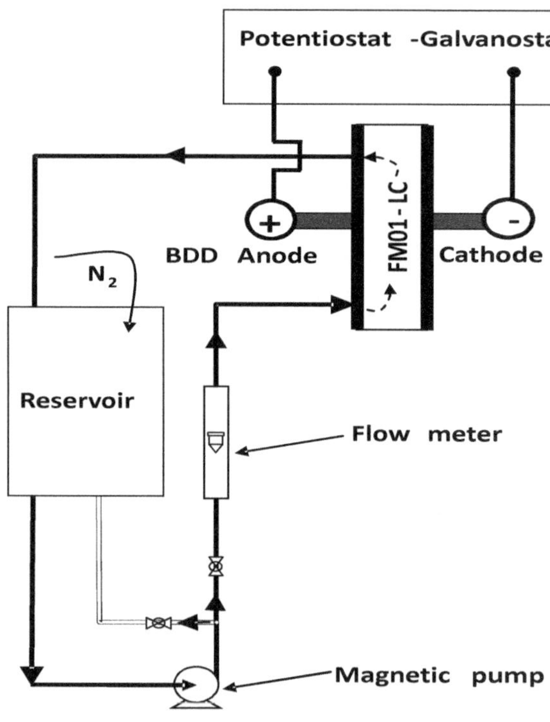

***Figure 1.** Electrical and flow circuit for the measurement of electrochemical incineration kinetics at FM01-LC electrolyzer.*

The undivided FM01-LC cell, with a single electrolyte compartment and the electrolyte flow circuit, is shown in Figure 1. The electrolyte was contained in a 1 L polycarbonate reservoir. A magnetically coupled pump of 1/15 hp March MFG, model MDX-MT-3 was used; the flow rates were measured by a variable area glass rotameter from Cole Palmer, model F44500. The electrolyte circuit was constructed from Master Flex tubing, C-Flex 6424-16, of 0.5 inch diameter. The valves and the three way connectors were made of PVC.

Scanning electron microscopy

Surface characterization of the metallic coating was performed using a SEM Carl Zeiss DSM 940A microscope. The energy of the primary electrons beam employed was 15 keV.

Methodology

Preparation of the DSA material

A ternary oxide (IrO_2-SnO_2–Sb_2O_5) film was deposited onto a Ti plate and mesh to be used in the three electrode cell and in the flow cell (Figure 1) by Pechini method using appropriate molar ratios of the oxide components. The precursor polymer solution was a mixture of citric acid (CA) in ethylene glycol (EG) at 60-70 °C. After total dissolution of the CA, $H_2IrCl_6xH_2O$, $SnCl_4$ and $SbCl_3$ were added to the mixture according to a molar composition of EG:CA:Ir:Sn:Sb as 16:0.12:0.0296:0.0296:0.0004, maintaining the temperature at 60-70 °C for 30 min. This mixture was then applied with a brush to both sides of the pre-treated Ti support. After the application of the coating, the electrode was heated at 100 °C for 5 min in a furnace in order to induce the polymerization of the precursor. This procedure was repeated eight times. After the final coating, the electrodes were maintained at 550 °C for 1 h in order to calcinate the polymer and form the ternary oxide (IrO_2-SnO_2–Sb_2O_5); XRD analysis confirmed at such at temperature these oxide phases are obtained [22]. The temperature did not exceed 600 °C to avoid the formation of TiO_2 that markedly reduces the electrocatalytic properties of the Ti/IrO_2-SnO_2–Sb_2O_5 coating due to passivation [23].

Microelectrolysis tests

Anodic polarization curves to determine the limits of potential and current density where the media is oxidized at Ti/IrO_2-SnO_2–Sb_2O_5 electrode were performed. These studies were carried out in the solution containing phosphate buffer (pH 7.4), and in the presence of 2×10^{-5} M RNO in the same buffer at room temperature (298 K). Anodic potential limit of 1.6 V *vs.* SHE was applied from open circuit potential (OCP) (0.82 V *vs.* SHE) using the linear sweep voltammetry technique at 50 mV s^{-1}. Based on these polarization curves the detection of hydroxyl radicals was performed.

Detection of hydroxyl radicals

In this paper RNO was used as spin trap for the detection of low concentration of $^{\bullet}OH$ radicals formed by water discharge at the Ti/IrO_2-SnO_2-Sb_2O_5 electrode and the bleaching of the yellow colour was measured during electrolysis [17]. RNO traps the $^{\bullet}OH$ radical by an addition reaction to produce a more stable radical (spin adduct), Eq. (3) [17].

$$\underset{\text{(spin trap)}}{\mathrm{RNO}} + {}^{\bullet}\mathrm{OH} \rightarrow \underset{\text{(spin adduct)}}{\mathrm{RN}\underset{\mathrm{OH}}{\underset{|}{\mathrm{O}}}{}^{\bullet}} \qquad (3)$$

It is important to mention that RNO is electrochemical inactive at Pt, SnO_2 and IrO_2 anodes [17,20]. A divided cell for the indirect detection of $^{\bullet}OH$ radicals was used (see microelectrolysis experiments section). Anodes screening tests were carried out in phosphate buffer (pH=7.4) containing 2×10^{-5} M RNO. Galvanostatic electrolyses at current densities of 5, 7 and 10 mA cm^{-2} applied to the Ti/IrO_2-SnO_2-Sb_2O_5 electrode were performed; at the same time the bleaching (*in-situ*) of the yellow color of RNO during electrolysis was followed. The same tests were performed using Pt plate (1 cm^2) as anode for which the surface •OH radical concentration is almost zero [17].

Electrochemical incineration in the filter-press flow cell.

Electrochemical incinerations of indigo were carried out in the FM01-LC cell equipped with mesh-(Ti/IrO_2-SnO_2-Sb_2O_5) at current density of 7 mA cm^{-2}, value determined from microelectrolysis studies, at different volumetric flow rates between 0.1–0.4 L $min.^{-1}$.

Incineration evolution was estimated by COD analysis of samples taken at different times. The COD values were determined by closed reflux dichromate titration method [24]. It is important to mention that estimating residual organic matter by COD analysis allowed eliminating any interference from chloride species. For this method, an excess of $HgSO_4$ was added and Ag_2SO_4 in the digestion and catalyst solutions, respectively, with the purpose of eliminating possible interferences from chloride species during the estimation of the residual organic matter from COD analysis [12].

The chloride concentration was evaluated by volumetric titration using a 0.5 M $AgNO_3$, confirmed by potentiometric measurements [12]. In addition, the color removal was determined by the decrease in absorbance at 639 nm, during electrolyzes.

Results and Discussion

Characterization of DSA

Figure 2 presents typical scanning electron micrographs for freshly prepared electrode Ti/IrO_2-SnO_2-Sb_2O_5. The surface morphology of the layer is characterized by the presence of crackers and plates. The presence of plates on the surface is probably due to the drastic heat treatment to which the sampled was submitted, that promoted the rapid exit of CO_2 gas originated from the decomposition of the organic polymer. EDX analyses focused on several plate structures show heterogeneous atomic percentage ratio of Sn and Ir (between 1.6 to 2.74), indicating that Sn segregates from other oxide to form a Sn rich deposit. Moreover, antimony was randomly detected along the electrode, showing that Sb is not homogeneously distributed along the electrode surface owing to its low content.

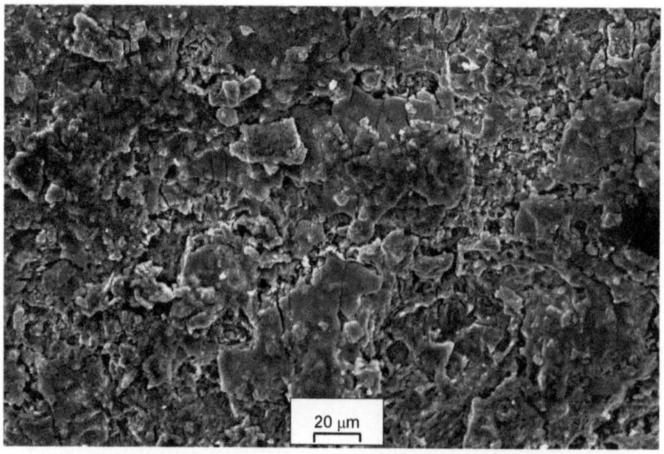

***Figure 2**. SEM images of Ti/IrO_2-SnO_2-Sb_2O_5.*

Figure 3 shows typical linear sweep voltammetries obtained on Ti/IrO_2-SnO_2-Sb_2O_5 electrode in the solution containing phosphate buffer (pH 7.4), and in the presence of 2×10^{-5} M RNO in the same buffer where no differences were detected. The fact that no changes were detected in both electrolytic solutions suggests the oxidation of water which is found in excess. Tafel slope performed on Ti/IrO_2-SnO_2-Sb_2O_5 from these curves (see inset), gives value of 190 mV dec^{-1}, which is different to that reported for Ti/IrO_2/SnO_2-Sb_2O_5 and Ti/Pt/SnO_2-Sb_2O_4, 120 and 204 mV dec^{-1} obtained at 298 K, respectively [12,25]; this difference is associated with the electrode composition and by the method of preparation.

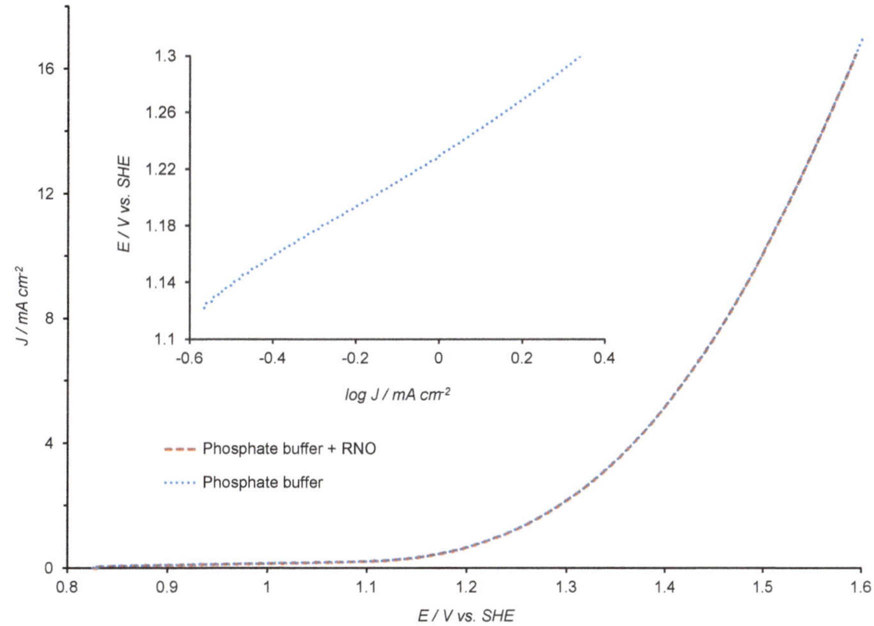

***Figure 3**. Typical linear sweep voltammetries on Ti/IrO_2-SnO_2-Sb_2O_5 anode. Electrolyte: phosphate buffer (pH 7.4), and phosphate buffer + 2×10^{-5} M RNO. The scan rate was 50 mV s^{-1}. The inset shows the Tafel plot for J-E curves for phosphate buffer.* A = *1 cm²*. T = *298 K.*

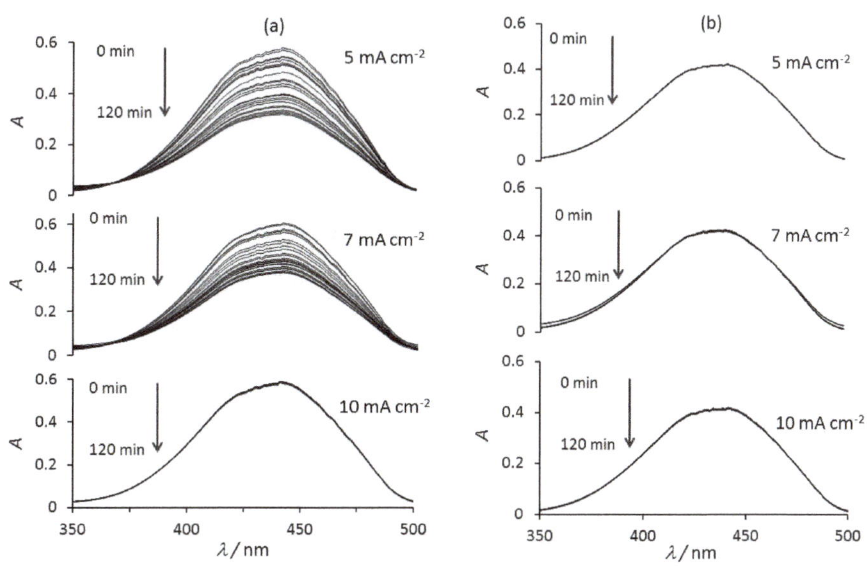

***Figure 4.** Absorbance spectra of RNO (2×10^{-5} M) in phosphate buffer (pH=7.4) obtained at 5 min intervals during galvanostatic electrolyses with Ti/IrO_2-SnO_2-Sb_2O_5 **(a)** and Pt **(b)** anodes.* A = *1 cm²*. T=*298 K.*

For screening tests of anodes we used RNO as spin trap of •OH radicals. Figure 4 shows the absorption spectrum of aqueous solution (2×10^{-5} M RNO) in phosphate buffer at pH 7.4 during galvanostatic electrolysis at 5, 7 and 10 mA cm^{-2} with Ti/IrO_2-SnO_2-Sb_2O_5 and Pt electrode. With Pt anode, there is no decrease in absorbance at 440 nm, at the three current densities, contrary to the Ti/IrO_2-SnO_2-Sb_2O_5 anode for which there is a rapid decrease in the absorbance at 5 and 7 mA cm^{-2}. These results show that there is accumulation of $^{\bullet}OH$ radicals at the Ti/IrO_2-SnO_2-Sb_2O_5 electrode surface contrary to Pt anode for which the surface $^{\bullet}OH$ radical is almost zero. The fact that the Ti/IrO_2-SnO_2-Sb_2O_5 anode at 10 mA cm^{-2} behaves similar to that Pt suggests that at such current density the accumulation of $^{\bullet}OH$ radicals is zero and the oxygen evolution reaction starts to appear. Therefore, according to the proposed reactions (Eqs. (1) and (2)) [14,17] the Ti/IrO_2-SnO_2-Sb_2O_5 will favor complete combustion of indigo textile dye at 5 and 7 mA cm^{-2}.

Electrochemical incineration of indigo textile dye in the FM01-LC using DSA electrode

Figures 5 (a) and (b) show the normalized color (detected at λ = 639 nm) and COD results obtained from experiments performed at constant current density (7 mA cm^{-2}) and variable volumetric flow rates. In these figures, the normalized color decreases faster than COD with the electrolysis time at different volumetric flow rates. COD kinetic was lower than that obtained for color decay owing to the slower combustion of by-products. However, color and COD depletion do not show marked improvement at the elevated volumetric flow rates.

Given that the presence of chloride ions (*i.e.*, 0.05 M in this study) is relevant due to the possible formation of active chlorine by oxidation at Ti/IrO_2-SnO_2-Sb_2O_5, the chloride consumption at the end of the electrolysis was measured (Figure 6), giving an average conversion between 15-40 %. This value did not show a marked dependence with hydrodynamics. This indicates that, despite the predominant role of Ti/IrO_2-SnO_2-Sb_2O_5 ($^{\bullet}OH$) as oxidant species, indigo and/or its by-products can be simultaneously destroyed by other oxidants such as dissolved chlorine gas, hypochlorous acid (HClO) and hypochlorite ion (ClO^-), as well as chlorate and perchlorate ions formed upon electro-oxidation with Ti/IrO_2-SnO_2-Sb_2O_5 electrode.

The complete combustion obtained here confirms that the $^{\bullet}OH$ radical, in addition to the other oxidants, are responsible for the oxidation of indigo, which does not occur on platinum electrodes, where the oxidation of indigo in chloride medium achieved 60 % in terms of COD [7]. The results obtained here are in agreement with other articles carried out by our group, where we achieved the complete combustion of indigo mediated by $OH^{\bullet}$ and active chlorine (produced on BDD in the same filter-press flow cell) [12,13].

The fact that hydrodynamics does not improve indigo oxidation and color removal may be associated with a complex mechanism of indigo degradation. HPLC studies would be helpful in the identification of possible indigo oxidation by-products; however, these were beyond the scope of the present work. It is important to point out that all of the electrolyses presented herein were carried out in the undivided FM01-LC cell, for which reason the degradation of indigo may also involve reactions at the cathode (Ti/Pt).

With the data obtained from COD for all of the electrolyses at their respective volumetric flow rates, integral current efficiency and energy consumption were analyzed as a function of percentage of indigo oxidation, for electrolyses performed at 7 mA cm^{-2}, Figure 7 (a)-(b). The estimation of integral current efficiency and energy consumption were determined using Equations (4) and (5) [12]:

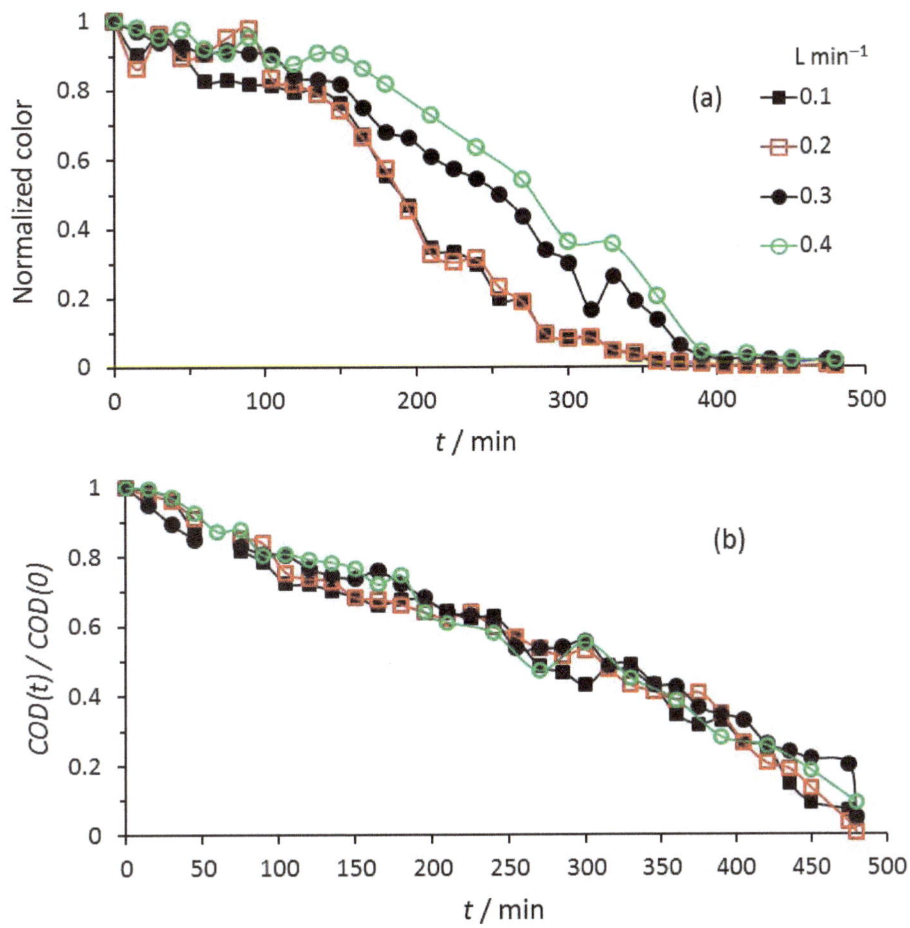

Figure 5. *Normalized color* (λ *= 639 nm)* ***(a)*** *and COD* ***(b)*** *decay during the electrolyses of indigo on (Ti/IrO$_2$-SnO$_2$-Sb$_2$O$_5$) in the FM01-LC electrolyzer. Electrolyte: 1 mM indigo in 0.05 M NaCl; this composition resembles a denim laundry wastewater.* A = *112 cm^2*, j = *7 mA cm^{-2}*, T = *298 K. Volumetric flow rates are shown in the figure.*

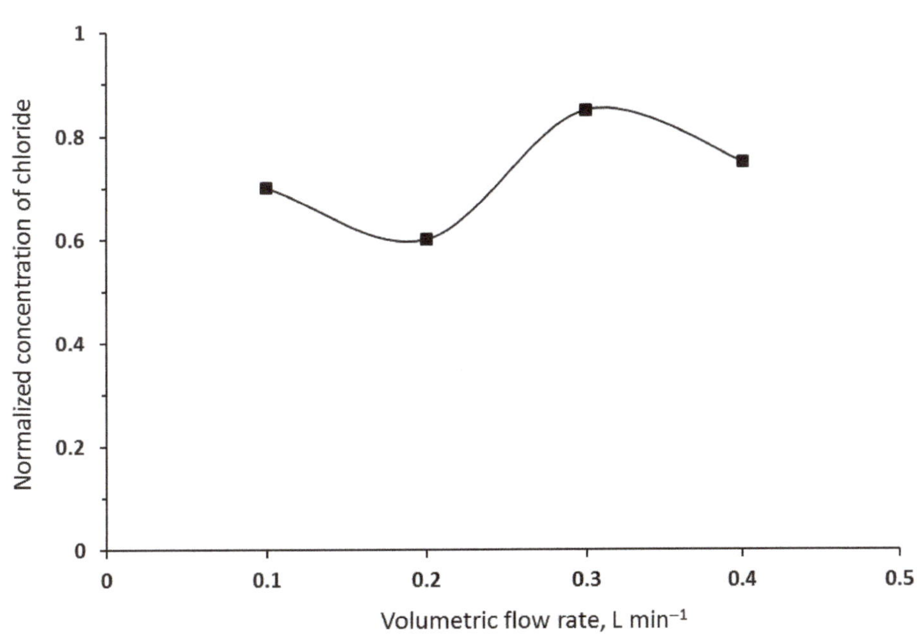

Figure 6*. Normalized concentration of chloride versus volumetric flow rates evaluated at the end of the electrolyses similar to those from Fig. 5(b). Electrolyte: 1 mM indigo in 0.05 M NaCl.* A = *112 cm^2*, j = *7 mA cm^{-2}*, T = *298 K. Volumetric flow rates are shown in the figure.*

$$\varphi = \frac{4FV[COD(0) - COD(t)]}{It} \tag{4}$$

$$E_c = \frac{4FE_{cell}}{\varphi V_m} \frac{1}{3.6} \tag{5}$$

where F is the Faraday constant, 96485 C mol^{-1}, V is the solution volume (cm^{-3}), COD*(0)* and COD*(t)* are the chemical oxygen demand initially and at time (t) of the electrolysis, in mol cm^{-3}, I is the applied current, in A, t is the time of electrolysis (s), E_{cell} is the cell potential in V, and V_m is the molar volume in cm^3 mol^{-1}. The value of 3.6 is a correction factor which converts E_c to units of KWh m^{-3}.

Figure 7(a) shows that current efficiency surpasses 100 % (theoretical value) at volumetric flow rates of 0.1 and 0.3 L min^{-1}, suggesting those indigo oxidation by-products and/or the processes taking place at the cathode enhance the degradation of indigo. A similar behavior was obtained in a previous communication carried out by our group [12], during indigo mineralization process in the same filter-press reactor. On the other hand, for the volumetric flow rates of 0.2 and 0.4 L min^{-1}, the current efficiencies were lower than that obtained for 0.1 and 0.3 L min^{-1}. It is important to remark that at the end of the electrolyses the current efficiency where 80 % for all volumetric flow rates studied, and there are no marked effects of the hydrodynamics on current efficiency in the set of electrolyses studied herein.

The analysis of Figure 7(b) shows that the energy consumption is not strongly influenced by hydrodynamics at 0.2-0.4 L min^{-1}. It is important to emphasize that the energy consumption is at least four times lower than those obtained in a previous paper, carried out by our group using the FM01-LC electrolyzer equipped with BDD electrodes in the same indigo solution [12]. This savings in energy consumption is due to the lower electrode polarization obtained using DSA (1.2 V) than the obtained on BDD (2.4 V), diminishing cell potential.

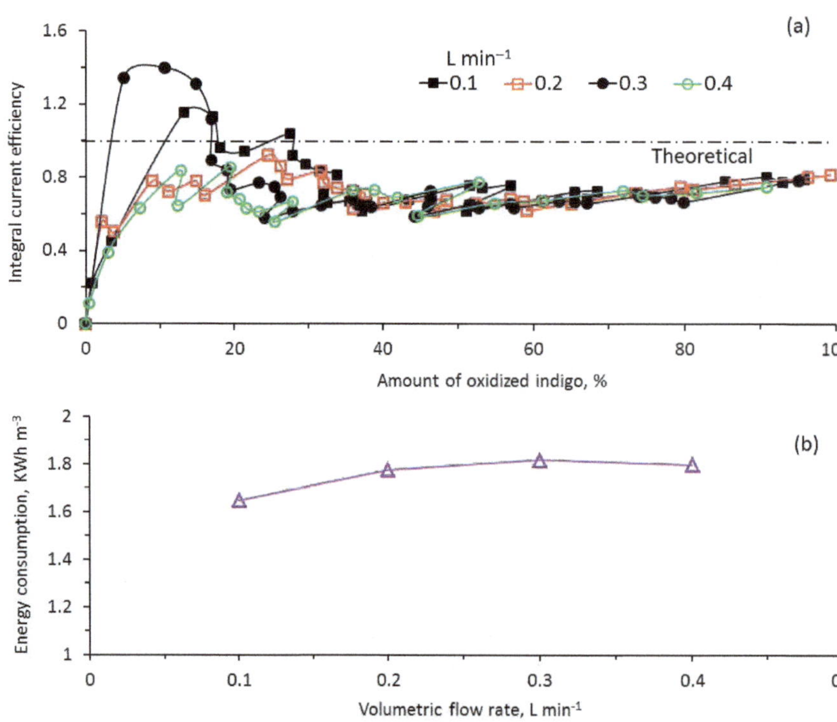

***Figure 7. (a)** Integral current efficiency versus percentage of oxidized indigo in the FM01-LC electrolyzer, evaluated from the electrolyses similar to those from Fig. 5(b). **(b)** Energy consumption versus volumetric flow rate evaluated at 88 % of degradation from the electrolyses similar to those from Fig. 5(b).*

The study presented here indicates that, despite the predominant role of Ti/IrO_2-SnO_2-Sb_2O_5(•OH) as oxidant species, indigo and/or its by-products can be simultaneously destroyed by other oxidants such as dissolved chlorine gas, hypochlorous acid (HClO) and hypochlorite ion (ClO^-), as well as chlorate and perchlorate ions formed upon electro-oxidation with Ti/IrO_2-SnO_2-Sb_2O_5 electrode.

Conclusions

The detection of •OH radicals formed by water discharge at Ti/IrO_2-SnO_2-Sb_2O_5 using N,N-dimethyl-*p*-nitrosoaniline (RNO) as a spin trap showed that exits an accumulation of •OH radical at Ti/IrO_2-SnO_2-Sb_2O_5 surface. Therefore, the Ti/IrO_2-SnO_2-Sb_2O_5 anode favors complete combustion of indigo by bulk electrolysis.

The galvanostatic tests using RNO as spin trap of •OH radicals confirmed that Ti/IrO_2-SnO_2-Sb_2O_5 will favor the hydroxyl radical formation at current densities between 5 and 7 mA cm^{-2}, while at current density of 10 mA cm^{-2} the oxygen evolution reaction occurs.

Electrolyses in a FM01-LC flow cell indicates, that despite the predominant role of Ti/IrO_2-SnO_2-Sb_2O_5 (•OH) as oxidant species, indigo and/or its by-products can be simultaneously destroyed by other oxidants such as dissolved chlorine gas, hypochlorous acid (HClO) and hypochlorite ion (ClO^-), as well as chlorate and perchlorate ions formed upon electro-oxidation with Ti/IrO_2-SnO_2-Sb_2O_5 electrode.

The mineralization of indigo carried out at 0.2 L min^{-1} and 7 mA cm^{-2} achieved values of 100 %, with current efficiencies 80 %, and energy consumption of 1.78 KWh m^{-3}. The FM01-LC equipped with mesh-(Ti/IrO_2-SnO_2-Sb_2O_5) improves space-time yield, allowing better interaction between mesh-(Ti/IrO_2-SnO_2-Sb_2O_5)(•OH) and organics, a phenomenon that increases organic mineralization efficiency.

In this manner, the complete mineralization of indigo with high current efficiency, obtained in this work is a notable improvement over those reported in the literature by using other DSA electrode. Additionally, the performance of the FM01-LC electrolyzer equipped with mesh-(Ti/IrO_2-SnO_2-Sb_2O_5) electrodes, demonstrate the convenience of using this electrochemical reactor as a pre-pilot cell for other water samples containing recalcitrant organic matter.

Acknowledgements: *María I. León and Zaira G. Aguilar thank CONACYT for the given grant. Authors are grateful to CONACYT and CONCYTEG for the economic support via the project FOMIX GTO-2012-C04-195057. Authors also acknowledge Universidad de Guanajuato for the financial support.*

References

[1] Y. Wong, J. Yu, *Water Research* **33** (1999) 3512-3520.
[2] P. Cañizares, F. Martínez, C. Jiménez, J. Lobato, M. A. Rodrigo, *Environmental Science & Technology* **40** (2006) 6418-6424.
[3] N. Mohan, N. Balasubramanian, V. Subramanian, Chemical *Engineering & Technology* **24** (2001) 749-753.
[4] M. Faouzi, P. Cañizares, A. Gadri, J. Lobato, B. Nasr, R. Paz, M. A. Rodrigo, C. Sáez, *Electrochimica Acta* **52** (2006) 325-331.
[5] X. Chen, G. Chen, *Separation and Purification Technology* **48** (2006) 45-49.
[6] X. Chen, F. Gao, G. Chen, *Journal of Applied Electrochemistry* **35** (2005) 185-191.
[7] D. Dogan, H. J. Turkdemir, *Journal of Chemical Technology and Biotechnology* **80** (2005) 916-923.

[8] D. Rajkumar, B. J. Song, J. G. Kim, *Dyes and Pigments* **72** (2007) 1-7.

[9] A. M. Faouzi, B. Nasr, G. Abbdellati, *Dyes and* Pigments **73** (2007) 86-89.

[10] P. Cañizares, A. Gadri, J. Lobato, B. Nasr, M. A. Rodrigo, C. Saez, *Industrial & Engineering Chemistry Research* **45** (2006) 3468-3473.

[11] F. H. Oliveira, M. E. Osugi, F. M. M. Paschoal, D. Profeti, P. Olivi, M. V. B. Zanoni, *Journal of Applied Electrochemistry* **37** (2007) 583-592.

[12] E. Butrón, M. E. Juárez, M. Solis, M. Teutli, I. González and J. L. Nava, *Electrochimica Acta* **52** (2007) 6888-6894.

[13] J. L. Nava, I. Sirés, E. Brillas. *Environmental Science and Pollution Research* **21** (2014) 8485-8492.

[14] C. L. P. S. Zanta, P. A. Michaud, C. Comninellis, A. R. de Andrade, J. F. C. Boodts, *Journal of Applied Electrochemistry* **33** (2003) 1211-1215.

[15] M. Tian, L. Bakovic, A. Chen, *Electrochimica Acta* **52** (2007) 6517-6524.

[16] N. Matyasovszky, M. Tian, A. Chen, *Journal of* Physical *Chemistry A* **113** (2009) 9348-9353.

[17] C. Comninellis, *Electrochimica Acta* **39** (1994) 1857-1862.

[18] B. Correa-Lozano, Ch. Comninellis, A. De Battisti, Journal *of Applied Electrochemistry* **27** (1997) 970-974.

[19] R. Chaiyont, C. Badoe, C. Ponce de León, J.L. Nava, J. Recio, I. Sirés, P. Herrasti, F.C. Walsh, *Chemical Engineering & Technology* **36** (2013) 123-129.

[20] J. Muff, L. R. Bennedsen, E. G. Søgaard, *Journal of* Applied *Electrochemistry* **41** (2011) 599–607.

[21] C. J. Brown, F. C. Walsh, D. Pletcher, *Transactions of the Institute of Chemical Engineers*, **73A**, (1994) 196–205.

[22] X. Qin, F. Gao, G. Chen, *Journal of Applied* Electrochemistry **40** (2010) 1797–1805.

[23] M. P. Pechini, N. Adams, US Patent 3 3,330,697 (1967).

[24] APHA, AWWA, WPCF, Standard methods for the examination of water and wastewater, New York, USA, 1995.

[25] D. Santos, A. Lopes, M. J. Pacheco, A. Gomes, L. Ciríaco, *Journal of the Electrochemical Society* **161** (2014) H564–H572.

19

Enhanced electrocoagulation: New approaches to improve the electrochemical process

Carlos E. Barrera-Díaz✉, Gabriela Roa-Morales, Patricia Balderas Hernández, Carmen María Fernandez-Marchante* and Manuel Andrés Rodrigo*

Centro Conjunto de Investigación en Química Sustentable UAEM – UNAM, Carretera Toluca-Atlacomulco, km 14.5, Unidad El Rosedal, C.P. 50200, Toluca, Estado de México, México
**Departartment of Chemical Engineering. Faculty of Chemical Sciences & Technologies, Enrique Costa Building, Campus Universitario s/n 13071 Ciudad Real, Spain*

✉Corresponding Author

Abstract
Electrocoagulation is a promising technology for the removal of colloids from different types of wastewater and it has also demonstrated good efficiencies for the breaking-up of emulsions. It consists of the dissolution of aluminum or iron anodes, promoting the formation of coagulant reagents in wastewater that helps to coagulate pollutants and the formation of bubbles that favors the mixing (electroflocculation) and the removal of suspended solids by flotation (electroflotation). During the recent years, the combination of this technology with other treatment technologies has become a hot topic looking for a synergistic improvement in the efficiencies. This work aims to review some of the more recent works regarding this topic, in particular the combination of electrocoagulation with ozonation, adsorption and ultrasound irradiation.

Keywords
Electrocoagulation; ozonation; adsorption; ultrasound irradiation; pulse application

Introduction

Electrochemical treatment techniques have attracted a great deal of attention because of their versatility and environmental compatibility. Electrochemical reactions take place at the anode and the cathode of an electrolytic cell when an external direct current voltage is applied. In fact, the main reagent is the electron, which is a "clean reagent"[1,2] and this fact helps to explain the lower production of wastes associated to these technologies. Applications studied in the recent years range from the oxidation of organic pollutants contained in wastewater to the electroremediation of soils.

Electrochemical methods have also been used as coagulation processes to remove color and cloudiness from turbid industrial wastewater. In this application, the electrochemical process generated numerous flocculates, achieving high efficiency in clearing the wastewater[3,4]. The term electrocoagulation involves the in situ generation of coagulants by electrolytic oxidation of an appropriate sacrificial anode (iron or aluminum), which causes the dissolution of electrode plates into the effluent. Metal ions, at an appropriate pH, can form wide range of coagulated species and metal hydroxides that destabilize and aggregate particles or precipitate and adsorb the dissolved contaminants. Main stages involved in the electrocoagulation process using aluminum anodes have been previously identified [5,6]. The anodic process involves the oxidative dissolution of aluminum into aqueous solution as reaction (1) indicates as well as the oxidative dissociation of water as reaction (2) shows.

$$Al \rightarrow Al^{3+} + 3e^{-} \quad (1)$$

$$2H_2O \rightarrow O_{2(g)} + 4H^{+} + 4e^{-} \quad (2)$$

In the case of iron or steel anodes, it is not iron (III) but iron (II) the main product of the electrochemical process (Eq.3) [7]. Then, oxygen is known to be involved for further Fe^{2+} oxidation into Fe^{3+} (Eq. 4)

$$Fe_{(s)} \rightarrow Fe^{2+} + 2e^{-} \quad (3)$$

$$4\,Fe^{2+} + 4H^{+} + O_{2(g)} \rightarrow 4Fe^{3+} + 2H_2O \quad (4)$$

Once dissolved iron and aluminum, can participate in many chemical reactions (Eqs. 5-10). In fact, speciation of iron and aluminum during electrocoagulation is very complex [8,9] and the description of the interactions between pollutants and the coagulant species is one of the most relevant topics nowadays in this field [10-13].

$$M(OH)_4^{-} + H^{+} \leftrightarrows M(OH)_3 + H_2O \quad (5)$$

$$M(OH)_3 + H^{+} \leftrightarrows M(OH)_2^{+} + H_2O \quad (6)$$

$$M(OH)_2^{+} + H^{+} \leftrightarrows M(OH)^{+2} + H_2O \quad (7)$$

$$M(OH)^{+2} + H^{+} \leftrightarrows M^{+3} + H_2O \quad (8)$$

$$M(OH)_3(s) \leftrightarrows M^{3+} + 3OH^{-} \quad (9)$$

It is interesting that in electrocoagulation papers little attention has been paid on cathodic reactions. Regardless of whether iron or aluminum is used, the main reaction that is reported is water reduction (Eq. 10).

$$2H_2O + 2e^{-} \rightarrow H_{2(g)} + 2OH^{-}_{(aq)} \quad (10)$$

However, this reaction has three important implications on the electrocoagulation technology:

a. provides hydroxyl ions which then react in bulk solution with iron or aluminum cations to form insoluble species and other coagulants (Eqs. 5 to 9);
b. hydrogen gas is produced increasing turbulence. This process contributes in the destabilization of colloidal particles leading to flocculation (so-called electro-flocculation process), and
c. contribution to electroflocculation which is a simple process that floats pollutants (or other substances) by their adhesion onto tiny the bubbles formed by the hydrogen evolution [14] (so-called electroflotation process)

As a consequence of this complex interaction, the electrochemical cell combines several processes at the same time in the same reactor and this becomes a significant advantage of this type of processes as compared with conventional coagulation treatments. In particular, from the economical point of view they compare favorably with coagulation processes [15-17] in many applications.

As for coagulation processes, electrocoagulation highly depends on the wastewater pH and it becomes a critical parameter in the performance of this technology. This parameter determines the speciation of aluminum and iron and hence the primary coagulation mechanisms occurring in the electrocoagulation cell. In fact, pH is one of the key differences between coagulation and electrocoagulation as conventional coagulation acidifies the treated wastewater due to the acidic properties of the typical coagulants dosed (iron chloride, aluminum sulfate, etc.), which are known to behave as Lewis acid. These properties make necessary the neutralization of wastewater after the coagulation treatment and this process implies an undesired increase in the salinity of the effluent. On the other hand, electrocoagulation typically buffers the pH during the treatment in values within the range 8-9, which should be a proper value even for direct discharge and no further neutralization is required [18].

Recent studies shows many promising applications of electrocoagulation in the treatment of lowland surface water [19], water [6,20,21], metal plating wastes [22], other types of industrial wastewater [5,23-30], urban wastewater [31-33] and even in disinfection [34,35]. In fact, it is one of the most promising environmental technologies based on electrochemical engineering [36,37]. Electrochemical methods offer two main advantages over traditional chemical treatment: less coagulant ion is required and less sludge is formed [19,22,31]. In the recent years, the potential of this technology is tried to be even further increased by the synergistic combination with other treatment technologies. The objective of the present manuscript is to review the potential of electrocoagulation for the treatment of industrial effluents coupling it with four types of processes:

- Electrocoagulation-ozone processes
- Electrocoagulation- adsorption processes
- Electrocoagulation-ultrasound processes
- Electrocoagulation-pulses processes

2. Electrocoagulation-ozone processes

Ozonation implies the use of ozone in the treatment of wastewater. Ozone is a strong oxidant that oxidizes organic pollutants via two pathways: direct oxidation with ozone molecules and/or the generation of free-radical intermediates, such as the •OH radical, which is a powerful, effective, and non-selective oxidizing agent [38]. The ozonation process has the advantage of being able to be applied when the flow rate and/or composition of the effluents are fluctuating. However, the high cost of equipment and maintenance, as well as energy required to supply the process, constitutes some of the disadvantages. Moreover, ozonation process requires the transfer of ozone molecules from gas phase to liquid phase, where the attack on the organic molecules occurs. Therefore, mass transfer limitations are also a relevant factor to be considered in the oxidation process involving ozone. In many cases, the ozone consumption rate per unit of volume can be so high that mass transfer is the limiting step, reducing the process efficiency and increasing the operating costs [39]. In addition, the ozonation performance is affected by the presence of organic matter, suspended solids, carbonate, bicarbonate and chlorine ions and also by pH and temperature [40]. Some studies using real industrial wastewater have pointed out that ozone by itself does not achieve high levels of pollutant removal [41]. In particular, the oxidation of wastewater from molasses fermentation with ozone results in an effective color removal but is less effective in removing organic matter [42]. Similar results were obtained when ozone was used to treat textile wastewater, where ozone treatment proves to be very effective for complete color removal but provides only partial reduction of the chemical oxygen demand (COD) [43]. Also, previous research on ozone-coupled methods indicates that the ozonation of anaerobically pretreated wastes enhances significantly the organic removal in comparison to the ozonation of unpretreated wastes, and substrate conversions in the range of 40–67 % are obtained [44].

This behavior in the reduction of COD can be ascribed to the initial pH value of wastewater, where the decomposition of ozone in water to form hydroxyl radicals occurs through the following mechanism [45], where hydroxide ions initiate the reaction (Eqs. 11-16):

$$O_3 + OH^- \rightarrow O_2 + HO_2^- \quad (11)$$

$$O_3 + HO_2^- \rightarrow HO_2^{\cdot} + O_3^{\cdot -} \quad (12)$$

$$HO_2^{\cdot} \rightarrow H^+ + O_2^{\cdot -} \quad (13)$$

$$O_2^{\cdot -} + O_3 \rightarrow O_2 + O_3^{\cdot -} \quad (14)$$

$$O_3^{\cdot -} + H^+ \rightarrow HO_3^{\cdot} \quad (15)$$

$$HO_3^{\cdot} \rightarrow OH^{\cdot} + O_2 \quad (16)$$

According to reactions (11) and (12) the initiation of ozone decomposition can be artificially accelerated by increasing the pH value. Side reaction (Eq. 17) is a fast process and plays an important role in waters with low dissolved organic carbon and alkalinity [46] since it can reduce the oxidative capacity of the system:

$$OH^{\cdot} + O_3 \rightarrow HO_2^{\cdot} + O_2 \quad (17)$$

Regarding the combined process, Table 1 summarizes the main papers found in the literature. Typically, the iron provided by the electrochemical reactor is not enough to remove all the pollutants present in aqueous solution. Thus, the ozone contributes importantly to improve the pollutant removal. Initially, the ozone contribution in the integrated process increases the oxidation of pollutants that are dissolved in the solution and that cannot be eliminated via electrocoagulation. An advantage of supplying ozone into the reactor is that it promotes the mixing between the reactants and also maximizes the organics oxidation, that results in the decreasing of COD and color [28,47,48]. Furthermore, the ozone provides good mixing thought the reactor which improves the mass transfer. The ozone action also contributes to reduce the amount of sludge produced.

In addition to processes coming from the combination of the effects of the single treatment technologies, the combined process involves an increased hydroxide radical production because Fe^{2+} catalyzes ozone decomposition to generate hydroxyl radicals (Eqs. 18-20) in the well-known Fenton process. This process helps to explain the synergistic effect of the combination of both technologies and the resulting high efficiencies.

$$Fe^{2+} + O_3 \rightarrow (FeO)^{2+} + O_2 \quad (18)$$

$$FeO^{2+} + H_2O \rightarrow Fe^{3+} + HO\cdot + OH^- \quad (19)$$

$$FeO^{2+} + Fe^{2+} + 2H^+ \rightarrow 2Fe^{3+} + H_2O \quad (20)$$

Table 1. *Pollutant removal using coupled electrocoagulation - ozone processes*

Wastewater	**Process Conditions**	**Poll. Removal**	**Ref.**
C.I. Reactive Yellow 84	Ozone flow rate 20 mL min^{-1}, Iron electrodes; current density 15 mA cm^{-2}	85 % TOC 100 % color	[49]
Reactive Blue 19	Ozone flow rate 20 mL min^{-1}, Iron electrodes; current density 10 mA cm^{-2}	80 % TOC 96 % Color	[50]
Reactive Black 5	Ozone flow rate 20 mL min^{-1}, Iron electrodes; current density 10 mA cm^{-2}	60 % COD 94 % Color	[51]
Distillery effluent	Ozone flow rate 15 L min^{-1}; initial pH 6 Iron electrodes; current density 3 Adm^{-2}	83 % COD 100 % Color	[52]
Industrial wastewater	Ozone flow rate 23 L min^{-1}; initial pH 7 Iron electrodes; current density 26 mA cm^{-2}	63 % COD 90 % Turbidity	[42]
Red MX-5B	Ozone flow rate 0.5 L min^{-1}; initial pH 6.1 Iron electrodes current density of 1.5 mA cm^{-2}	100 % Color	[53]
Boat pressure washing wastewater	Iron and aluminium electrodes current density 17 mA cm^{-2}	88.46 %,TOC 76.28 % COD	[54]
Acid Orange 6 azo Dye	Ozone concentration 36 mg L^{-1}; initial pH 4.5 Iron electrodes current density 88.6 mA cm^{-2}	50 % TOC 80 % Color	[55]

Main challenge for this technology is the scale-up. Most of the studies are at the lab-scale or the bench-scale and typically efficiencies can be greatly improved if a proper scale up is carried out. The design of the reactor seems to be a critical point because it fixes the flow patterns and hence the interaction of the species formed by electrocoagulation with ozone. Another challenge for this technology is the production of ozone by simultaneous anodic oxidation, taking advantage of the possibilities of electrochemical technology to produce oxidants[56]. A good possibility could be the use of cells equipped with bipolar cells such as the recently proposed by Llanos *et al.* [35]

3. Combined electrocoagulation- adsorption processes

Adsorption is a very well-known water and wastewater treatment process, which is gaining prominence as a means of reducing metal ion and organic concentrations in industrial effluxents [57]. The biosorbents derived from dead biomass, are considered the cheapest and most abundant environmentally friendly option [58,59]. Nowadays, the development of inexpensive adsorbents for the treatment of wastewater is an important area in the environmental sciences [60,61].

The use of an electrochemical treatment in combination with adsorption as a pre-treatment step to enhance adsorption capability of biosorbents has been assessed in many cases. However, the applications must be carefully evaluated, because technical incompatibilities may arise. This combined technology demonstrates a very good efficiency in the removal of many different pollutants as it is shown in Table 2. The filtering capacity of the sorbent bed is an efficient treatment to remove the suspended solids produced by the electrocoagulation process while simultaneously it helps to remove all soluble pollutants that were not effectively trapped by the flocs. Most of the studies select aluminum instead of iron as anode because aluminum coagulants promotes neutralization coagulation processes instead of enmeshment into growing precipitates which helps avoiding operational problems in the filtering system.

Table 2. *Pollutant removal using coupled electrocoagulation-adsorption processes*

Wastewater	Process Conditions	Pollutant Removal	Ref.
Cr(VI)	Al electrodes, sorbent red onion skin, pH 3-6	97 % Cr	[62]
Cardboard paper mill effluents	Al electrodes, current density 4.41 mA cm^{-2} sorbent granular activated carbon, pH 5.3	99 % COD	[63]
Marine Blue Erionyl MR	Al electrodes, sorbent granular activated carbon, pH 6.0	100 % dye	[64]
Reactive Black 5	current density 277 A m^{-2} sorbent granular activated carbon, pH 7	100 % dye, 100 % COD, 100% Toxicity	[65]
Cr(VI)	Al-Fe, current density 26.7 mA cm^{-2} Sorbent granular activated carbon	99 % Cr(VI)	[66]
Indigo carmine	Al electrodes Sorbent granular activated carbon	99 % Colorant	[67]
Nakdong River	Al electrodes, Al-fiber filter	65 % TOC	[20]
Industrial Wastewater	Al electrodes, current density 45.45 A m^{-2} Sorbent Ectodermis of Opuntia, pH 8	84 % COD, 78 %, BOD5, 97 % color, 98 % turbidity, 99 % fecal coliforms	[68]

Thus, the coupling of electrochemical and adsorption processes might prove a judicious choice for treating industrial wastewater with mixtures of different types of pollutants including both organic and inorganic pollutants. This technology has also been studied in systems in which an adsorbent bed used for the fast removal of pollution from wastewater is continuously regenerated using electrolysis [69,70]. Efficiencies obtained are high enough to consider this technology as a promising choice in the treatment of many effluents polluted with organic species. Most studied found in the literature are carried out at the lab or bench-scale. As for the combination of electrocoagulation with ozone, it is expected that with a proper scale-up, which develop an efficient cell from the view point of the filtering, adsorption and electrochemical processes, efficiencies obtained would be even higher.

4. Combined electrocoagulation-ultrasound processes

The treatment of wastewater in an electrolytic cell by ultrasound irradiation is expected to improve significantly the kinetics and the effectiveness of the electrode processes taking place in the cell [71-73]. A number of favorable impacts of using ultrasound in electrocoagulation are the following:

- Destruction of the compact layer formed at the electrode surfaces by the products of electrode reactions.
- Decrease in the thickness of the diffuse part of the electrical double layer created at the electrode surface.
- Direct activation of the ions in the reaction zone at the electrodes by ultra-sound waves.
- Activation of the electrode surfaces by means of generation of defects in the crystal lattices of the electrodes.
- Local augmentation of the temperature at the electrode surfaces as a result of friction between the liquid and the surfaces.

However, the ultrasound used may cause a few negative effects directly related to the purification process, such as the following:

- Destruction of a part of the obtained colloidal hydroxides by the action of the acoustic waves. This means a diminution of the solid phase that takes part in the adsorption process and a diminution of the removed contaminations respectively.

- Destruction of a part of the formed adsorption layer at the surface of the colloidal particles and possible return of the adsorbed ions to the liquid phase.
- Disorganization of the migration processes in the medium by the ultra-sonic waves.

Main studies found in the literature regarding this combined technology are shown in Table 3.

Table 3. *Pollutant removal using coupled electrocoagulation- US irradiation processes*

Pollutant	Process Conditions	Pollutant Removal, %	Ref.
Cl - SO_4^{-2}	treatment time 60 min; Ultrasonic low-frequency Electrocoagulation Fe, current density 40 mA cm^{-2} Media Cl- 500 ppm pH 3.8; Media SO_4^{-2} 500 ppm pH 2.8; Mine water	Very important removals with an increase in the amount of sludge at 25Hz.	[71]
Cr(VI)	Ultrasonic and sludge obtained for Electrocoagulation Fe of pharmaceutical wastewater ; EC: Conditions: rpm = 150, pH = 7.0 and sludge = 10 g L^{-1}.; Sono-EC Conditions: frequency 30 kHz, pH = 7.0 and time = 100 min.	100 % of removal at 275 min when used 200 mg L^{-1} of Cr(VI); 100 % of removal at 190 min when used 200 mg L^{-1} of Cr(VI)	[72]
Cu(II)	Current 1.0 A. Electrolysis 8 h In the sonicated field, voltage and temperature were constantly increased, in order to maintain the same thermal conditions for non-sonicated solutions. Temperature was adjusted to match those during the sonicated process.	Electrolysis 100 and 200 mg L^{-1} removed 55 an 63 % of Cu(II) increasing concentration of Cu the removal was of 93 %; Sono-Electrolysis 100 and 200 mg L^{-1} removed 94.6 an 95.5 % of Cu(II) increaseing concentration of Cu the removal	[73]
Non-ionic surfactants (SA)	Current density 0.5–2.5 A dm^{-2}, treatment time 5–40 min, ultrasonic power density 0.5–3.0 W cm^{-2}. Frequency 22 ± 1 kHz. Treatment time: 10 min	68 % of AS only CE EC with US 90 %	[74]
car-washing wastewater	I=1.2 A pH= 6 treatment time 20 min	COD 68.77 % and turbidity 96.27 %	[75]

Main results of these studies show that the combined process promotes the flocculation through vigorous mixing and the oxidation through the formation of radicals that contribute to the enhancement of the efficiency of electrocoagulation processes by chemical polishing of the surface of the flocs and by the oxidation of soluble pollutants in the bulk. This fact helps to explain the high efficiencies reached. In this case, performance of iron electrodes is better than that of aluminum electrodes. This fact can be explained in terms of the enhanced performance of the enmeshment of the pollutant into growing metal hydroxide flocs which is much more important for iron than for aluminum coagulants.

Hence, sono-electrocoagulation treatment has demonstrated superior performances in treatment of industrial effluents than single electrocoagulation. However, and as it was described for the other two previous technologies, scale-up should be considered as a major challenge.

5. Electrocoagulation-pulsed processes

Pulsed electrocoagulation technology is a novel method for wastewater treatment. It uses the interactions of electrochemical technology and polarity reversal in an electrical field to induce dipole formation in nonpolar particles in the wastewater, thus enabling the formation of micro-aggregates of insoluble substances.

The aggregates formed are further assisted in forming macro-aggregates. Charge neutralization of ions or charged materials also takes place in the electrochemical reactor, turning them into insoluble, suspended substances in the wastewater. The neutralization process enhances the efficiency of removing electrical conductivity [76]. In Table 4, there are some examples of using electrocoagulation pulsed treatments, which show that has been done with wastewater of different origins. As it can be observed, it has proven effective in the treatment of urban wastewater and different types of industrial wastes.

Table 4. *Pollutant removal using coupled electrocoagulation-pulsed processes*

Wastewater	Process Conditions	Pollutant Removal	Ref.
Higher Cr(VI) concentrations	Cr(VI) initial concentrations (50- 1000 mg L^{-1}) Electrical energy consumption (EEC) range: 4-58 kWh m^{-3} wastewater, current density (CD): 56–222 A m^{-2}, operating time: 20–110 min, pH 3–9 ($pH_{optimum}$ 5), voltage: 15–25 V.	99 %	[47]
Synthetic solutions containing mercury(II)	Hg (II) 2×10^{-5} M,distance between the electrodes was 3 cm, current density ranging from 2.5 to 3.1 A dm^{-2}; charge loading 9.33- 15.55 F m^{-3}, iron and aluminum electrodes, 3 - 7.	99.9 %. With iron, 15 min of electrolysis was sufficient to reach the highest removal; aluminum required 25 min for the same result.	[77]
Solutions of a dye Dianix Yellow CC (DY) and Procion Yellow (PY)	Range pH (4-8), Current density (40-120 A m^{-2}) Frequency (200-900 Hz^{-1}), Operation time (100 min)	99 %	[78]
Industrial and municipal wastewater	Pilot plant of electrochemical treatment system (0.3 m^3h^{-1}). Ti/RuO_2–TiO_2 anode was larger than with a platinum anode	The removal of T-N, T-P, NH_4–N and COD was approximately 90 %	[79]
Berberine hydrochloride (BH) wastewater	Fe electrodes and Al electrodes. The optimal conditions of reaction time of 3.5 h, pulse duty cycle of 0.3, pulse frequency of 1.0 kHz, current density of 19.44 mA cm^{-2}, and electrode distance of 2.0 cm	90.1 % BH and 62.6 % COD	[80]
Dye wastewater	Fe electrodes and 1000 mg L^{-1} Dye solution in a 15 mins electrolyzing time	99.62% of color removal and 91.15% of COD	[81]
Old corrugated containerboard (OCC)-based Paper Mill Wastewater	Current density of 0 to 240 A m^{-1}, a hydraulic retention time of 8 to 16 min and a coagulant (anionic polyacrylamide) dosage of 0 to 30 mg L^{-1}	Electrical conductivity: 47.7 %; Suspended. Solids: 99.3 %;COD: 75 %	[76]
Cooking oil (1800 mg/L, scour (1000 mg/L) and sodium sulfate (1g/L)	Al electrodes, dimensions of 50×110×2 mm; AC power (SMD 30)	Passivation of Al electrodes is not observed	[82]
Electroplating wastewater	Having a pH of 4, voltage 2.5 V, hydraulic retention time of 15 minutes, current density of 25 A m^{-2}	99.5 %	[83]
Oil wastewater	Electrode distance of 3.3 cm, pH of 4, current density of 49.38 $mAcm^{-2}$, reaction time of 15 min and pole switching time of 10 s.	96.21%	[84]

Conclusions

Electrocoagulation has demonstrated to be a promising technology in the removal of pollutants from different types of wastewater. However its combination with other technologies can help to increase efficiency due to synergistic effects such as those derived from the formation of radicals in the ozonation (by interaction of ozone with iron (II)) or in the US irradiation. Results depend on the particular application (technology combined and type of wastewater) and should be evaluated carefully. Scale-up is the major challenge of this technology for the next years, although the very positive results obtained at the lab and bench scales make these studies very promising.

***Acknowledgements**: The authors wish to acknowledge the support given by the UAEM trough the project 3409/2013 M and financial support from CONACYT in project 153828 is greatly appreciated.*

References

[1] G. Roa-Morales, E. Campos-Medina, J. Aguilera-Cotero, B. Bilyeu, C. Barrera-Díaz, *Separation and Purification Technology* **54** (2007) 124-129.

[2] L. J. J. Janssen, L. Koene, *Chemical Engineering Journal* **85** (2002) 137-146.

[3] A. B. Ribeiro, E. P. Mateus, L. M. Ottosen, G. Bech-Nielsen, *Environmental Science & Technology* **34** (2000) 784-788.

[4] O. T. Can, M. Bayramoglu, M. Kobya, *Industrial & Engineering Chemistry Research* **42** (2003) 3391-3396.

[5] O. T. Can, M. Kobya, E. Demirbas, M. Bayramoglu, *Chemosphere* **62** (2006) 181-187.

[6] P. K. Holt, G. W. Barton, M. Wark, C. A. Mitchell, *Colloids and Surfaces A: Physicochemical and Engineering Aspects* **211** (2002) 233-248.

[7] K. Rajeshwar, J. Ibañez, Environmental Electrochemistry; fundamentals and applications in Pollution Abatement, Academic Press, United States of America, 1997.

[8] P. Canizares, F. Martinez, C. Jimenez, J. Lobato, M.A. Rodrigo, *Industrial & Engineering Chemistry Research* **45** (2006) 8749-8756.

[9] P. Canizares, C. Jimenez, F. Martinez, C. Saez, M.A. Rodrigo, *Industrial & Engineering Chemistry Research* **46** (2007) 6189-6195.

[10] P. Canizares, F. Martinez, C. Jimenez, J. Lobato, M.A. Rodrigo, *Environmental Science & Technology* **40** (2006) 6418-6424.

[11] E. Lacasa, P. Canizares, C. Saez, F. Martinez, M. A. Rodrigo, *Separation and Purification Technology* **107** (2013) 219-227.

[12] P. Canizares, F. Martinez, M. A. Rodrigo, C. Jimenez, C. Saez, J. Lobato, *Separation and Purification Technology* **60** (2008) 155-161.

[13] P. Canizares, F. Martinez, M. A. Rodrigo, C. Jimenez, C. Saez, J. Lobato, *Separation and Purification Technology* **60** (2008) 147-154.

[14] R. G. Casqueira, M. L. Torem, H. M. Kohler, *Minerals Engineering* **19** (2006) 1388-1392.

[15] P. Canizares, F. Martinez, C. Jimenez, C. Saez, M. A. Rodrigo, *Journal of Chemical Technology and Biotechnology* **84** (2009) 702-710.

[16] E. Yuksel, E. Gurbulak, M. Eyvaz, *Environmental Progress & Sustainable Energy* **31** (2012) 524-535.

[17] M. Bayramoglu, M. Kobya, O. T. Can, M. Sozbir, *Separation and Purification Technology* **37** (2004) 117-125.

[18] P. Canizares, C. Jimenez, F. Martinez, M. A. Rodrigo, C. Saez, *Journal of Hazardous Materials* **163** (2009) 158-164.

[19] J.-Q. Jiang, N. Graham, C. André, G.H. Kelsall, N. Brandon, *Water Research* **36** (2002) 4064-4078.

[20] Y. Cho, G. Ji, P. Yoo, C. Kim, K. Han, *Korean Journal of Chemical Engineering* **25** (2008) 1326-1330.

[21] E. Lacasa, C. Saez, P. Canizares, F. J. Fernandez, M. A. Rodrigo, *Separation Science and Technology* **48** (2013) 508-514.

[22] C. Barrera-Díaz, M. Palomar-Pardavé, M. Romero-Romo, S. Martínez, *Journal of Applied Electrochemistry* **33** (2003) 61-71.

[23] N. Abdessamad, H. Akrout, G. Hamdaoui, K. Elghniji, M. Ksibi, L. Bousselmi, *Chemosphere* **93** (2013) 1309-1316.

[24] X. Chen, G. Chen, P.L. Yue, *Separation and Purification Technology* **19** (2000) 65-76.

[25] B. Merzouk, B. Gourich, A. Sekki, K. Madani, C. Vial, M. Barkaoui, *Chemical Engineering Journal* **149** (2009) 207-214.
[26] B. Merzouk, M. Yakoubi, I. Zongo, J.P. Leclerc, G. Paternotte, S. Pontvianne, F. Lapicque, *Desalination* **275** (2011) 181-186.
[27] C. Barrera-Díaz, F. Ureña-Nuñez, E. Campos, M. Palomar-Pardavé, M. Romero-Romo, *Radiation Physics and Chemistry* **67** (2003) 657-663.
[28] C. Barrera-Díaz, G. Roa-Morales, L. Ávila-Córdoba, T. Pavón-Silva, B. Bilyeu, *Industrial & Engineering Chemistry Research* **45** (2005) 34-38.
[29] S. Zodi, O. Potier, C. Michon, H. Poirot, G. Valentin, J. P. Leclerc and F. Lapicque, *J. Electrochem. Sci. Eng.* **1(1)** (2011) 55-65.
[30] M. Kraljić Roković, M. Čubrić and O. Wittine, *J. Electrochem. Sci. Eng.* **4(4)** (2014) 215-225.
[31] E.A. Vik, D.A. Carlson, A.S. Eikum, E.T. Gjessing, *Water Research* **18** (1984) 1355-1360.
[32] L. Zaleschi, C. Teodosiu, I. Cretescu, M. Andres Rodrigo, *Environmental Engineering and Management Journal* **11** (2012) 1517-1525.
[33] M.A. Rodrigo, P. Canizares, C. Buitron, C. Saez, *Electrochimica Acta* **55** (2010) 8160-8164.
[34] S. Cotillas, J. Llanos, P. Canizares, S. Mateo, M. A. Rodrigo, *Water Research* **47** (2013) 1741-1750.
[35] J. Llanos, S. Cotillas, P. Canizares, M. A. Rodrigo, *Water Research* **53** (2014) 329-338.
[36] S. Bebelis, K. Bouzek, A. Cornell, M. G. S. Ferreira, G. H. Kelsall, F. Lapicque, C. P. de Leon, M. A. Rodrigo, F. C. Walsh, *Chemical Engineering Research & Design* **91** (2013) 1998-2020.
[37] I. Sires, E. Brillas, M. A. Oturan, M. A. Rodrigo, M. Panizza, *Environmental Science and Pollution Research* **21** (2014) 8336-8367.
[38] M. A. García-Morales, G. Roa-Morales, C. Barrera-Díaz, B. Bilyeu, M. A. Rodrigo, *Electrochemistry Communications* **27** (2013) 34-37.
[39] J. M. Britto, M. d. C. Rangel, *Química Nova* **31** (2008) 114-122.
[40] U. von Gunten, *Water Research* **37** (2003) 1443-1467.
[41] P. Cañizares, M. Hernández-Ortega, M. A. Rodrigo, C. E. Barrera-Díaz, G. Roa-Morales, C. Sáez, *Journal of Hazardous Materials* **164** (2009) 120-125.
[42] M. Hernández-Ortega, T. Ponziak, C. Barrera-Díaz, M. A. Rodrigo, G. Roa-Morales, B. Bilyeu, *Desalination* **250** (2010) 144-149.
[43] R. Rosal, A. Rodríguez, J. A. Perdigón-Melón, A. Petre, E. García-Calvo, *Chemical Engineering Journal.* **149** (2009) 311-318.
[44] P. Cañizares, R. Paz, C. Sáez, M. A. Rodrigo, *Journal of Environmental Management* **90** (2009) 410-420.
[45] R. Andreozzi, V. Caprio, A. Insola, R. Marotta*, Catalysis Today* **53** (1999) 51-59.
[46] E. Guinea, E. Brillas, F. Centellas, P. Cañizares, M. A. Rodrigo, C. Sáez, *Water Research.* **43** (2009) 2131-2138.
[47] E. Keshmirizadeh, S. Yousefi, M.K. Rofouei, *Journal of Hazardous Materials.* **190** (2011) 119-124.
[48] H. M. Menapace, N. Diaz, S. Weiss, *Journal of Environmental Science and Health, Part A* **43** (2008) 961-968.
[49] Z. Q. He, S. Song, J. P. Qiu, J. Yao, X. Y. Cao, Y. Q. Hu, J. M. Chen, *Environmental Technology* **28** (2007) 1257-1263.
[50] S. Song, J. Yao, Z. He, J. Qiu, J. Chen, *Journal of Hazardous Materials* **152** (2008) 204-210.
[51] S. Song, Z. He, J. Qiu, L. Xu, J. Chen, *Separation and Purification Technology* **55** (2007) 238-245.
[52] P. Asaithambi, M. Susree, R. Saravanathamizhan, M. Matheswaran, *Desalination* **297** (2012) 1-7.
[53] C.-H. Wu, C.-L. Chang, C.-Y. Kuo, *Dyes and Pigments* **76** (2008) 187-194.

[54] V. Orescanin, R. Kollar, K. Nad, *Journal of Environmental Science and Health, Part A* **46** (2011) 1338-1345.

[55] H.-J. Hsing, P.-C. Chiang, E. E. Chang, M.-Y. Chen, *Journal of Hazardous Materials* **141** (2007) 8-16.

[56] P. Canizares, C. Saez, A. Sanchez-Carretero, M. A. Rodrigo, *Journal of Applied Electrochemistry* **39** (2009) 2143-2149.

[57] K. K. H. Choy, J. F. Porter, G. McKay, *Langmuir* **20** (2004) 9646-9656.

[58] B. Nasernejad, T. E. Zadeh, B. B. Pour, M. E. Bygi, A. Zamani, *Process Biochemistry* **40** (2005) 1319-1322.

[59] L. K. Cabatingan, R. C. Agapay, J. L. L. Rakels, M. Ottens, L. A. M. van der Wielen, *Industrial & Engineering Chemistry Research* **40** (2001) 2302-2309.

[60] K. Wagner, S. Schulz, *Journal of Chemical & Engineering Data* **46** (2001) 322-330.

[61] A. K. Jain, V. K. Gupta, S. Jain, Suhas, *Environmental Science & Technology* **38** (2004) 1195-1200.

[62] Y. Ait Ouaissa, M. Chabani, A. Amrane, A. Bensmaili, *Chemical Engineering & Technology* **36** (2013) 147-155.

[63] S. Bellebia, S. Kacha, A.Z. Bouyakoub, Z. Derriche, *Environmental Progress & Sustainable Energy* **31** (2012) 361-370.

[64] S. Bellebia, S. Kacha, Z. Bouberka, A.Z. Bouyakoub, Z. Derriche, *Water Environment Research* **81** (2009) 382-393.

[65] S.-H. Chang, K.-S. Wang, H.-H. Liang, H.-Y. Chen, H.-C. Li, T.-H. Peng, Y.-C. Su, C.-Y. Chang, *Journal of Hazardous Materials* **175** (2010) 850-857.

[66] N. Vivek Narayanan, M. Ganesan, *Journal of Hazardous Materials* **161** (2009) 575-580.

[67] M. S. Secula, B. Cagnon, T. F. de Oliveira, O. Chedeville, H. Fauduet, *Journal of the Taiwan Institute of Chemical Engineers* **43** (2012) 767-775.

[68] I. Linares-Hernández, C. Barrera-Díaz, G. Roa-Morales, B. Bilyeu, F. Ureña-Núñez, *Journal of Hazardous Materials* **144** (2007) 240-248.

[69] P. Canizares, J. Lobato, J. Garcia-Gomez, M. A. Rodrigo, *Journal of Applied Electrochemistry* **34** (2004) 111-117.

[70] P. Canizares, J. A. Dominguez, M. A. Rodrigo, J. Villasenor, J. Rodriguez, *Industrial & Engineering Chemistry Research* **38** (1999) 3779-3785.

[71] V. K. Kovatcheva, M. D. Parlapanski, *Colloids and Surfaces A: Physicochemical and Engineering Aspects* **149** (1999) 603-608.

[72] M. N. Kathiravan, K. Muthukumar, *Environmental Technology* **32** (2011) 1523-1531.

[73] R. Farooq, Y. Wang, F. Lin, S. F. Shaukat, J. Donaldson, A. J. Chouhdary, *Water Research* **36** (2002) 3165-3169.

[74] V. G. Sister, E. V. Kirshankova, *Chemical and Petroleum Engineering* **41** (2005) 553-556.

[75] J. Y. Chu, Y. R. Li, N. Li, W. H. Huang, *Advanced Materials Research* **433** (2012) 227-232.

[76] P. Yuan-Shing, W. E. I-Chen, Y. Shih-Tsung, C. An-Yi, S. Chi-Yuan, *Water quality research Journal of Canada* **42** (2007) 63-71.

[77] C. P. Nanseu-Njiki, S. R. Tchamango, P. C. Ngom, A. Darchen, E. Ngameni, *Journal of Hazardous Materials* **168** (2009) 1430-1436.

[78] M. Eyvaz, M. Kirlaroglu, T. S. Aktas, E. Yuksel, *Chemical Engineering Journal* **153** (2009) 16-22.

[79] C. Feng, N. Sugiura, S. Shimada, T. Maekawa, *Journal of Hazardous Materials* **103** (2003) 65-78.

[80] M. Ren, Y. Song, S. Xiao, P. Zeng, J. Peng, *Chemical Engineering Journal* **169** (2011) 84-90.

[81] Z. Cao, L. Sun, X. Cao, Y. He, *Advanced Materials Research* **233-235** (2011) 444-451.

[82] X. Mao, S. Hong, H. Zhu, H. Lin, L. Wei, F. Gan, *Journal of Wuhan University of Technology-Mater. Sci. Ed.* **23** (2008) 239-241.
[83] Q. Yuan, S. Yongqi, Z. Xiangyang, L. Desheng, K. Wu, *Chinese Journal of Environmental Engineering* **6** (2009) 1029-1032.
[84] Y.-F. Xiang, Z.-M. Xie, Y. Zou, *Journal of Chongqing University*, **9(1)** (2010) 41-46.

20

Electrochemical studies on zirconium phosphoborate based heterogeneous membranes

SANDEEP KAUSHAL, PRITPAL SINGH* AND SUSHEEL K. MITTAL**,✉

Department of Applied Sciences and Humanities, Punjab Technical University, Jalandhar, India
**Sri Guru Granth Sahib World University, Fatehgarh Sahib (Pb), India*
***School of Chemistry and Biochemistry, Thapar University, Patiala 147004, India*

✉Corresponding Author

Abstract

Electrode potential measurements have been applied to study electrical characteristics like transport numbers, permselectivity & fixed charged density of zirconium phosphoborate ion exchange membranes. The potential measurements were made across the cation exchange membrane maintained at 27±0.1 °C, using halide and nitrate salts of alkali and alkaline earth metals as electrolytes. The membrane potentials, transport numbers and permselectivity values increase with increase in average concentration from 0.0055 M to 0.0495 M for 1:1 and 1:2 electrolytes. With the increase in concentration of the electrolyte, the number of counter ions interacting with the membrane surface increases leading to enhanced Donnan exclusion responsible for the increase of transport numbers. Fixed charge density of the membrane (X) for 1:2 electrolytes is higher in magnitude than for 1:1 electrolytes indicating that the cation exchange is taking place as hydrated species. This hypothesis is supported by higher transport numbers for alkaline earth metal ions than alkali metal ions throughout the range of concentration.

Keywords

Electrochemical studies, Ion-exchange membrane, transport numbers, alkali metal ions, ZrPB

Introduction

Inorganic ion-exchangers are stable towards chemical degradation and are more selective for metal ions over their organic analogs [1]. The ion-exchange membranes have diverse applications, from desalting of brackish water to treating industrial effluents as well as processing biological effluents. Due to the development of new ion-exchange membranes with better selectivities, lower electrical resistance and improved electrochemical and chemical properties, the ion exchange membranes find applications in food, drugs, chemical processes and biotechnology industries [2,3]. The great interest in the ion exchange membranes is mainly due to their exceptional electrochemical properties.

The ion exchange membranes combine the ability to act as a separation wall between the two solutions, with chemical and electrochemical properties of ion exchanger. The most important of these are the pronounced difference in permeability for counter ions, co-ions and neutral molecules and their high electrical conductivity. When in contact with electrolyte solutions of low or moderate concentrations, the membrane contains a large number of counter ions but relatively few co-ions, due to Donnan exclusion. The membrane is perm selective for counter ions. The perm selectivity is reflected not only in differences in permeability, but also in electrical potential difference which arises between the two solutions (membrane potential).

The effect of membrane potential and adsorption on the permeability of ions, electrical conductance, ion exchange capacity and perm selectivity behavior on diffusion phenomena in a large number of inorganic membranes was studied by Malik *et. al.* [4-9]. Gnusin *et. al.* [10] investigated the concentration dependence of a wide range of transport characteristics of inorganic ion exchange membranes. To assess the suitability of any membrane for any specific application, two vital parameters i.e. the transport characteristics and its structural properties must be evaluated. The behavior of ion exchange membranes (IEMs) depends on the physiochemical properties, in addition to the nature of electrolyte solutions used. It is also beneficial to predict the behavior of prepared membranes on the basis of their structural properties. Few reports are available on investigation of the effect of nature of electrolyte solutions' on IEMs properties despite the fact that they have significant influences [11-16].

In this paper, electrode potential measurements have been used to study electrical characteristics like transport numbers, perm selectivity and fixed charge density across zirconium phosphoborate based heterogeneous ion exchange membranes. Halide and nitrate salts of alkali and alkaline earth metals were used as electrolytes. The effect of electrolyte concentration on transport numbers and perm selectivity on the ion exchange characteristics of zirconium phosphoborate membrane has also been studied. The membrane behaved as cation selective under ambient experimental conditions. The counter ion transport number, membrane perm selectivity and concentration of fixed ionic sites in the zirconium phosphoborate membrane are estimated by membrane potential data. Fixed ionic concentration data have been analyzed in the light of non-thermodynamic principle for its correlation with membrane structure and permeability behavior. The proposed membrane system has been used to study the selective behavior for the alkali/ alkaline earth metal ions.

Experimental

Reagents

Zirconyl oxychloride (Loba Chemie, India), boric acid (S.D. Fine Chem., India) and phosphoric acid (S.D. Fine Chem., India) were used for synthesis. All other chemicals used were of A.R. grade. Standard solutions were prepared by direct weighing of AR grade reagents using double distilled water.

Synthesis of zirconium phosphoborate

The exchanger was prepared by adding zirconyl oxychloride (0.1 M) solution to a continuously stirred mixture of boric acid solution (0.1 M) and phosphoric acid solution (0.1 M) at 60 °C, in the volume ratio 2:1:1. The gel produced was allowed in round bottomed flask to stand overnight. Then the gel was repeatedly washed with distilled water to remove chlorides from the mother liquor. The absence of chlorides in the mother liquor was tested with $AgNO_3$ solution. After the gel became free from chlorides, it was filtered through Whatmann No.1 filter paper using Buchner funnel and suction pump. The gel was transferred from Buchner funnel to petri dish. The precipitates in the Petri dish were dried in an air oven at 40 °C. When the gel dried completely, distilled water was added. Small granules of the ion-exchanger were formed with cracking sound.

Determination of ion-exchange capacity

Ion-exchange capacity was determined by column operation. Exchanger in the H^+-form was placed in a column containing a glass wool support. Sodium nitrate solution (1.0 M) was used as an eluent and about 400 ml of it was passed at a rate of 8-10 drops per minute through the ion exchanger column containing 1 g of exchanger. H^+ eluted from the column was determined titrimetrically against standard solution of sodium hydroxide.

Preparation of Membrane

A desired quantity of the finely ground zirconium(IV) phosphoborate (ZrPB) was added to epoxy resin in the ratio ZrPB : epoxy resin = 80 : 20 (w/w) with constant stirring till a homogeneous slurry was obtained. This slurry was spread between the folds of a filter paper and dried in air to get the membrane of ≅ 1 mm thickness. The dried membrane was dipped in distilled water to remove the filter paper. The sheet of the membrane thus obtained was cut with a sharp knife into circular discs of about 18 mm diameter. The membrane discs with good surface qualities were selected for further investigations

Electrode Assembly

The membrane was pasted on one end of the electrode assembly using araldite. The membrane was fixed from backside to other part of the electrode assembly. The electrode chambers were filled with 1.0 M solution of each of the electrolytes such as lithium chloride, sodium chloride, potassium chloride, magnesium chloride and barium chloride for 16 hours, to convert it into appropriate ionic form. After equilibration, the electrode assembly was washed with demineralized water (DMW). The membrane was kept immersed in DMW for 2 hours to remove the excess of electrolyte solution. When not in use, the electrode chambers were filled with demineralized water. The electrode assembly was kept immersed in water thermostat maintained at 27 ± 0.1 °C. Membrane potential measurements were made using saturated calomel electrodes as reference electrodes.

$$Ag/AgCl(s), Cl^- || solution\ C_1 | Membrane | solution\ C_2 || Cl^-, AgCl\ (s) | Ag$$

Potential measurements were made for different concentrations of the same electrolyte on two sides of the membrane in such a way that the concentration ratio $C_2/C_1 = 10$. The potential difference developed across the membrane was measured against Ag/AgCl reference electrode on either side of membrane, using a digital potentiometer. The electrode chambers were rinsed with electrolyte solution of next higher concentration and then filled with the same solution. The membrane was allowed to equilibrate for 2 hours and the new potential difference was then noted.

The membrane potentials across zirconium phosphoborate membrane were determined using some 1:1 electrolytes such as lithium chloride, sodium chloride, potassium chloride, sodium nitrate, lithium nitrate and potassium nitrate and some 1:2 electrolytes such as magnesium chloride, calcium chloride, strontium chloride, barium chloride, magnesium nitrate, calcium nitrate strontium nitrate and barium nitrate, in overall concentration range of 0.001 M to 0.1 M. The membrane potential measurements were reproducible to ± 0.1 mV.

Results and Discussion

Membrane Potentials

A membrane separating electrolyte solutions of unequal concentrations exhibits a difference in electrical potential due to unequal ionic mobility and is equal to liquid junction potential in a non-selective membrane. In an ideally selective membrane, according to the TMS theory [17], the membrane potential values are related to the activities of the electrolyte on the two sides of membrane, a_1 and a_2:

$$E_m = 2.303\frac{RT}{nF}\left(2\bar{t}_+ - 1\right)\log\frac{a_2}{a_1} \quad (1)$$

where E_m is membrane potential across the membrane.

The maximum electrical potential is given by:

$$[(E_m)_{t=0}]_{max} = \frac{RT}{nF}\ln\frac{a_2}{a_1} \quad (2)$$

provided the solutions are dilute. E_m changes with change in mean concentration of the electrolyte. From Table 1 and 2, it is evident that higher membrane potentials are observed at higher concentrations of electrolytes and membrane potential increases with increase in concentration of the electrolyte. For 1:1 halide electrolytes, membrane potentials are in order

$Li^+ > K^+ > Na^+$

Table 1: *Membrane potential values of 1:1 electrolytes for zirconium phosphoborate*

Concentration of electrolyte, M	Membrane potential, V					
	LiCl	NaCl	KCl	$LiNO_3$	$NaNO_3$	KNO_3
0.001-0.01	0.02	0.012	0.02	0.01	0.026	0.016
0.002-0.02	0.03	0.016	0.02	0.016	0.026	0.018
0.003-0.03	0.04	0.017	0.02	0.018	0.027	0.022
0.004-0.04	0.04	0.019	0.02	0.019	0.031	0.023
0.005-0.05	0.05	0.02	0.02	0.02	0.033	0.025
0.006-0.06	0.05	0.02	0.03	0.021	0.035	0.028
0.007-0.07	0.05	0.021	0.02	0.024	0.036	0.027
0.008-0.08	0.05	0.021	0.02	0.025	0.036	0.028
0.009-0.09	0.05	0.022	0.02	0.025	0.037	0.029
0.01-0.1	0.05	0.026	0.01	0.025	0.038	0.028

This order of the membrane potentials prevails in the concentration range 0.01 to 0.06 M, while for 1:1 nitrates, membrane potentials are in the order:

$K^+ > Na^+ > Li^+$

Table 2. *Membrane potential values of 1:2 electrolytes for zirconium phosphoborate*

Concentration of electrolyte, M	Membrane potential, V							
	$MgCl_2$	$CaCl_2$	$SrCl_2$	$BaCl_2$	$Mg(NO_3)_2$	$Ca(NO_3)_2$	$Sr(NO_3)_2$	$Ba(NO_3)_2$
0.001-0.01	0.021	0.025	0.017	0.034	0.019	0.020	0.0165	0.019
0.002-0.02	0.028	0.032	0.026	0.041	0.027	0.026	0.027	0.026
0.003-0.03	0.033	0.035	0.031	0.042	0.032	0.031	0.032	0.030
0.004-0.04	0.036	0.038	0.034	0.044	0.035	0.034	0.035	0.031
0.005-0.05	0.038	0.040	0.036	0.045	0.038	0.035	0.036	0.032
0.006-0.06	0.040	0.041	0.038	0.046	0.040	0.036	0.037	0.033
0.007-0.07	0.041	0.042	0.039	0.046	0.042	0.037	0.038	0.034
0.008-0.08	0.042	0.043	0.040	0.047	0.042	0.038	0.040	0.034
0.009-0.09	0.042	0.043	0.40	0.047	0.042	0.038	0.040	0.035
0.01-0.1	0.043	0.042	0.038	0.048	0.041	0.039	0.039	0.036

The ionic perm selectivity of membrane is also expressed quantitatively, based on migration of counter ions through cation exchange membrane [11,16,18-26]:

$$P_s = (\bar{t}_+ - t_+)/(1 - t_+)$$

Where $\bar{t}_+$ refers to the value of transport number in the membrane and t_+ is the transport number of counter ions in solution [27].

Ion exchange capacity of ZrPB

The ion exchange capacity of zirconium phosphoborate was determined for some monovalent and bivalent cations like Na^+, K^+, Mg^{2+} and Ca^{2+} cations and is given in Table 3.

Table 3. *Ion exchange capacity of zirconium phosphoborate for some alkali and alkaline earth metals*

Sr. No.	Metal ion	Salt solution used	Ion-exchange capacity, eq kg^{-1}
1	Na^+	NaCl	0.29
2	K^+	KCl	0.58
3	Ca^{2+}	$CaCl_2$	0.52
4	Mg^{2+}	$MgCl_2$	0.32

The ion exchange capacity of ZrPB has been found in the order $K^+ > Na^+$ and $Ca^{2+} > Mg^{2+}$. These results suggested that the ion exchange capacity decreases as the radii of hydrated metal ions increase.

The magnitude of membrane potential depends on:

1. Adsorption of anions of diffusing electrolyte on membrane surface
2. Exchangeability of cations
3. Size of cations
4. Diffusion potential across the membrane

Higher membrane potential for lithium is because of its small size and high positive field around it, hence, it establishes higher potential across the charged membrane. Higher the potential, higher is the permeability [28] as observed in our experiments in the concentration range (0.01 to

0.06 M). The trend in the membrane potential and hence greatest permselectivity for lithium ions is because ion mass transfer coefficient through the ion-exchange membrane increases as a function of ion size [29].

Ionic radii and hydrated ionic radii

Smaller the size of the alkali metal less is the ionic radius. As the size increases from top to bottom, ionic radii increase. The extent of hydration depends upon the size of the ion. Smaller the size of the ion, more highly it is hydrated and greater is its hydrated ionic radius and less is its ionic mobility (conductance). The radii and mobility of alkali metal ions are given in table 4.

Table 4. *Values of ionic radii, hydration numbers, hydrated ionic radii and ionic mobility of alkali metal ions*

Metal ions	Li^+	Na^+	K^+	Rb^+	Cs^+
Ionic Radius, pm	76	102	138	152	167
Hydration Number	25.3	16.6	10.5	10.0	9.9
Hydrated Radius, pm	340	276	232	228	228
Ionic Mobility	33.5	43.5	64.5	67.5	68

Transport Numbers

The transport number in exchanger phase is calculated from the slope of equation (1). The transport numbers and perm selectivity values of 1:1 halides and nitrates, and the transport numbers of 1:2 halides and nitrates are given in the Table 5 and 6, respectively.

Table 5. *Transport numbers and perm selectivity values at mean concentration for 1:1 electrolytes*

Mean concentration of electrolyte, M	Transport No. ($\bar{t}_+$)						Perm selectivity (P_s)					
	Metal halide electrolyte			Metal nitrate electrolyte			Metal halide electrolyte			Metal nitrate electrolyte		
	LiCl	NaCl	KCl	$LiNO_3$	$NaNO_3$	KNO_3	LiCl	NaCl	KCl	$LiNO_3$	$NaNO_3$	KNO_3
0.0055	1.08	0.53	0.75	0.45	1.15	0.8	1.12	0.23	0.51	0.179	1.247	0.609
0.011	1.44	0.71	0.67	0.74	1.17	0.85	1.65	0.52	0.35	0.613	1.280	0.706
0.0165	1.66	0.75	0.89	0.81	1.21	0.92	1.98	0.59	0.71	0.717	1.344	0.843
0.022	1.85	0.86	0.93	0.85	1.4	0.94	2.26	0.77	0.86	0.777	1.656	0.882
0.0275	2.02	0.89	0.97	0.92	1.47	0.97	2.51	0.82	0.94	0.881	1.771	0.941
0.033	2.06	0.91	0.99	0.94	1.55	1.03	2.56	0.85	0.98	0.911	1.899	1.059
0.0385	2.09	0.93	0.98	1.1	1.6	1.01	2.6	0.89	0.96	1.147	1.979	1.02
0.044	2.14	0.95	1.01	1.1	1.62	1.04	2.67	0.91	1.02	1.147	2.01	1.078
0.0495	2.16	0.96	1.02	1.12	1.66	1.05	2.7	0.93	1.04	1.176	2.075	1.098
0.055	2.22	1.2	0.39	1.13	1.71	1.03	2.78	0.32	0.19	1.190	2.145	1.059

It is observed that the transport numbers increase with increase in concentration of the electrolytes. This may be due to the fact that with increase in average concentration of the electrolyte, the number of counter ions interacting with the membrane surface increase leading to enhanced Donnan exclusion responsible for increase of transport numbers. The obtained results are in contrast with the Donnan equilibrium theory. The transport numbers tend to stabilize up to a mean concentration of 0.045 M. Thereafter, the values of transport numbers and permselectivity tend to stay constant. This is due to concentration polarization phenomenon at high concentration resulting in increased co-ion percolation and hence resisting further increase in transport number of cations.

Table 6. *Transport numbers at mean concentration for 1:2 electrolytes*

Mean concentration of electrolyte, M	Transport No. $\bar{t}_+$							
	$MgCl_2$	$CaCl_2$	$SrCl_2$	$BaCl_2$	$Mg(NO_3)_2$	$Ca(NO_3)_2$	$Sr(NO_3)_2$	$Ba(NO_3)_2$
0.0055	1.43	1.60	1.25	2.01	1.34	1.39	1.25	1.34
0.011	1.74	1.92	1.65	2.32	1.70	1.65	1.65	1.65
0.0165	1.96	2.05	1.87	2.37	1.90	1.87	1.87	1.83
0.022	2.09	2.2	2.0	2.45	2.0	2.0	2.00	1.87
0.0275	2.18	2.27	2.09	2.50	2.18	2.05	2.09	1.92
0.033	2.27	2.31	2.18	2.54	2.27	2.09	2.18	1.96
0.0385	2.31	2.36	2.23	2.54	2.35	2.14	2.23	2.0
0.044	2.36	2.40	2.27	2.59	2.35	2.18	2.27	2.0
0.0495	2.36	2.40	2.27	2.59	2.35	2.18	2.27	2.05
0.055	2.35	2.36	2.18	2.56	2.31	2.16	2.18	2.02

Fixed charge density

The electrical character of a membrane is expressed in terms of fixed charge density. This fixed charged density of zirconium phosphoborate membrane for 1:1 electrolytes has been evaluated by using Kobatake's equation [30] and is given in Table 7.

Table 7. *Values of parameters* α, β *and* $\bar{X}$ *for zirconium phosphoborate membrane*

Electrolyte	α	β	$\bar{x}$
LiCl	0.80	2.04	0.0075
NaCl	0.56	4.79	0.0078
KCl	0.60	3.83	0.0047
$LiNO_3$	0.56	4.42	0.0072
$NaNO_3$	0.68	2.55	0.0024
KNO_3	0.62	4.34	0.0011
$MgCl_2$	0.67	0.34	1.51
$CaCl_2$	0.69	0.23	2.22
$SrCl_2$	0.70	0.26	1.94
$BaCl_2$	0.75	0.60	0.61
$Mg(NO_3)_2$	0.57	6.22	.0036
$Ca(NO_3)_2$	0.57	1.15	0.49
$Ba(NO_3)_2$	0.62	0.76	0.67
$Sr(NO_3)_2$	0.61	0.13	4.69

When negatively charged membrane separates solutions of electrolyte of different concentrations, the membrane potential is given by:

$$E_m = -\frac{RT}{F}\left[\frac{1}{\beta}\ln\frac{C_2}{C_1} - \left(1+\frac{1}{\beta}-2\alpha\right)\ln\frac{C_2+\alpha\beta\bar{X}}{C_1+\alpha\beta\bar{X}}\right] \tag{3}$$

where $\alpha = \dfrac{u}{u+v}$ and $\beta = 1 + \dfrac{KF\overline{X}}{u}$

E_m = membrane potential difference,

u and v = molar mobilities of cation and anion, respectively,

K = a constant depending on the solution viscosity,

$\overline{X}$ = fixed charged density,

F = faraday constant.

In order to evaluate α, β and $\overline{X}$, Kobatake has derived two useful limiting forms, (a) and (b) of equation (3)

a) When $C_2 << \delta$, equation (3) may be written as:

$$\left|E_m^{\sigma}\right| = \frac{1}{\beta}\ln\delta - \left(\frac{\delta-1}{\alpha\beta\delta}\right)\left(1+\frac{1}{\beta}-2\alpha\right)\frac{C_2}{\overline{X}} \tag{3a}$$

where, $\left|E_m^{\sigma}\right|$ = absolute value of membrane potential given by

$$\left|E_m^{\sigma}\right| = \frac{FE_m}{RT} \text{ and } \delta = \frac{C_2}{C_1}$$

From the plot of $\left|E_m^{\sigma}\right|$ *vs.* C_2 in low concentration region, a straight line with an intercept equal to $1/\beta \ln \delta$ is obtained (Figure 1). Thus β can be calculated.

b) At fixed δ, inverse of apparent transport number $1/\overline{t}_{+app}$ for a coion species in a negatively charged membrane, varies linearly with the inverse of concentration C_2 at higher electrolyte concentrations, where $1/\overline{t}_{+app}$ is defined as

$$\left|E_m^{\sigma}\right| = (1-2\overline{t}_{+app})\ln\delta$$

Substituting for E_m^{σ} in equation (3) and expanding resultant for expression for $1/\overline{t}_{+app}$ in power of $1/C_2$ gives

$$\frac{1}{\overline{t}_{+app}} = \frac{1}{1-\alpha} + \frac{(1+\beta-2\alpha\beta)(\delta-1)\alpha}{2(1-\alpha)^2\ln\delta}\left(\frac{\overline{X}}{C_2}\right) \tag{3b}$$

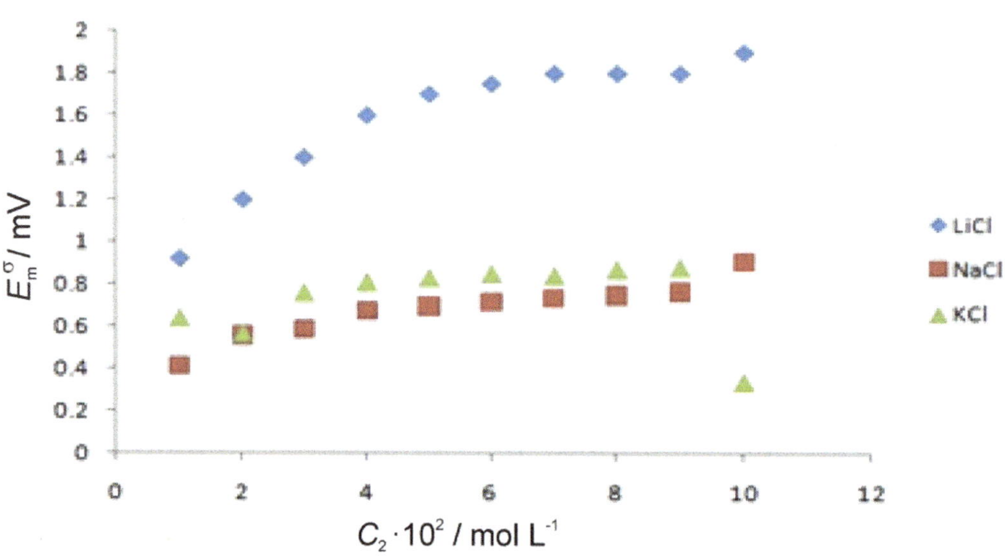

Figure 1. *Variation of membrane potential with the concentration of electrolyte*

From equation (3b), it is clear that a plot of $1/\bar{t}_{+app}$ *vs.* $1/C_2$ (Figure 2) at fixed δ value should be a straight line with an intercept equal to $1/(1-\alpha)$, from where α can be calculated.

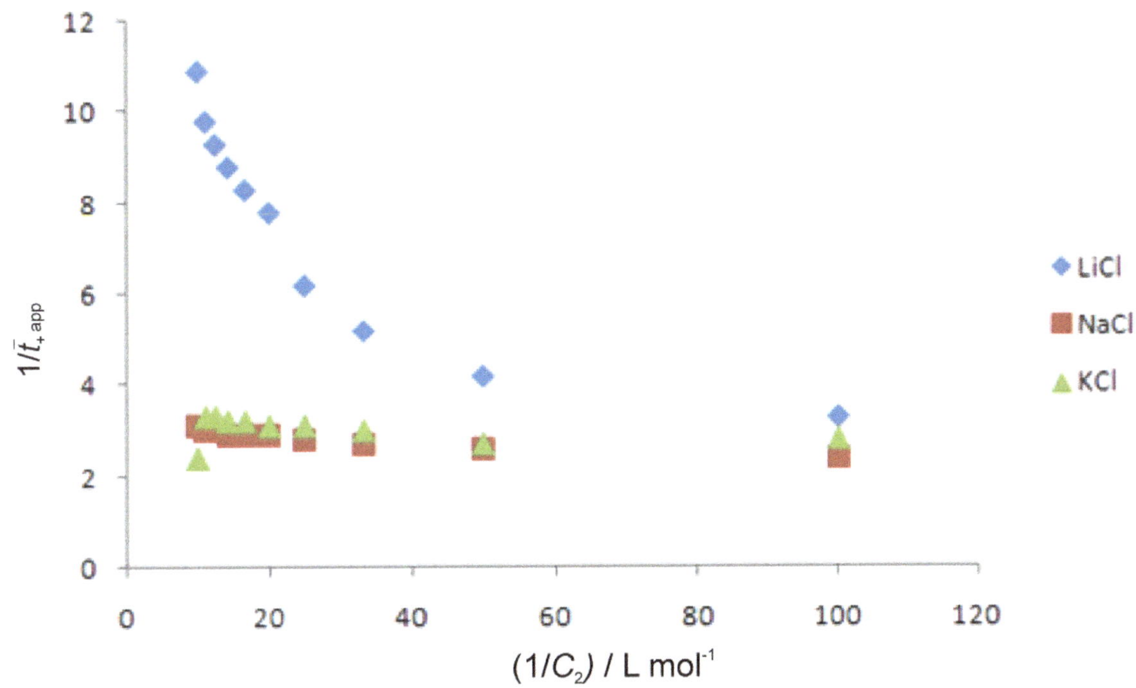

Figure 2. *Variation of apparent transport number with reciprocal of concentration*

For determining the fixed charge density $\bar{X}$ in dilute concentration range, the value of slope determined from the plot of $|E_m^\sigma|$ *vs.* C_2 plot was equated with the slope of equation (3a). α and β being known earlier, $\bar{X}$ can be calculated.

$$\text{Slope} = \frac{\delta-1}{\alpha\beta\delta}\left(1+\frac{1}{\beta}-2\alpha\right)\frac{1}{\bar{X}} \qquad (4)$$

Apparent transport number $1/\bar{t}_{+app}$ indicates transport number of a metal ion in the exchanger whereas the transport number of the same metal ion in solution phase is represented by t_+. As cited in the literature [31], transport numbers of alkali metal ions in an aqueous system generally increase sharply with initial increase in concentration in low concentration range and then become constant at higher concentration. This is true for all alkali metal ions including Li^+, Na^+ and K^+. As shown in Figure 2, the apparent transport numbers for Na^+ and K^+ do not change at all, whereas for Li^+, an appreciable linear increase in trend is observed, which indicates that the ion exchanger matrix is selective for some metal ions. The selective behavior of the ion exchange membrane for Li^+ ions may be either due to steric or electronic reasons. This can't be due to steric reasons because hydrated radii of all the alkali metal ions are almost of the same size. Li^+ in dehydrated form has large charge to radius ratio as compared to that of Na^+ and K^+.

The different trends of transport number in solution and in membrane phase confirm that the membrane is highly selective for Li^+ over Na^+ and K^+. This property of the membrane can also be generalized for transition metal ions as well because metal ions of lanthanide series are also different from one another due to their electronic properties while their ionic radii (steric factor) do not change much.

It is observed that $\bar{t}_{+app}$ decreases with increase in mean concentration of the lithium chloride electrolyte. These membrane permeate interactions indicate crystalline morphology [32], being more in amorphous and less in crystalline membranes. The low values indicate very low degree of crystallinity of exchanger material.

The observed values (Table 7) of fixed charge density $\bar{X}$ are much lower than those expected from the fixed charge concentration of the exchanger. It indicates that larger part of internal fixed charge remains inactive. It may be due to the reason that active fixed charges in these membranes are essentially those of external surface of grain. As observed in Table 7, the magnitude of $\bar{X}$ values for alkali metal nitrates is lower than for alkali metal chlorides. Hence, the trend in fixed charge density observed for halide and nitrate salts indicates that Donnan exclusion is more applicable for halide salts than for nitrate salts

Conclusions

The present investigation shows that ion exchange capacity of the ion exchanger decreases with the increase in the radii of hydrated metal ions. The transport numbers increase with increase in average concentration of the electrolyte due to enhanced Donnan exclusion at low electrolyte concentration. The transport numbers decrease at high concentration due to concentration polarization phenomenon. Hence, zirconium phosphoborate membrane shows better characteristics at lower concentrations (up to 0.045 M) beyond which no appreciable change in activity of the membrane is noticed and remains almost constant. The proposed ion exchange membrane behaves much more selectively for alkaline earth metal ions than alkali metal ions as observed from their respective fixed charge density values.

Acknowledgement: *SK and PPS gratefully acknowledge Punjab Technical University (PTU), Jalandhar for permission to work on the project. SKM is thankful to Director, Thapar University, Patiala for the support.*

References

[1] K. G. Varshney, U. Gupta, *Bull. Chem. Soc. (Japan)* **63** (1990) 1915-1921.
[2] R. D. Noble, S. A. Stern, Membrane Separations Technology: Principles and Applications (Elsevier, Amsterdam), 1995, Chap 7.
[3] M. Ulbricht, K. Richau, H. Kamusewitz, *Coll. Surf. A. Physicochem. Eng. Aspects* **138** (1998) 353-366.
[4] W. U. Malik, S. A. Ali, *Kolloid Z.* **175** (1961) 139-144.
[5] W. U. Malik, F. A. Siddiqi, *J. Colloid Sci.* **18** (1963) 161-175.
[6] W. U. Malik, M. S. Anwar, *Indian J. Chem*. **3** (1965) 491-496.
[7] W. U. Malik, H. Arif, F. A. Siddiqi, *Bull. Chem. Soc. (Japan)* **40** (1967) 1746-1753.
[8] W. U. Malik, S. K. Srivastava, P. N. Razdan, S. Kumar, *J. Electroanal. Chem*. **72** (1976) 111-116.
[9] W. U. Malik, S. K. Srivastava, S. P. Arora, R. K. Gulati, *J. Electrochem. Soc. (India)* **26** (1977) 72-78.
[10] N. P. Gnusin, N. P. Berezina, O. A. Demina, N. A. Kononeko, *Russ. J. Electrochem.* **32** (1996) 154-163.
[11] S.M. Hosseini, S.S. Madaeni, A.R. Khodabakhshi, *Sep. Sci. Technol.* **45** (2010) 2308-2321.
[12] P. Dlugolecki, B. Anet, S.J. Metz, K. Nijmejjer, M. J. Wessling, *J. Member. Sci.* **346** (2010) 163-171.
[13] P. Dlugolecki, K. Nymeijer, S. J. Metz, M. J. Wessling, *Membr. Sci.* **319** (2008) 214-222.
[14] R. K. Nagarale, G. S. Gohil, V. K. Sahni, R. Rangranjan, *Colloids Surf. A* **251** (2004) 133-140.

[15] K. Urano, Y. Masaki, M. Kawabata, *Desalination* **58** (1986) 171-176.
[16] S. M. Hosseini, S. S. Madaeni, A. R. Khodabakhshi, A. Zendehnam *J. Member. Sci*, **365** (2010) 438-446.
[17] N. Lakshinarayanaih, *Membrane Electrodes*, Academic Press, New York, 1976, pp. 50-94.
[18] S. M. Hosseini, S. S. Madaeni, A. R. Khodabakhshi, *Sep. Sci. Technol*. **46** (2011) 794-808.
[19] S. M. Hosseini, S. S. Madaeni, A. R. Khodabakhshi, *J. Member. Sci*. **362** (2010) 550-559.
[20] S. M. Hosseini, S. S. Madaeni, A. R. Khodabakhshi, *J. Member. Sci*. **351** (2010) 178-188.
[21] S. M. Hosseini, S. S. Madaeni, A. R. Khodabakhshi, *J. Appl. Polym. Sci.* **118** (2010) 33713383.
[22] R .K. Nagarale, V. K. Sahi, S. K. Thampy, R. Rangranjan, *React. Funct. Polym.* **61** (2004) 131-138.
[23] V. K. Shahi, S.K. Thampy, R. Rangaranjan, *J. Membr. Sci*. **158** (1999) 77-83.
[24] R. K. Nagarale, V. K. Shahi, R. Ragranjan, *J. Membr. Sci*. **248** (2005) 37-44.
[25] G. S. Gohil, V. V. Bishnu, V. K. Shahi, *J. Membr.Sci*. **280** (2006) 210-218.
[26] J. Schauer, V. Kuleda, K. Richau, R. Mohr, *Desalination* **198** (2006) 256-264.
[27] D. R. Lide, *CRC Handbook of Chemistry and Physics*, 87th Ed., CRC press, Taylor & Francis Group, Florida (2006-2007).
[28] A. Cimen, M. Ersoz, S. Yildiz, *Gazi Uni. J. Sci.* **25** (2012) 355-362.
[29] L. Li, J. Dong, T. M. Nenoff, *Sep. Purif. Technol.* **53** (2007) 42-48.
[30] Y. Kobatake, N. T. Toyoshima, H. Fuzita, *J. Phys. Chem*. **69** (1965) 3981-3988.
[31] R. L. Balokhra, A. Nag, *Indian J. Chem*. **32A** (1993) 610-615.
[32] R. K. Nagarale, G. S. Gohil, V. K. Shahi, R. Rangrajan, *Colloids Surf. A* **251** (2004) 133-140.

Permissions

All chapters in this book were first published in JESE, by International Association of Physical Chemists (IAPC); hereby published with permission under the Creative Commons Attribution License or equivalent. Every chapter published in this book has been scrutinized by our experts. Their significance has been extensively debated. The topics covered herein carry significant findings which will fuel the growth of the discipline. They may even be implemented as practical applications or may be referred to as a beginning point for another development.

The contributors of this book come from diverse backgrounds, making this book a truly international effort. This book will bring forth new frontiers with its revolutionizing research information and detailed analysis of the nascent developments around the world.

We would like to thank all the contributing authors for lending their expertise to make the book truly unique. They have played a crucial role in the development of this book. Without their invaluable contributions this book wouldn't have been possible. They have made vital efforts to compile up to date information on the varied aspects of this subject to make this book a valuable addition to the collection of many professionals and students.

This book was conceptualized with the vision of imparting up-to-date information and advanced data in this field. To ensure the same, a matchless editorial board was set up. Every individual on the board went through rigorous rounds of assessment to prove their worth. After which they invested a large part of their time researching and compiling the most relevant data for our readers.

The editorial board has been involved in producing this book since its inception. They have spent rigorous hours researching and exploring the diverse topics which have resulted in the successful publishing of this book. They have passed on their knowledge of decades through this book. To expedite this challenging task, the publisher supported the team at every step. A small team of assistant editors was also appointed to further simplify the editing procedure and attain best results for the readers.

Apart from the editorial board, the designing team has also invested a significant amount of their time in understanding the subject and creating the most relevant covers. They scrutinized every image to scout for the most suitable representation of the subject and create an appropriate cover for the book.

The publishing team has been an ardent support to the editorial, designing and production team. Their endless efforts to recruit the best for this project, has resulted in the accomplishment of this book. They are a veteran in the field of academics and their pool of knowledge is as vast as their experience in printing. Their expertise and guidance has proved useful at every step. Their uncompromising quality standards have made this book an exceptional effort. Their encouragement from time to time has been an inspiration for everyone.

The publisher and the editorial board hope that this book will prove to be a valuable piece of knowledge for researchers, students, practitioners and scholars across the globe.

List of Contributors

SUZANA SOPČIĆ
University of Zagreb, Faculty of Chemical Engineering and Technology, Marulićev trg 19, HR-10000 Zagreb, Croatia

MARIJANA KRALJIĆ ROKOVIĆ
University of Zagreb, Faculty of Chemical Engineering and Technology, Marulićev trg 19, HR-10000 Zagreb, Croatia

ZORAN MANDIĆ
University of Zagreb, Faculty of Chemical Engineering and Technology, Marulićev trg 19, HR-10000 Zagreb, Croatia

VIJAYA GOPAL SRIBHARATHY
Corrosion Research Centre, PG and Research Department of Chemistry, GTN Arts College, Dindigul-624005, Tamil Nadu, India

SUSAI RAJENDRAN
Department of Chemistry, RVS School of Engineering and Technology, Dindigul-624005, Tamil Nadu, India

AJAR KAMAL
Department of Chemistry, UGC-Centre for Advanced Studies, Guru Nanak Dev University, Amritsar- 143005, India

RAJIV KUMAR PURI
Department of Chemistry, UGC-Centre for Advanced Studies, Guru Nanak Dev University, Amritsar- 143005, India

RAGHU RAJ
Department of Chemistry, UGC-Centre for Advanced Studies, Guru Nanak Dev University, Amritsar- 143005, India

VIPAN KUMAR
Department of Chemistry, UGC-Centre for Advanced Studies, Guru Nanak Dev University, Amritsar- 143005, India

RAKESH KUMAR MAHAJAN
Department of Chemistry, UGC-Centre for Advanced Studies, Guru Nanak Dev University, Amritsar- 143005, India

ANTHONY SAMY SAHAYA RAJA
PG and Research Department of Chemistry, GTN Arts College, Dindigul-624005, Tamil Nadu, India

SUSAI RAJENDRAN
Department of Chemistry, RVS School of Engineering and Technology, Dindigul-624005, Tamil Nadu, India

Ana Carolina O. Santana
Escola de Química, Universidade Federal do Rio de Janeiro, Av. Horácio Macedo, 2.030, Centro de Tecnologia, Bloco E, E-203, Cidade Universitária,CEP 21941-909, Rio de Janeiro, Brasil

Erica F. Southgate
Escola de Química, Universidade Federal do Rio de Janeiro, Av. Horácio Macedo, 2.030, Centro de Tecnologia, Bloco E, E-203, Cidade Universitária,CEP 21941-909, Rio de Janeiro, Brasil

João Paulo B. G. Mendes
Instituto de Química, Universidade Federal do Rio de Janeiro, Av. Horácio Macedo, 2.030, Centro de Tecnologia, Bloco A, A-302, Cidade Universitária, CEP 21941-909, Rio de Janeiro, Brasil

Jo Dweck
Escola de Química, Universidade Federal do Rio de Janeiro, Av. Horácio Macedo, 2.030, Centro de Tecnologia, Bloco E, E-203, Cidade Universitária,CEP 21941-909, Rio de Janeiro, Brasil

Eliana Mosse Alhadeff
Escola de Química, Universidade Federal do Rio de Janeiro, Av. Horácio Macedo, 2.030, Centro de Tecnologia, Bloco E, E-203, Cidade Universitária,CEP 21941-909, Rio de Janeiro, Brasil

Ninoska Isabel Bojorge Ramirez
Universidade Federal Fluminense, Dep. Engenharia Química e de Petróleo, R. Passo da Pátria, 156, Bl E-226, São Domingos , Niterói, CEP 24210-240, Rio de Janeiro, Brasil

BOULBABA ELADEB
Laboratoire Réactions et Génie des Procédés, CNRS-Université de Lorraine, ENSIC, 1 rue Grandville, F-54001 Nancy, France

CAROLINE BONNET
Laboratoire Réactions et Génie des Procédés, CNRS-Université de Lorraine, ENSIC, 1 rue Grandville, F-54001 Nancy, France

ERIC FAVRE
Laboratoire Réactions et Génie des Procédés, CNRS-Université de Lorraine, ENSIC, 1 rue Grandville, F-54001 Nancy, France

FRANÇOIS LAPICQUE
Laboratoire Réactions et Génie des Procédés, CNRS-Université de Lorraine, ENSIC, 1 rue Grandville, F-54001 Nancy, France

LUO JIANCHENG
School of Chemistry and Environment, South China Normal University, Guangzhou 510006, China

YANG JIE
School of Chemistry and Environment, South China Normal University, Guangzhou 510006, China

LI WEISHAN
School of Chemistry and Environment, South China Normal University, Guangzhou 510006, China
Key Laboratory of Electrochemical Technology on Energy Storage and Power Generation of Guangdong Higher Education Institutes, South China Normal University, Guangzhou 510006, China
Engineering Research Center of Materials and Technology for Electrochemical Energy Storage
(MOE), South China Normal University, Guangzhou 510006, China

HUANG QIMING
School of Chemistry and Environment, South China Normal University, Guangzhou 510006, China
Key Laboratory of Electrochemical Technology on Energy Storage and Power Generation of Guangdong Higher Education Institutes, South China Normal University, Guangzhou 510006, China
Engineering Research Center of Materials and Technology for Electrochemical Energy Storage
(MOE), South China Normal University, Guangzhou 510006, China

XU HONGKANG
Dongguan Hongjie Environmental Technologies Ltd, Dongguan 523039, China

SVETLOZAR IVANOV
Rostislaw Kaischew Institute of Physical Chemistry, Bulgarian Academy of Sciences, Sofia 1113, Bulgaria

IOANNA MINTSOULI
Department of Chemistry, Aristotle University of Thessaloniki, Thessaloniki 54124, Greece

JENIA GEORGIEVA
Rostislaw Kaischew Institute of Physical Chemistry, Bulgarian Academy of Sciences, Sofia 1113, Bulgaria

STEPHAN ARMYANOV
Rostislaw Kaischew Institute of Physical Chemistry, Bulgarian Academy of Sciences, Sofia 1113, Bulgaria

EUGENIA VALOVA
Rostislaw Kaischew Institute of Physical Chemistry, Bulgarian Academy of Sciences, Sofia 1113, Bulgaria

GEORGIOS KOKKINIDIS
Department of Chemistry, Aristotle University of Thessaloniki, Thessaloniki 54124, Greece

IOANNIS POULIOS
Department of Chemistry, Aristotle University of Thessaloniki, Thessaloniki 54124, Greece

SOTIRIS SOTIROPOULOS
Department of Chemistry, Aristotle University of Thessaloniki, Thessaloniki 54124, Greece

Camilo González-Vargas
Laboratorio de Electroquímica Medio Ambiental, LEQMA, Departamento de Ciencias del Ambiente, Facultad de Química y Biología, Universidad de Santiago de Chile, USACh, Casilla 40, Correo 33, Santiago, Chile

Ricardo Salazar
Laboratorio de Electroquímica Medio Ambiental, LEQMA, Departamento de Ciencias del Ambiente, Facultad de Química y Biología, Universidad de Santiago de Chile, USACh, Casilla 40, Correo 33, Santiago, Chile

Ignasi Sirés
Laboratori d'Electroquímica dels Materials i del Medi Ambient, Departament de Química Física, Facultat de Química, Universitat de Barcelona, Martí i Franquès 1-11, 08028 Barcelona, Spain

FADHIL M. MOHAMMED
School of Chemical Engineering and Analytical Science, University of Manchester. The Mill, Oxford Road, Manchester M13 9PL, UK
Ministry of Science and Technology, Baghdad, Iraq

EDWARD P. L. ROBERTS
School of Chemical Engineering and Analytical Science, University of Manchester. The Mill, Oxford Road, Manchester M13 9PL, UK
Ministry of Science and Technology, Baghdad, Iraq

ANDREW K. CAMPEN
Arvia Technology Ltd, Daresbury Innovation Centre, Keckwick Lane, Daresbury, Cheshire WA4 4FS, UK

NIGEL W. BROWN
Arvia Technology Ltd, Daresbury Innovation Centre, Keckwick Lane, Daresbury, Cheshire WA4 4FS, UK

PERCIO AUGUSTO MARDINI FARIAS
Department of Chemistry, Pontifícia Universidade Católica do Rio de Janeiro, 22451-900, Rio de Janeiro, RJ, Brazil

ARNALDO AGUIAR CASTRO
Fac. Química, Química Analítica, Universidad de La Habana, Cuba

ANA ISA PEREZ CORDOVES
Fac. Química, Química Analítica, Universidad de La Habana, Cuba

RICHARD HONG PENG LIANG
Raffles Science Institute, Raffles Institution, One Raffles Institution Lane, Singapore 575954

TANGSHENG ZOU
Raffles Science Institute, Raffles Institution, One Raffles Institution Lane, Singapore 575954

KARTHIK SOMASUNDARAM
Department of Mechanical Engineering, National University of Singapore, Singapore 117576

WEI TONG
Department of Mechanical Engineering, National University of Singapore, Singapore 117576

ERIK BIRGERSSON
Department of Chemical and Biomolecular Engineering, National University of Singapore, Singapore 117576

ALIYA TOLEUOVA
Electrochemical Innovation Lab, Department of Chemical Engineering, University College London, Torrington Place, London WC1E 7JE, UK
School of Engineering, Nazarbayev University, 53 Kabanbay Batyr Ave, Astana 010000, Republic of Kazakhstan

VLADIMIR YUFIT
Department of Earth Science and Engineering, Imperial College London, London SW7 2AZ, UK

STEFAAN SIMONS
Electrochemical Innovation Lab, Department of Chemical Engineering, University College London, Torrington Place, London WC1E 7JE, UK
International Energy Policy Institute, UCL Australia, 220 Victoria Square, Adelaide, Australia

WILLIAM C. MASKELL
Electrochemical Innovation Lab, Department of Chemical Engineering, University College London, Torrington Place, London WC1E 7JE, UK
Department of Earth Science and Engineering, Imperial College London, London SW7 2AZ, UK

DANIEL J. L. BRETT
Electrochemical Innovation Lab, Department of Chemical Engineering, University College London, Torrington Place, London WC1E 7JE, UK

BRUNA PEKEC
Institute of Chemistry, Department of Analytical Chemistry, Karl-Franzens University, Graz, Austria
Institute of Pharmaceutical Science, Department of Pharmaceutical Chemistry, Karl-Franzens University, Graz, Austria

BIRGIT FEKETEFÖLDI
Institute of Chemistry, Department of Analytical Chemistry, Karl-Franzens University, Graz, Austria

VOLKER RIBITSCH
Institute of Chemistry, Department of Analytical Chemistry, Karl-Franzens University, Graz, Austria

ASTRID ORTNER
Institute of Pharmaceutical Science, Department of Pharmaceutical Chemistry, Karl-Franzens University, Graz, Austria

KURT KALCHER
Institute of Chemistry, Department of Analytical Chemistry, Karl-Franzens University, Graz, Austria

QUENTIN MEYER
Electrochemical Innovation Laboratory, Centre for CO2 Technology, Department of Chemical Engineering, University College London, UK

SIMON BARASS
Electrochemical Innovation Laboratory, Centre for CO2 Technology, Department of Chemical Engineering, University College London, UK

OLIVER CURNICK
Intelligent Energy Ltd, Loughborough, Leicestershire, UK

TOBIAS REISCH
Intelligent Energy Ltd, Loughborough, Leicestershire, UK

DANIEL J. L. BRETT
Electrochemical Innovation Laboratory, Centre for CO2 Technology, Department of Chemical Engineering, University College London, UK

VELIZAR STANKOVIĆ
Department of Metallurgical Engineering, Technical Faculty Bor, University of Belgrade, Serbia

XIAOZHONG ZHOU
Key Laboratory of Eco-Environment-Related Polymer Materials of Ministry of Education, Key Laboratory of Polymer Materials of Gansu Province, College of Chemistry and Chemical Engineering, Northwest Normal University, Lanzhou 730070, Gansu Province, P. R. China

SHUHUA HUA
Key Laboratory of Eco-Environment-Related Polymer Materials of Ministry of Education, Key Laboratory of Polymer Materials of Gansu Province, College of Chemistry and Chemical Engineering, Northwest Normal University, Lanzhou 730070, Gansu Province, P. R. China

LIANHUA BAI
Key Laboratory of Eco-Environment-Related Polymer Materials of Ministry of Education, Key Laboratory of Polymer Materials of Gansu Province, College of Chemistry and Chemical Engineering, Northwest Normal University, Lanzhou 730070, Gansu Province, P. R. China

DONG YU
Key Laboratory of Eco-Environment-Related Polymer Materials of Ministry of Education, Key Laboratory of Polymer Materials of Gansu Province, College of Chemistry and Chemical Engineering, Northwest Normal University, Lanzhou 730070, Gansu Province, P. R. China

María I. León
Departamento de Ingeniería Geomática e Hidráulica, Universidad de Guanajuato, Av. Juárez 77, Zona Centro, C.P. 36000, Guanajuato, Guanajuato, Mexico

Zaira G. Aguilar
Departamento de Ingeniería Geomática e Hidráulica, Universidad de Guanajuato, Av. Juárez 77, Zona Centro, C.P. 36000, Guanajuato, Guanajuato, Mexico

José L. Nava
Departamento de Ingeniería Geomática e Hidráulica, Universidad de Guanajuato, Av. Juárez 77, Zona Centro, C.P. 36000, Guanajuato, Guanajuato, Mexico

Carlos E. Barrera-Díaz
Centro Conjunto de Investigación en Química Sustentable UAEM - UNAM, Carretera Toluca-Atlacomulco, km 14.5, Unidad El Rosedal, C.P. 50200, Toluca, Estado de México, México

Gabriela Roa-Morales
Centro Conjunto de Investigación en Química Sustentable UAEM - UNAM, Carretera Toluca-Atlacomulco, km 14.5, Unidad El Rosedal, C.P. 50200, Toluca, Estado de México, México

Patricia Balderas Hernández
Centro Conjunto de Investigación en Química Sustentable UAEM - UNAM, Carretera Toluca-Atlacomulco, km 14.5, Unidad El Rosedal, C.P. 50200, Toluca, Estado de México, México

Carmen María Fernandez-Marchante
Departartment of Chemical Engineering. Faculty of Chemical Sciences & Technologies, Enrique Costa Building, Campus Universitario s/n 13071 Ciudad Real, Spain

Manuel Andrés Rodrigo
Departartment of Chemical Engineering. Faculty of Chemical Sciences & Technologies, Enrique Costa Building, Campus Universitario s/n 13071 Ciudad Real, Spain

SANDEEP KAUSHAL
Department of Applied Sciences and Humanities, Punjab Technical University, Jalandhar, India

PRITPAL SINGH
Sri Guru Granth Sahib World University, Fatehgarh Sahib (Pb), India

SUSHEEL K. MITTAL
School of Chemistry and Biochemistry, Thapar University, Patiala 147004, India